INTERNATIONAL ENVIRONMENTAL

International Environmental Economics

A Survey of the Issues

Edited by

GÜNTHER G. SCHULZE
HEINRICH W. URSPRUNG

OXFORD
UNIVERSITY PRESS

OXFORD

UNIVERSITY PRESS

Great Clarendon Street, Oxford OX2 6DP

Oxford University Press is a department of the University of Oxford.
It furthers the University's objective of excellence in research, scholarship,
and education by publishing worldwide in

Oxford New York

Auckland Bangkok Buenos Aires Cape Town Chennai
Dar es Salaam Delhi Hong Kong Istanbul Karachi Kolkata
Kuala Lumpur Madrid Melbourne Mexico City Mumbai Nairobi
São Paulo Shanghai Taipei Tokyo Toronto

Oxford is a registered trade mark of Oxford University Press
in the UK and in certain other countries

Published in the United States
by Oxford University Press Inc., New York

British Library Cataloguing in Publication Data

Data available

Library of Congress Cataloging in Publication Data

International environmental economics: a survey of the issues /
edited by Günther Schulze, Heinrich W. Ursprung.
p. cm.
"Contains ten surveys of special aspects of the literature on international environmental
economics"– Pref.
Includes bibliographical references and index.
1. Environmental policy – Economic aspects.   2. International trade – Environmental
aspects.   3. Environmental law, International.   I. Schulze, Günther G.
II. Ursprung, Heinrich W.
HC79.E5 I5355 2001 333.7–dc21 2001016398
ISBN 0–19–829766–1 (hbk.)
ISBN 0–19–926111–3 (pbk.)

1 3 5 7 9 10 8 6 4 2

Typeset by Newgen Imaging Systems (P) Ltd., Chennai, India
Printed in Great Britain
on acid free paper by
T. J. International Ltd., Padstow. Cornwall

Λ 10029716 28

Preface

Environmental problems and environmental policies have an obvious international dimension. Many of the most pressing environmental problems are truly global: tropical deforestation, global warming, depletion of the ozone layer, to name just a few. These problems can only be solved by means of internationally concerted efforts. Other environmental problems are transboundary or even local, but these problems still have an international dimension since the pattern of pollution depends on the international division of labor which, in turn, is predicated, among other factors, on the individual countries' environmental regulation. A crucial role in this context plays the strictness of a country's environmental regulation *vis-à-vis* the environmental policies conduced by her neighbors. This is because individual countries compete with each other on a global level for market shares and for mobile factors; they might, therefore, be tempted to use environmental policies to gain a competitive edge. The environment-trade nexus and the interaction of environmental and international trade policies are thus blatantly obvious and therefore, need to be scrutinized to gain a profound understanding of environmental and trade policy formation.

These issues are, however, not only an academic challenge. The international dimension of environmental policy and the implications of international trade for the environment have in recent years become the focus of politically vigorous nongovernment organizations. The ongoing globalization of the economy has raised fears that the attendant increase of economic activity and erosion of the regulatory power of the nation states will see the environment on the receiving end. Especially the most prominent international organizations have, as a consequence, been exposed to intense criticism ranging from well-meaning do-it-yourself economists to rioting street-fighters. The World Trade Organization's (WTO) third Ministerial Meeting in Seattle (30 November–3 December, 1999), for example, took place under some kind of siege. Whatever one may think about the events which are now called the "battle of Seattle," these events clearly document that the globalization of the economy in general and international trade in particular represent a key issue in the agenda of environmentalists. It would also be a grave mistake to lay aside the "battle of Seattle" as an isolated and typical American incident. Shortly after the WTO Meeting, the inconspicuous World Economic Form Annual Meeting (27 January–1 February, 2000) in Davos, Switzerland, was targeted by more or less the same groups and with a similar venom, and in summer 2000 preparations for large-scale protests against the Annual Meeting of the Boards of Governors of the International Monetary Fund (IMF) and the

World Bank Group held in September 2000 in Prague ran under the slogan "Turn Prague into Seattle!"

All this is not to say that the trade–environment nexus has been neglected by serious economists. On the contrary: nobody denies that many if not most of the toughest environmental problems of today exhibit crucial international or even global components and that these problems can only be solved by explicitly addressing and investigating the international dimension of the problems at hand. The pertinent questions comprise the following: Is a liberal world economic order detrimental to a sustainable quality of the environment? What does it take to render international environmental agreements feasible and stable? How are environmentally sustainable growth paths to be designed? What characterizes sustainable forestry and agriculture? Do lenient environmental policies in fact attract foreign investment and thereby constitute an incentive for the apprehended race to the bottom? These and many other questions are addressed in this volume in a systematic way. The volume contains ten surveys of special aspects of the literature on international environmental economics written by well-established scholars in the respective fields. All of these surveys are strongly guided by theoretical considerations, but they are nontechnical and issue-oriented and therefore easily accessible; they are concise yet comprehensive in their coverage of the most important contributions to the literature. The book balances theoretical findings with empirical evidence in order to assess the relevance of the former by the latter. One of the objectives of the volume is to allow the reader to critically assess the various allegations made in the political debate about the nature of the interdependence of international economic activity on the one hand and environmental quality on the other.

The subject matter will be of interest to scholars and graduate students in economics, political science, and environmental studies who want to familiarize themselves with the main issues and contributions to the various aspects of international environmental economics. The book may serve as an introduction to the literature or rather the various strands of the literature on international environmental economics; it may also be used as a compendium for researchers in the field.

In making this book we drew on many people's help and advice. First and foremost, we are grateful to the authors themselves who not only contributed their own work but helped us tremendously in making this book into a coherent set of surveys by commenting on each others' work. Karl-Josef Koch, Arik Levinson, James Markusen, and Ronnie Schöb, among others, provided critical comments and very valuable advice. Lisa Green, Susanne Holder and Emil Totchkov helped us to edit the book. For all these contributions we express our sincere thanks.

Günther G. Schulze
Heinrich W. Ursprung

Contents

Contents

Figures

Tables

Abbreviations

AoA	Agreement on Agriculture
APHIS	Animal and Plant Health Inspection Service
CAP	Common Agriculture Policy
CEC	WTO Commission for Environmental Cooperation
CFC	Chloro Fluoro Carbons
CGE	Computable General Equilibrium
CIFOR	Centre for International Forestry Research
CITES	Convention on International Trade in Endangered Species of Wild Fauna and Flora
CRP	Conservation Reserve Program
CTE	WTO Committee on Trade and Environment
ECC	Environmental Control Costs
EIA	Environmental Investigation Agency
EKC	Environmental Kuznets Curve
EPA	US Environmental Protection Agency
ETM	Environmental Trade Measures
EU	European Union
FAIR	Federal Agricultural Improvement and Reform
FAO	Food and Agriculture Organization (United Nations)
FCC	Framework Convention on Climate Change
FDI	Foreign Direct Investment
FPE	Factor Price Equalization
FPMO	Forest Protection and Management Obligations
FQPA	Food Quality Protection Act
FRA	Forest Resources Assessment
FRG	Federal Republic of Germany
GATT	General Agreement on Tariffs & Trade
GDP	Gross Domestic Product
GDR	German Democratic Republic
GEF	Global Environment Facility
GEMS	Global Environment Monitoring System
GMO	Genetically Modified Organisms
HNIS	Harmful Non-Indigenous Species
IEA	International Environmental Agreement
ITTA	International Tropical Timber Agreement
ITTO	International Tropical Timber Organization
IUCN	International Union for Conservation of Nature and Natural Resources

LDC	Less Developed Countries
LUS	Lesser Used Species
NAAQS	National Ambient Air Quality Standards
NAFTA	North American Free Trade Agreement
NEI	Netherlands Economic Institute
NFAP	National Forest Action Programme
NIMBY	Not in my Backyard
OAU	Organization of African Unity
OECD	Organisation for Economic Cooperation and Development
OFI	Oxford Forestry Institute
OPEC	Organization of Petroleum Exporting Countries
OTA	Office of Technology Assessment (no longer exists)
PPP	Polluter-Pays Principle
R&D	Research and Development
SITC	Standard International Trade Classification
SPS	Sanitary and Phytosanitary
SWF	Social Welfare Function
TBT	Technical Barriers to Trade
TRI	Toxic Release Inventory
UNCED	United Nations Conference on Environment and Development
UNCTAD	United Nations Conference on Trade & Development
UNECE	United Nations Economic Commission for Europe
URA	Uruguay Round Agreement
USDA	United States Department of Agriculture
VAT	Value Added Tax
WRI	World Resources Institute
WWF	World Wildlife Fund
WTO	World Trade Organization

Notes on Contributors

Edward B. Barbier, Department of Economics and Finance, University of Wyoming, Laramie, WY 82071, USA.

Lucas Bretschger, Ernst-Moritz-Arndt-University of Greifswald, Friedrich-Loefflerstr. 70, D-17487 Greifswald, Germany.

Roger D. Congleton, Center for Study of Public Choice; George Mason University, Fairfax, VA 22030, USA.

Hannes Egli, Ernst-Moritz-Arndt University of Greifswald, Friedrich-Loefflerstr. 70, D-17487 Greifswald, Germany.

David E. Ervin, Environmental Sciences and Resources, Portland State University, 2411 Cramer, Portland, OR 97207, USA.

Michael Rauscher, University of Rostock, Economics Department, Ulmenstr. 69, D-18051 Rostock, Germany.

Carsten Schmidt, Arthur Andersen Frankfurt am Main, Erzweg 22, 61118 Bad Vilbel, Germany.

Günther G. Schulze, Department of Economics, Albert-Ludwig University of Freiburg, Platz der Alten Synagoge, D-79085 Freiburg i. Br., Germany.

Sjak Smulders, Department of Economics and Center for Economic Research, Tilburg University, P.O. Box 90153, 5000 LE Tilburg, The Netherlands.

Heinrich W. Ursprung, Department of Economics, University of Konstanz, Box D-138, D-78457 Konstanz, Germany.

1

Introduction

GÜNTHER G. SCHULZE AND HEINRICH W. URSPRUNG

Closed-economy models have dominated environmental economic analysis for a long time. Most monographs and omnibus volumes on environmental economics still reflect this closed-economy approach (see, for example, Baumol and Oates (1998)), with the consequence that international aspects of environmental problems are quite often not covered at all, or dealt with only as an afterthought and in a rather perfunctory manner. Yet we live in a world in which there is awareness of environmental problems that spill over national borders. Moreover, environmental problems are generally perceived to be linked to the globalization of the economy in a straightforward and adverse manner. The November 1999 "battle of Seattle" in which environmental groups played a significant role documents this popular perception most impressively. A national economy picture is therefore clearly unsatisfactory for an overall investigation of environmental externality problems. Rather, in an integrated world, environmental policy decisions need to take into consideration environmental consequences of international trade, international factor movements, and international policy interdependence. This calls for an international view and an accompanying open-economy approach to analyzing environmental problems and policies. Such an approach has been referred to as "International Environmental Economics" (Walter 1976).

International Environmental Economics made its appearance in the early 1970s. At that time the issue of impending global environmental plight was much talked about (the stirring report for the club of Rome's project on the predicament of mankind appeared in 1972; see Meadows *et al.* (1972)). This discussion gave rise to a first round of investigations of the link between international trade and the environment (see Baumol 1971; Siebert 1974; Markusen 1975; Walter 1975; Pethig 1976). A second round of investigations was spurred in the early 1990s, when trade liberalization clashed with the interests of the environmental groups who, in the meantime, had developed into political actors to be reckoned with (for the organization of environmental interests in the European Union, see Felke 1998: 88–93). The negotiations that led to the North American Free Trade Agreement (NAFTA) were fiercely attacked by environmentalists in the United States. Somewhat later, the "infamous" decision of a dispute settlement panel, which ruled that the US dolphin-safe tuna fishing legislation violated the General Agreement on Tariffs and Trade (GATT), became a worldwide focal point in the controversy between advocates of free trade and environmental protection. The second-generation literature on trade and the environment (see, for example, the well-received omnibus volumes by Anderson and Blackhurst (1992), Low (1992b), and Carraro (1994)) had a more substantial impact on the economics profession than the earlier literature of the 1970s, which was, to a large extent, more narrowly focused on abstract trade theory concepts.

The international dimension of environmental policymaking was singled out by Maureen Cropper and Wallace Oates in their famous 1992 survey to represent the most important future extension of the then existing literature on environmental economics (p. 731). After a decade of intensive research in the area of international environmental economics, the time has come to take stock of the state-of-the-art knowledge in this field. It is the objective of this volume to provide such a review; the following chapters survey specific aspects of the literature. While no single survey attempts to provide an integrating view of the entirety of issues involved, together the surveys cover the international economic and environmental linkages that have been the focus of the discussion.

International Environmental Economics explicitly recognizes the existence of national borders and nation states. The following section 1.1 elaborates on the significance of the international dimension of environmental policymaking. To understand the reluctance of the narrow mainstream of economists to integrate international aspects into the canon of environmental economics, we then continue in section 1.2 by briefly elaborating on the view presented by mainstream economics of national government, and on how this view then bounds the perspective on environmental economics in general, and the analysis of international linkages in particular. After the discussion of paradigms, we review the issues encountered in international environmental economics via an overview of the chapters of this volume in section 1.3.

1.1. INTERNATIONAL ENVIRONMENTAL ECONOMICS

What makes International Environmental Economics such an important sub-field of environmental economics, or of international economics for that matter? From a theoretical viewpoint environmental economics essentially deals with a special kind of externality; normative environmental economics investigates how these externalities are best internalized, while the positive approach explains why the observed policy measures are actually taken. There are three characteristics that impart environmental economics a genuinely international dimension.

First, environmental problems often are *transboundary* by their very nature. While national borders delineate the limits of the coercive power of nation states, national jurisdictions coincide only by chance with the areas that are affected by pollution of some sort. Environmental problems may be local, regional or even global. Examples for the latter two types are manifold: upstream pollution of a river may take place in one country while the negative effects of the polluting activity may become evident downstream in another country, a chemical plant emitting toxic releases in the air may affect the neighboring country if the plant is located close enough to the border, CO_2 emissions in one country affect the global climate and thereby all other countries, etc. In other words, countries are affected by polluting economic activities over which they do not have control. Those countries that do have control, that is, the countries in which the pollution originates, may not and typically will not have an incentive to fully internalize these externalities because they would have to bear the full costs of internalization but would reap only a part of the benefits, the remaining part being realized abroad. Independent national environmental policy is clearly suboptimal since it disregards international repercussions; optimal environmental policy calls for international cooperation. The issue thus arises as to how international cooperation schemes need to be designed if they are to be feasible (i.e., generally enacted), stable and efficient in the sense that they optimally internalize the environmental externality at hand.

Second, even in the case where externalities are only local, environmental economics has a significant international dimension if the *source of the externality can be shifted internationally* by way of economic transactions. If, for example, a country has a comparative advantage in polluting industries, increasing international economic integration may give rise to higher pollution in that country. This specialization in polluting activities can be brought about by *internal* relocation of factors or by *international* capital movements. Here the interesting question is to what extent international integration influences environmental quality in different countries via international trade in goods and international factor trade. In particular, do tariff-ridden developing countries suffer more from pollution than their less distorted counterparts and do international capital movements from industrialized countries to less developed countries exacerbate the environmental problems in the "South" or do they alleviate the environmental burden by virtue of the transfer of less environment-intensive production technologies? Also, if environmental policies are suboptimal (as is probably the case in agriculture and forestry), to what extent does international trade contribute to this policy inactivity? How, finally, does international trade in hazardous waste effect environmental quality and welfare?

Closely related to the previous issues is the third issue, the *interaction of environmental policy with other international policies*. If trade policy influences environmental quality through its impact on the pattern of international trade, does this imply that the time-honored free trade policy is suboptimal from a more encompassing welfare perspective? Conversely, if environmental regulation, by influencing a country's endowment with environment as a factor of production, redirects international trade and capital flows, does this imply that environmental policy can and will be used for trade policy ends? After all, if international capital flows react to differences in environmental standards, some countries might be tempted to lower their standards strategically to attract capital ("pollution haven hypothesis"), or they might lower their *production* standards simply to gain a competitive edge in the respective industries ("ecodumping"). Some countries might also be tempted to use environmental *product* standards to protect domestic industries from foreign competition which might find it more difficult to meet the specified standards ("ecoprotectionism"). The interaction between environmental and trade policy could, of course, also prove to be beneficial; if environmental taxes increase government revenue, these extra funds can, in principle, be used to lower trade taxes (tariffs), thereby possibly providing a "double dividend" in terms of internalization of environmental externalities plus a reduction in tariff distortions.

It thus becomes clear that environmental policy has an important international dimension that must not be disregarded. Especially in today's integrating world economy the burden of proof for relevance lies with purely national environmental economics, not with international environmental economics. Although the literature on international environmental economics is developing dynamically and now constitutes a considerable body of literature, no comprehensive presentation of the material exists so far.[1] The surveys collected in this volume were written with the intention to provide such a presentation.

Before reviewing the issues covered by the individual surveys in section 1.3, we now proceed to elaborate on the reasons for the perfunctory treatment international environmental economics has received from the narrow mainstream of economics for such a long time. This inquiry will also shed some light on the paradigmatic differences between traditional environmental economics and international environmental economics.

1.2. ENVIRONMENTAL ECONOMICS AND THE NATION STATE

Given the rather conservative bend of mainstream economics, it is maybe somewhat surprising that environmental economics managed to become a constituent part of economic theory within a decade (the 1970s). When the issue of environmental pollution came to the attention of the policymakers and the general public, the economics profession, for once, "was ready and waiting" (see Cropper and Oates 1992: 675). The phenomenon of environmental pollution was readily identified as a special kind of externality which gives rise to market failure. The orthodox approach to economic policy analysis in such cases calls for government intervention to remedy the market failure by applying policy instruments that internalize the troublesome externalities. The narrow mainstream of economics did not hesitate to propose what they

---

[1] Surveys on selected aspects are available, see e.g. Rauscher (1997) and Bommer (1998).

regarded as appropriate measures for government to engage in environmental regulation and management.

The swift assimilation of environmental economics in academia, however, has not been accompanied by the anticipated impact on actual environmental policymaking. The policy advice advanced by academic economists did not find a sympathetic ear with the political and administrative agents responsible for environmental legislation. This was due to two different kinds of reasons. The first reason has to do with the economic way of reasoning as such. In the early days of environmental economics, the economic calculus as applied to the measurement of the benefits and costs of environmental programs was still rather undifferentiated and based on *ad hoc* assumptions. The economic analysis was, therefore, subject to ready criticism. Moreover, since direct and indisputable market valuations of the benefits of pollution control are not usually available, the general public and policymakers showed reluctance, if not opposition on principle, to accept any monetary measures of environmental impact (see Schulze and Ursprung 2000). These objections have lost some of their initial poignancy. There has been considerable progress in the development of (environmental) valuation techniques and the general political setting appears to have become much more receptive to market-economy approaches to solving social problems (see Cropper and Oates 1992: section VI).

The second reason for the neglect of economic expertise is related to the role that narrow mainstream economic theory assigns to government activities, and to the ways in which political processes are accordingly portrayed. Proceeding from the well-known problems of private provision of nonexcludable public goods, orthodox economic policy analysis derives an immediate justification for the existence of an intervening state. The role for government is, however, not backed up by an accompanying realistic approach to modeling the behavior of the political agents. On the contrary, narrow orthodox economic theory shies away from the political economy of redistribution contests fought out via the political process. There is too much controversy in the recognition of the redistributive contests. Whatever the reason for the refusal to come to terms with the coercive redistributional state (for an explanation see Hillman 1998), the purely normative mode of argumentation of orthodox economics comes at a cost in terms of explaining the events that take place in the world in which we live. By confining policy recommendations exclusively to efficiency considerations, the narrow orthodox advice becomes either appropriate only for those few cases in which no redistribution-motivated resistance to efficiency enhancing policy measures is present. Or the advice is irrelevant because the presence of redistributional consequences is assumed away by the assumption of politically infeasible lump-sum compensation.

When the political process is subsumed in a so-called "benevolent dictator", economists are spared involvement in controversial and inherently unsolvable political disputes. Nothing is, however, gained from this ostrich-like approach to economic theorizing. The proposal to assign the provision of public goods – such as environmental quality – to the state in response to the free-rider problem of voluntary provision does not resolve any basic problems. The problems do not disappear but reemerge in the form of a coordination problem of "rationally ignorant" voters who do not monitor public-office holders. As a consequence, politicians have discretion to grant privileges to special interest groups in return for political support. By assigning the provision of environmental quality to the domain of the state, market failure is thus transformed, at least to some extent, into political failure. In a recent paper on pollution control Dwight Lee summarizes this transformation process as follows:

Politicians realize that the biggest payoff comes from providing programs that concentrate benefits on politically active groups organized around narrow interests. The public's general desire for public goods justifies granting government the power to tax and regulate to provide them, but the private interest of organized groups shape the details of how those public goods are provided. And these organized groups, and their political agents, have a lot of latitude to secure private benefits by reducing the efficiency with which public goods are provided. Because the cost of the reduced efficiency is spread over the general public, any effort to mobilize opposition to the cost is frustrated by the free-rider problem. As with market decisions, private interest is the dominant force behind political decisions. (Lee 1999: 126)

When these political-economic relationships are not taken into account, environmental policy advice advanced by academic economists will have no impact on actual legislation; and this, as we have pointed out above, has been the fate of much of environmental economics in the past. Environmental economists simply need to accept the evident fact that economic policy is endogenous to the political-economic system. This insight, trivial as it might be for people familiar with actual policy decision making in the world, implies that policy advice should not be targeted at the big stakeholders in the ongoing political process. If environmental economics can influence the ongoing process of environmental legislation at all, it is by informing the general public, and not by a vain attempt to "enlighten" the directly affected interests who are pre-committed to the policies that best serve their personal interests.

In any case, economic expertise stands a better chance of exerting a significant impact in the long run, by helping to design political institutions that will result in more satisfactory outcomes of the environmental policy process (see Vanberg and Buchanan 1989; Buchanan 1994). Such a (normative) constitutional economic program explicitly acknowledges that environmental pollution calls for state-imposed coordination of individual behavior and is firmly based on a political-economic analysis of endogenous environmental-policy formation. Developing an efficacious normative theory of environmental policy thus requires a preceding positive, that is political-economic, analysis of the environmental policy process that results from alternative institutional settings governing the interaction between the political agents.

Some environmental economists have taken up the challenge of putting politics back into the economic analysis of environmental policy (see, for example, the collection of papers in Congleton (1996) and the recent monograph by Dijkstra (1999)). Some of the contributions to the still small but continuously growing literature on the political economy of environmental policy rank with the most novel and exciting studies on environmental economics published during the past few years.

This is not to say, of course, that traditional environmental policy analysis is redundant. Abstracting from the existence of self-interested political actors and postulating instead a benevolent and omnipotent state agency may provide a convenient benchmark for assessing the behavior of alternative institutions governing the political process. Confusing the traditional welfare–theoretic approach to economic policy analysis with a positive investigation into the workings of observed political-economic processes, does, however, give rise to results far removed from anything resembling the problems encountered in the real world. Since economists do admit to having difficulties in taking the narrow efficiency-focused "explanations" seriously (see Clower 1993), we should not be surprised with the observation that the "policy environment (is) characterized by a real ambivalence (and in some instances, an active hostility) to a central role for economics in environmental decision making" (Cropper and Oates 1992: 729). If the assumption of a benevolent and omnipotent state agency yields impracticable policy proposals in a national, that is closed-economy, context (in which a public agency indeed exists – even if it is neither omnipotent nor necessarily benevolent), this fiction leads to a pathetic caricature in an international policy-coordination context. This is so because interaction at the international level is characterized, if not defined, by the absence of any supranational agency endowed with coercive powers. Even the most ardent supporters of the traditional welfare–theoretic approach to economic policy analysis concede that national policy makers have objectives that are narrower than *global* welfare maximization; the narrow orthodox approach is thus clearly not applicable to the issues arising in the field of international environmental economics. It is undisputed that national borders are responsible for many severe environmental problems, and it is therefore paramount to analyze national policymakers' objectives and their behavior since, in economic terms, national borders are nothing but the embodiment of different government policies.

We thus come to the conclusion that pollution externalities and the non-excludability of environmental quality give rise to market failure calling (because of absence of supranational government structures) for regulatory intervention of the established nation states. However, simply calling upon the state does not resolve the problem.

First of all, the regulating agency, whatever the motivation of its members, is incompletely informed about the benefits and costs of environmental programs. This is a tough issue, which has received a great deal of attention in environmental economics over the last 30 years. It would not be a gross exaggeration to claim that this issue, to a large extent, constitutes environmental economics as perceived by the textbook canonization.

Second, and more fundamentally, invoking the (national) state always generates problems of its own, since the political agents are liable to use the discretion provided by the given political institutions to pursue their own private interests. The regulatory policies therefore may not serve the interest of the general public. This kind of political failure is the subject of political-economic investigations, which trace observed policies back to the preferences of the political agents and the prevailing political institutions. The problems that arise when environmental management is delegated to national states are confounded by the international dimension. Even if political failure on the national level were not a serious problem, the economic and environmental interdependence of nation states with their diverse interests would still be expected to cause failures of coordinating national environmental policies.

The contributions collected in this volume are concerned with environmental problems arising from political failure at the international level. Two of the contributions acknowledge the fact that political failure at the national level cannot be divorced from political failure on the international level and so explicitly adopt a political-economic viewpoint (see Chapters 4 and 10). The other contributions, in order to focus the analysis on the international aspect, abstract from serious malfunctioning of the political process in the national arenas. This choice is, however, clearly a matter of analytical convenience and not of principle. International environmental economics looks – by necessity – at the dark side of government intervention, and deals, therefore, with an inherently politicized subject matter.

Lest we be misinterpreted as denigrating the role of national governments in environmental protection, we emphasize here the double-edged mode of government action: government action is required to remedy market failures in environmental protection, but it comes at the cost of political failure. Denying the cost side of government action does not make environmental policy more satisfactory. The only consequence of this course is to estrange the expertise of the economics profession from actual environmental policymaking.

1.3. STRUCTURE OF THE BOOK

The division of the world into sovereign nation states brings about serious environmental problems arising from the friction of the individual national interests at the international level. The contributions collected in this volume survey this highly politicized subject matter from various viewpoints. The volume, in particular, integrates political-economic views with traditional normative analyses based on a social welfare perspective. Both approaches contribute to our understanding of international environmental economics; they complement each other in the sense that a realistic portrait of the political process needs to build on a thorough analysis of the economic relationships which the traditional models have focused on. The volume also provides a balanced account of the theoretical and empirical work on international environmental economics. An even-handed presentation of theory and empirical analysis allows to assess the relevance of the derived theoretical results on the one hand, and to provide guidelines for econometric modeling on the other hand. The volume is structured as follows.

The first three chapters present the basic relationships between international goods and factor trade and the environment and, consequently, also between the policies that influence the former and the latter. Chapter 2 by Günther Schulze and Heinrich Ursprung systematically surveys the sizeable welfare theoretical literature on trade, investment and the environment. In this chapter we seek in particular answers to the following questions: How does economic integration, that is, an increase in international trade, affect environmental quality, environmental policy, and welfare? To what extent is this relationship influenced by the size of the integrating countries and the prevailing trade pattern, considering that the trade pattern is determined by the countries' relative factor endowments which, in the case of the factor "environment," in turn, depends on the adopted environmental policy? How do the predictions change when pollution is transboundary rather than local and to what extent do the predictions depend on international differences in preferences for environmental quality? Conversely, since the endowment with "environment" is policy-dependent, how do environmental policies shape trade patterns and the international allocation of capital? In particular, is there scope for a competitive reduction of production standards in order to attract mobile capital and to gain a comparative advantage in pollution-intensive industries ("eco-dumping")? To what extent can environmental policy be substituted for trade policy in order to influence the terms of trade or to support domestic producers in imperfectly competitive markets ("strategic environmental policy")? These issues are addressed by classifying the numerous theoretical models in a systematic manner. We describe, evaluate and summarize the various results and demonstrate how these results follow from the adopted modeling approaches.

The studies surveyed in Chapter 3 confront the theoretical results elaborated upon in Chapter 2 with the available empirical evidence. After some methodological remarks on the measurement of pollution intensity and the stringency of environmental policy measures, we address the three most important empirical issues in this context. We begin by assessing the influence of environmental policy on the pattern of international trade. Subsequently we present the empirical evidence on the impact of the trade regime on environmental quality. The chapter is rounded off by a survey of the empirical literature investigating the question whether differences in environmental standards have actually led to the relocation of pollution-intensive industries to countries with more lenient standards ("pollution-haven hypothesis").

While the literature surveyed in Chapters 2 and 3 portrays governments as unitary actors maximizing national welfare, Chapter 4 presents the political economy approach to analyzing environmental and trade policy formation in an international context. This literature is still in its infancy and we provide the first survey of this exciting new research agenda. The interaction of environmental and international trade policies is at the heart of Chapter 4. Both policies are, of course, endogenous to the political process – and both policy fields are, in particular, strongly influenced by pressure group interests (namely the interests of environmental and protectionist lobbies). We first present the literature which assumes low international capital mobility and therefore focuses on commodity trade. One strand of this literature investigates to what extent environmental interests are liable to side with trade policy interest and whether environmental policies are abused for protectionist ends. A second strand of the literature investigates the endogenous reaction of environmental policy formation to trade liberalization. We then turn to the regime of high international capital mobility and survey the political-economic studies investigating the industrial-flight hypothesis, that is the question whether regulation induced relocation of industries represents, in actual fact, a serious threat to a globally efficient environmental policy. We conclude this chapter by presenting the scant empirical evidence relating to the political-economic view.

After having looked at the basic issues, Chapters 5–7 focus on three specific sectors which figure very prominently in the political debate about the nexus between international trade and the environment. Chapter 5 by David Ervin studies the link between international trade and agriculture which, by virtue of its large land and water requirements, usually affects a greater share of a nation's natural resources than all other sectors. The central question here is: How can trade expansion and environmental protection be managed such that the pursuit of one objective does not render the attainment of the other impossible? Ervin's analysis proceeds in four steps: after introducing the reader to the specific problems of the environment–trade nexus in the agricultural context, he analyzes the impact of the liberalization of trade in agricultural goods on environmental quality and subsequently the reverse relationship, that is, the impact of agri-environmental programs on international trade in agricultural goods. The survey covers theoretical as well as empirical studies and follows in many respects the structure of the more general Chapters 2 and 3. By focusing on the specific circumstances and institutions of the agricultural sector this chapter constitutes a natural and indispensable extension of the more general analyses presented in Chapters 2–4. Ervin concludes by drawing up a set of principles for designing agri-environmental programs and by summarizing his main findings.

Closely related to the agricultural sector is the forestry sector. The top issue in the forestry context is certainly the alarming tropical deforestation which is often caused not so much by international demand for forestry products but by expanding agriculture. The link between international trade and deforestation, in particular the agricultural conversion of forests, is the subject of Chapter 6 by Edward Barbier. Barbier's main focus is on the link between trade in forestry products and the sustainable management of forests. After providing an overview of world production, the supporting resource base and international trade in forestry products, he investigates market access, trade barriers and illegal logging for trade purposes. Special attention is given to the issue whether eco-labeling and certification represent suitable mechanisms to ensure that traded timber originates from sustainable sources. Implementation of public policies geared towards sustainable forestry and the question as to how such measures are to be financed (possibly through trade taxes) constitute the focus of the second part of his article.

In Chapter 7 Michael Rauscher deals with the topical issue of international trade in waste which represents one of the most obvious links between international trade and the environment. Like international trade in agricultural and forestry products, international trade in hazardous waste trade is a very important special topic of the environment–trade nexus. It deserves special attention since the general principles laid out in Chapters 2–4 need to be applied with explicit consideration of the particularities of hazardous goods. Unlike the two issues covered in the previous chapters, however, international trade in hazardous waste has not been given sufficient attention in the economic literature. This lack of interest is somewhat surprising given the frequent political upheaval which exports of hazardous waste have caused especially in developed countries. Michael Rauscher's survey is the first of this kind. He sets out by describing the institutional framework (the Basle Convention) and the volume and pattern of international trade in hazardous waste. He then proceeds to analyze the determinants of this trade and its welfare implications. Subsequently he elaborates on the optimal regulation of the waste sector in open economies and describes the effect of monopoly power on the part of the exporter of waste. Lastly he addresses the question whether restrictions on international trade in hazardous waste can be justified from a welfare maximizing perspective.

Sjak Smulders analyzes in Chapter 8 the interaction of trade and environmental policies in a second best world. While many of the theoretical papers surveyed in Chapter 2 and in later chapters assume the environmental externality to be the only distortion, it is clear that in actual fact economies are plagued by many distortions, among which tax-induced distortions feature prominently. This recognition has led to a substantial literature on the so-called "double dividend" of environmental taxation in tax-distorted economies (see Bovenberg 1999). Sjak Smulders applies this literature, which, up to now, has mainly focused on closed economy settings, to the international context. He addresses the question whether it is really possible to reap a double dividend by levying an environmental tax – the double dividend consisting of an improvement in environmental quality and an additional welfare gain associated with a revenue-neutral reduction in distorting tariffs. In the first part of the survey the double dividend literature applying to small open economies is reviewed, the second part then deals with large open economies. The last part of the survey focuses on the distribution of welfare gains from an environmental-for-labor tax swap in an interdependent multi-region world.

The notion of sustainable environmental quality does not play any explicit role in the chapters covered so far since all of them survey studies which are essentially conducted in a comparative-static mode. It is, however, extremely important to understand how the sustainability of environmental quality relates to the sustainability of economic growth and how this relationship is affected by the increasing globalization of the national economies. How can environmentally sustainable growth paths be reached and how are they characterized in an integrated world economy? These issues are dealt with in Chapter 9 by Lucas Bretschger and Hannes Egli. Although the relevance of these issues is obvious, only very few contributions which directly analyze environmentally sustainable growth in open economies exist so far. To be sure, a sizeable literature exists which deals with international trade and the environment (see Chapters 2–4), with economic growth and the environment in a closed economy context, and with international trade and economic growth; yet, the intersection of the three issues is almost empty although this intersection arguably represents the most relevant subset of the issues involved. This sorry state of affairs provides the structure for this survey (the first of this kind). First, Lucas Bretschger and Hannes Egli introduce the concept of sustainability. Afterwards it is explained how endogenous growth theory identifies sustainable development paths with the help of different modeling approaches which explicitly consider the influence of natural resources on (sustainable) growth (growth–environment nexus). The authors then discuss growth in open economies (i.e., the trade–growth nexus) and only then they bring the various strands together in a section on trade, growth and the environment. Finally, the relevant empirical evidence is presented.

The volume is rounded off by two chapters on international environmental agreements, the analysis of which is one of the most dynamically developing subfields of International Environmental Economics. Starting out from the recognition that many of today's most severe environmental problems cannot be solved by way of unilateral (or bilateral) national solutions, the literature has focused on the political and economic conditions for the feasibility and stability of voluntary multilateral environmental agreements. The importance of International Environmental Agreements (IEA) in the face of *global* environmental problems such as the depletion of the ozone layer (global warming), the depletion of water and energy resources, and the deforestation of the tropical rain forest are so blatantly obvious that the analysis of IEAs has become one of the best researched topics in international environmental economics.

In Chapter 10, Carsten Schmidt analyzes how international environmental agreements must be designed in order to provide the involved countries with sufficiently strong incentives to participate and to comply with the agreement. This issue of incentive compatibility is crucial since no supranational authority can coerce sovereign states to comply with the agreed upon restrictions. Insufficient incentives for voluntary participation and compliance emerge, for example, if the gains from the agreement are asymmetrically distributed or if free rider opportunities are present. In the first part of his survey Carsten Schmidt discusses how the choice and specific design of the instruments to internalize the environmental externality affects the incentives to participate in an international environmental agreement and to comply with its rules. In the second part he analyzes in a repeated game context how international environmental agreements can be made self-enforceable (internally stable) in the sense that a breach of the agreement by one country does not pay if she duly considers the reactions of the other countries. International environmental agreements can also be externally stabilized. Carsten Schmidt discusses in this context compensation schemes, issue linkage and trade sanctions. In the last part of his survey he describes how unilateral measures taken by individual nation states or subgroups of states influence the participation and compliance incentives of other countries and how long-term agreements can be flexibly designed to allow for changing circumstances.

Chapter 11 by Roger Congleton complements the previous analysis by introducing the political-economic dimension into the analysis of international environmental agreements. His model assumes realistic motives on the part of the main actors and drives home the point that the domestic political process is crucial for the understanding of the content of environmental treaties as well as the pattern of signatories. For instance, since international environmental agreements as a rule take effect only after a substantial delay, the formal agreement is often used by the negotiating politicians to signal future environmental policy shifts. Roger Congleton, therefore, distinguishes symbolic and substantive treaties and explains the observed pattern of international environmental agreements in three steps. First, he provides an overview of the environmental treaties signed during the last 50 years. He then explains why in the past nonsubstantive treaties have been signed more frequently than substantive ones. Finally he characterizes the type of nations most likely to sign international environmental agreements and reviews the empirical evidence on the pattern of agreed-upon rules and signatory states.

The theoretical, empirical and applied literature surveyed in this volume covers, indeed, a substantial amount of ground and constitutes a comprehensive outline of the state-of-the-art in international environmental economics. It is evident that progress in our attempts to protect the environment depends very much on our understanding of the link between pollution and the political fragmentation in a world undergoing a transition towards economic globalization. While the contributions collected in this volume clearly show that we are still far from a complete understanding of the intricacies of the relevant relationships, they also document that in the last decade substantial progress has been made. Environmental economics has finally become a dynamically developing discipline which squarely faces the fact that it deals with a highly politicized subject matter. This withdrawal from the ivory tower gives rise to the hope that in the future the economics profession will have a more significant impact on environmental policymaking than it has had to date.

2

International Trade, Investment, and the Environment: Theoretical Issues

GÜNTHER G. SCHULZE AND HEINRICH W. URSPRUNG

2.1. INTRODUCTION

This chapter addresses the current debate on the interdependence of global economic integration and environmental policymaking by critically surveying the economic literature on international trade, international factor movements, and the environment. We seek in particular answers to two sets of questions: First, how does global economic integration influence the environmental policy and the environment? Does trade liberalization and the abolition of capital controls give rise to a deterioration of the environment? To what extent does the answer to this question depend on the country's size and per capita income, and what is the influence of the prevailing trade pattern between the countries concerned? If it is the case that the environment really deteriorates in the course of intensified global competition, do the respective liberalization steps result in a *welfare* reduction? Is it possible to design mechanisms which resolve the prisoners' dilemma in which the governments are caught when deciding on environmental policy?

The second related set of questions asks to what extent environmental policy is likely to have an impact on economic integration: How are trade patterns, profits and capital allocations influenced by environmental policy? These questions are especially relevant in the context of global economic integration since governments tend to forgo the use of traditional trade policy instruments such as tariffs and quotas in the course of economic integration, and may therefore be tempted to resort to other available instruments – environmental policy being a prominent example. The main concern in this context is whether the debilitation of the coercive power of the nation states, which is incidental upon global economic integration, gives rise to a downward competition of environmental standards with the consequence of global environmental plight

We are grateful to Rolf Bommer, James Markusen, Michael Rauscher and Carsten Schmidt for helpful comments. The usual disclaimer applies.

(ecodumping). Countries may strategically lower environmental standards to gain an edge over their competitors or to attract foreign capital. Should then, as a consequence, environmental policy be harmonized?

All these questions can be answered from the traditional normative perspective by assuming a social welfare maximizing government, as well as from the political-economic viewpoint. The answers, of course, will depend on the adopted approach. The normative perspective, which we will adopt in this chapter, allows us to assess how a government would act if the government were not constrained by the political process and committed only to the well-being of the people it ruled. Under these circumstances the government would maximize a social welfare function (SWF). Underlying the SWF approach is, of course, the (unrealistic) assumption that individuals are either sufficiently homogeneous or that the gainers from a certain policy always compensate the losers, so that only the net effect needs to be considered. This approach, unrealistic as it may be, serves as a useful reference point and thus as a point of departure for more realistic political-economic analyses. It compels us to assess how a *given* (trade or environmental) policy affects the environment, the allocation of resources, and the pattern of trade. Without such a solid analysis it is impossible to identify the gainers and losers from the policies under consideration. In other words, we need to understand the economic consequences of different policies in order to identify the relevant political interests and to study how these interests pursue their goals under different political institutions. The distributional and – to a lesser extent – efficiency effects of a given range of environmental and trade policies determine the political stance of the voters and the relevant interest groups and thus the government's policy choice. The observed policies are then interpreted as the outcome of optimization processes in the economic and political sphere under the prevailing market configurations and institutional settings. The models surveyed in this chapter therefore provide the foundation not only for normative analyses of environmental and trade policies but also for the positive, political-economic approach to explaining the observed policies.

Up to now, most papers on the trade and environment nexus have been written from the traditional SWF perspective; the more recent political-economic approach on trade and environment is still in its infancy (see Chapter 4). We will provide an overview of the by now extensive body of normative literature by classifying the models in section 2.2. Then we will systematically survey the contributions; section 2.3 discusses the models, which do not portray factor mobility, and section 2.4 discusses the models in which factor movements play a prominent role. Section 2.5 summarizes the main results.

2.2. CLASSIFICATION OF MODELS

The normative literature on the interdependence of economic integration and the environment is fairly extensive and cannot be covered in its entirety in an article length survey. We therefore classify the literature and survey the most important contributions belonging to each class. The relevant models can be distinguished according to (i) the transmission channels through which one country's policy affect the other countries, (ii) the size, (iii) the market characteristics of the portrayed economies, and (iv) whether the models are partial or general equilibrium. Moreover, models differ with respect to the available policy instrument(s), that is, whether the government pursues some kind of environmental policy or trade policy (one of the two exogenously fixed), or both; and whether the environmental policy relies on taxes, quotas, or tradable permits. Lastly, the models may assume that the source of externality is in production or in consumption.

Let us now look at the distinguishing features in some more detail before we turn to the models themselves. We begin with the *international transmission channels* of national policies. Countries can be linked solely through commodity trade or through commodity and factor trade.[1] In the former case, trade policy will affect the sectoral production pattern as well as the consumption pattern and thereby also the level of pollution since pollution-intensities of production or consumption differ across sectors. Environmental policies have the same effect because they also change relative prices; in addition, environmental policies might have a direct effect on pollution levels since they restrict overall pollution (or pollution per unit of output). If also factors are mobile internationally, environmental and trade policies will not only impact on the intersectoral allocation of resources and thereby their remuneration (via the Stolper–Samuelson effect), they will also affect the international allocation of factors. The arbitrage condition of equal gross prices for goods is supplemented by a second arbitrage condition of equal gross prices for internationally mobile factors. The third transmission channel consists of transboundary pollution. If transboundary pollution is present environmental and trade policies affect the level of domestic pollution not only via their effect on domestic production but also via their effect on foreign production.

Another criterion is the *country size*; small open economies take international factor and commodity prices as given, whereas large countries influence world prices through their trade and environmental policies (optimal tariff argument and its analog for environmental regulation). While trade models are typically *general equilibrium* models, which take the interaction of factor and goods markets and especially income effects into account, there are also *partial equilibrium* approaches to analyse trade and environment. Partial equilibrium models which neglect spillover effects to other markets and the endogeneity of all prices might be justified if the market is small; this kind of analysis often

---

[1] In this survey we deliberately ignore models in which goods or factor trade is intersectoral, but not transboundary, such as Forster (1977) and Yohe (1979). These models deserve, however, credit for providing the building blocs for the models portraying international trade, factor movements, and the environment.

allows us to derive sharper results than those generated in a general equilibrium framework. Models can also be distinguished according to their market characteristics: perfect competition yields different results than imperfect competition. In a general equilibrium framework, monopolistic competition models for example can portray the role of scale economies and product diversity. Strategic trade policy considerations are typically modeled in a duopolistic partial equilibrium setup, where governments try to shift profits to "their" domestic firm with the help of policy intervention. Monopolistic competition and duopoly models have both been applied to environmental policy. The majority of models, however, still assumes perfect competition.[2] We begin our survey with the models, which assume that goods are mobile, but factors are not (section 2.3). The second part of the survey then deals with models assuming factor mobility (section 2.4). In each part, we start with partial equilibrium analyses and subsequently investigate how the results are altered when general equilibrium feedback effects are taken into account. Also we will differentiate between small country and large country models and discuss the few imperfect competition models separately. Moreover, we distinguish a special class of general equilibrium models which deals with pollution in a North–South context by explicitly assuming a special kind of asymmetry of the countries involved. Unless indicated otherwise, pollution is assumed to be caused in production. For each model it will become clear what the available policy instruments are. A taxonomy of the models surveyed in this chapter is presented in table 2.1.

2.3. COMMODITY TRADE AND ENVIRONMENT

2.3.1. Welfare Implications of Trade Liberalization when Environmental Policy is Exogenous

We first analyze the influence of trade policy, notably trade liberalization, on welfare in the presence of an environmental externality when environmental policy is exogenous. Pethig (1976) is a good point of departure. He analyzes a negative externality that occurs in production and does not generate any transboundary effects. These are standard assumptions which are shared by the other papers reported below unless indicated otherwise. Pethig shows in a simple and rather specialized general equilibrium model that traditional propositions of international trade theory, *viz.* the law of comparative advantage and the factor price equalization theorem, continue to hold when environmental externalities are present. The endowment with the factor "environment"

---

[2] Another possible classification distinguishes between models in which the externality occurs in consumption or in production (see table 2.2), and between models in which the externality affects only individual utility oral so productivity (e.g. the tourism industry will be less productive if pollution is high). Unless indicated otherwise, we will assume that the externality occurs in production and that it will affect only individual utility.

Table 2.1. *The traditional literature on trade and environment*

		Trade models		Factor mobility (and trade)	
Partial equilibrium models	SMOPEC	Anderson (1992)			
	LOPEC	Krutilla (1991) Anderson (1992) Ludema & Wooton (1994)			
	Imperfect Comp.	Barrett (1994) Conrad (1993, 1995, 1999) Kennedy (1994)[a] Ulph (1994, 1996, 1997)	Imperfect comp.	Rauscher (1995) Hoel (1997) Ulph & Valentini (1997)	
General equilibrium models	SMOPEC	Markusen (1975a,b) McGuire (1982) Ulph (1994)[a] Copeland (1994) Schweinberger (1997)	SMOPEC	Oates & Schwab (1988) Rauscher (1993) Copeland (1994)	
	LOPEC	Pethig (1976) Siebert (1977) Asako (1979) McGuire (1982) Rauscher (1991a, 1994) Ulph (1994)[a]	LOPEC	Long & Siebert (1991) McGuire (1982) Merrifield (1988)[a] Rauscher (1991b,[a] 1992,[a] 1993[a])	
	Imperfect Comp.	Gürtzgen & Rauscher (1999) Haupt (1999)	Imperfect comp.	Markusen, Morey, Olewiler (1993, 1995) Pflüger (1996) Markusen (1997)	
	North–South	Chichilnisky (1994) Copeland & Taylor (1994, 1995,[a] 1997)	North–South	Chichilnisky & Di Mattheo (1996) Copeland & Taylor (1997)	

[a] Includes (also) transboundary pollution, otherwise only local pollution.

Notes: LOPEC: large open economy; SMOPEC: small open economy; North–South models all assume perfect competition.

depends on environmental regulation; the endowment ratios are therefore policy-induced and so is the pattern of trade. Yet, it is no longer certain that both countries will gain from free trade. The country that exports the pollution-intensive good will suffer from higher pollution after opening up to trade, which might offset the traditional gains from trade. The country that exports the cleaner good on the other hand, will unambiguously be better off. These results, however, depend on the assumption that no environmental policy is in place. In general, with optimally adjusted environmental policies, both countries will gain from free trade as we will see below. Asako (1979) obtains a similar result: a small departure from autarky will make the exporter of the pollution-intensive good worse off; for supermarginal deviations from autarky the net effect of higher pollution and higher consumption is ambiguous.[3]

Anderson (1992) corroborates these results using the traditional graphical analysis in a partial equilibrium setting. A small open economy with no environmental policy gains from trade liberalization if liberalization increases the *imports* of the pollution-intensive commodity. Not only does the country realize the traditional gains from trade (i.e., with a world price below the autarky price, the liberalization induced increase in consumer rent exceeds the reduction in producer rent); the overall gain from trade liberalization, which includes the welfare effects of the externality, is larger than in the absence of the externality because the externality is negative and it is reduced. The reason for this is that with no environmental policy for correcting the externality, the marginal social costs of pollution exceed the marginal private costs. Thus, reducing domestic production and increasing imports yields an additional gain amounting to this marginal cost difference integrated over the reduction in output. Conversely, if the world market price for the dirty good is larger than the autarky price, increased production for exports raises the pollution level, which runs counter the traditional gains from trade: the net effect can have either sign. This result is formally proven in a short appendix at the end of this chapter.

Matters are different if the externality occurs in consumption. Trade liberalization makes imports cheaper and thus increases under normal circumstances consumption. Thus, if the import good causes pollution, trade liberalization will reduce environmental quality. If the consumption of the export good causes pollution, environmental quality is only reduced if consumption of the export good increases in the course of the trade liberalization. This is the case if the income effect deriving from gains from trade more than offsets the relative price effect from trade liberalization. The effects are summarized in table 2.2.

---

[3] Siebert (1977) obtains a very similar set of results. For the sake of brevity we refrain from reporting details of his setup.

Table 2.2. *The effect of trade liberalization in the absence of environmental policy (or environmental policy adjustment)*

	Pollution occurs in production	Pollution occurs in consumption
The polluting good is the export good	Free trade increases pollution	Free trade may increase pollution if income effect is dominant
The polluting good is the import good	Free trade decreases pollution	Free trade increases pollution

2.3.2. Partial Equilibrium Analyses

Let us now assume that an optimal environmental policy is in place in this case.[4] Then the trade liberalization will unambiguously enhance welfare also in the case of pollution-intensive exports.[5] A production tax internalizes the negative externality and limits the increase in production after trade has been liberalized. Pollution increases, and so does consumption; the changes are such that the social benefit and the social costs from exports are equalized at the margin. Anderson (1992) goes on to show that trade taxes could also be used to reduce environmental degradation, but that this instrument would be inefficient. The first best solution is of course a tax on the activity that causes the negative externality.[6]

The results carry over to the large country case, altered only by a terms-of-trade effect. If a large country opens its markets and thereby increases imports, welfare increases for the reasons stated above. Yet the welfare increase is smaller than in the small country case, because the world market price rises due to the additional demand for the imported good. This price increase reduces the consumer surplus. Liberalizing trade if the exports are pollution-intensive is still beneficial if an optimal environmental policy (i.e., a production tax on dirty exportables) is applied. But the additional supply on world markets reduces the world market price and thereby the increase in national welfare. Anderson notes that if pollution is transboundary the welfare gain may even be larger than in the small country case because production and thus pollution abroad is reduced. This additional gain could overcompensate the negative terms of trade effect. Whether spillovers are present or not, domestic production and pollution rise, but the increased pollution is more than offset by increased consumption possibilities due to the optimal environmental taxation. In the large country case, the taxation of polluting exportables not only

[4] In contrast to the previous subsection we assume the policy can be endogenously adjusted to changing market situations as trade policy changes.

[5] Recall that if not indicated otherwise, pollution is assumed to occur in production.

[6] To be even more precise, the first best solution is a tax on the emission itself, the second best is a tax on pollution-intensive production, and trade taxes are at most third best.

internalizes the negative externality, it also improves the terms of trade as it curbs production.

The interdependence of terms of trade and environmental considerations is also the focus of Krutilla (1991) who models a large open economy in a partial equilibrium setting. It is clear that a first-best policy consists of a standard Pigouvian tax and the standard optimal tariff (see Markusen 1975a).[7] If however the government is circumscribed in the use of tariffs, because, for example, it is limited to zero tariffs by a free trade agreement, the optimal tax will deviate from the Pigouvian tax. In the case of a production externality, the second best tax on production will exceed (fall short of) the Pigouvian tax, if the country is a net exporter (importer) of the respective good. If the production of the export good is taxed, the tax reduces export supply and gives rise to a monopolistic surplus in addition to the environmental benefit caused by lower production. A production tax in the import-competing sector not only reduces domestic pollution, but also increases world demand and therefore deteriorates the terms of trade. The optimal tax is thus lower than the Pigouvian tax; it is set such that the marginal loss of additional environmental damage is offset by the marginal benefit of reduced import costs. If the negative externality occurs in consumption, the optimal tax is smaller than the Pigouvian tax if the country exports and higher if the country imports the polluting good. The reasoning is the same as before, only the tax has the opposite impact: a tax on consumption reduces the demand for imports and improves the terms of trade, but it increases the export supply, which deteriorates the terms of trade. If the tariff is positive, but cannot be changed, the deviation from the Pigouvian tax is smaller in all cases since a reduction in international trade reduces the tariff revenue.

In a rather special partial equilibrium model, Ludema and Wooton (1994) take one step further by incorporating strategic interaction between the countries. The polluting good is only consumed at home, but production takes place in both countries. The externality is via consumption of the foreign, that is, imported, good only. Perfect competition prevails; Home levies an import tariff and Foreign an export tax. Welfare is defined as the sum of consumer and producer surplus plus tariff revenue less the externality for the home country and the sum of tax revenue and producer surplus for the foreign country. In the Nash equilibrium the sum of the trade taxes exceeds the world-optimal level (which would exactly internalize the externality) for the standard terms-of-trade reasons. The externality gives Home a reason to increase its tariff beyond the optimal level; as a consequence, Foreign reduces its export tax and is worse off. If abatement is possible at the production stage and Foreign can tax production of the polluting good, it will adopt a combination of export and pollution taxes. Abatement is costly for the foreigners who do not suffer from the externality, but abatement reduces the level of the domestic tariff and thereby

---

[7] This argument, of course, disregards retaliation, which might (but need not) annihilate the gains from setting optimal tariffs (see Johnson 1953).

increases foreign welfare. Convex abatement costs produce an inner solution. If trade taxes are restricted to zero, as would be the case in a free trade agreement, Foreign levies a production tax to improve its terms of trade, thereby improving the wasteful abatement costs. The results are a straightforward extension of Markusen (1975a) and Krutilla (1991) and nicely demonstrate the principles at work; yet the special setup seems very restrictive. It is hard to justify that the externality should result only from domestic consumption of imports. If the externality occurred, also by consuming the domestically produced good, the optimal policy would be a consumption tax cum import tariff (see Krutilla 1991). The results are analogous for production externalities.

2.3.3. General Equilibrium Analyses

The above results were derived in partial equilibrium settings, which do not account for the feedback effects in other sectors. In particular, they disregard changes in factor allocation and factor remuneration throughout the economy, as well as changes in production and trade patterns as a consequence of environmental control. We now turn to general equilibrium models, which do not suffer from this drawback.

Markusen (1975b) is an early and seminal paper for international environmental economics: many results which have been established in various model setups later on, can already be found in a precursory version in this paper. Markusen models two countries producing many goods, one of which causes pollution via production. The portrayed pollution is a pure public "bad" for both countries, regardless of the source of pollution. Countries may tax (or subsidize) production and trade. Markusen shows that for a small open country the noncooperative national optimum is achieved through free trade and a specific production tax in the polluting industry which is equal to the marginal damage in that country caused by domestic production. Clearly, this optimal production tax depends negatively on the other country's production tax since pollution is assumed to be transboundary. A marginal increase in the production of the polluting good in one country will reduce welfare in the other country since it increases pollution. Moreover, it reduces this good's production in the other country. Markusen shows that the resulting Cournot equilibrium is not a Pareto optimum from the perspective of the world as a whole, the reason being that each country takes only the marginal damage in its own territory into consideration. In the international optimum, each country's production tax equals the sum of the marginal damages resulting from domestic production in *both* countries. The Pareto optimum may or may not be reached by way of *simple cooperation*: simple cooperation implies that the countries raise their production taxes above the level set in the noncooperative regime up to the point where the marginal cost from doing so (curtailing production beyond the noncooperative values) equals the marginal benefit received in exchange, from reduced production of the polluting good abroad.

This may not be identical to the *joint maximization of welfare*, where the domestic marginal cost of domestic production equals the marginal cost of pollution imposed on the foreign country and thus does not necessarily equal the marginal benefit received from the foreign country.[8] Transfer payments, however, will ensure that the cooperative solution will also be Pareto-optimal. Markusen also derives optimal noncooperative policies for large open economies with market power on commodity markets. These policies consist of free trade, optimum production taxes and transfer payments; the qualitative results therefore remain unchanged.

McGuire (1982) studies the effects of environmental regulation in a two sector Heckscher–Ohlin model. Both commodities are produced with capital and labor; one sector is polluting and uses the environment as the third factor of production. One way of modeling environmental control is to levy a tax on the use of the environment, which raises the price of the environment and thereby its use. Another alternative is to issue tradable pollution permits which can be auctioned off. Optimality requires that the value marginal product of the environment is equal to its (shadow) price. McGuire shows that for linear homogeneous production functions with equal pairwise elasticities of substitution σ_{KE}, σ_{LE}, increased regulation is equivalent to negative neutral technical progress. Factor prices and factor allocation change if environmental control becomes stricter. In a small open economy, the factor used intensively in the regulated (polluting) sector will be worse off while the other factor unambiguously gains. To see this, assume that the polluting sector is relatively labor-intensive. Reduced environmental input caused by stricter control drives the marginal productivities of labor and capital in this sector down (given the assumption of equal and positive elasticities of substitution for all factors). This leads to an outflow of capital and labor into the other sector until the marginal productivities are equalized again; the production of the polluting sector shrinks, the other sector's production increases and both production processes have become more labor-intensive. Labor's (capital's) marginal physical product declines (rises), and so do their prices since commodity prices are given for a small open economy. If environmental regulation differs across countries, factor price equalization no longer holds and factors are used in different proportions across countries in the production of a given commodity. In the case of a large open economy tightening environmental control leads to an increase in the world price for the polluting commodity and thereby benefits the factor used intensively in the production of the polluting good in the rest of the world (Stolper–Samuelson effect).

Rauscher (1991a) analyzes optimal environmental policy in a simple Heckscher–Ohlin model, where one sector produces with capital only and the

---

[8] This is most obvious in the case where one country specializes in the production of the polluting good. The other country cannot offer the first (polluting) country benefits in terms of pollution reduction and hence there is no scope for cooperation.

other sector uses capital and the environment as inputs; the welfare function is additively separable in consumption and environmental quality. He shows that it is optimal for a small open economy to increase environmental standards (reduce emissions) in the course of trade liberalization if the country exports the nonpolluting commodity, because the domestic demand for the dirty good can also be satisfied by imports. The result is reverse if the country exports the pollution-intensive good.[9] For a large open economy, Rauscher (1991a) shows how environmental policy affects the terms of trade. He corroborates Anderson's and Krutilla's finding in a general equilibrium setting. If the home country exports the clean good, it will increase its emissions in order to reduce its imports and thereby raise the relative price of its exports. Conversely, the exporter of the dirty good will reduce its emissions to improve its terms of trade. These results assume that the foreign country leaves its emission level constant. This will not be the case, however, if the foreign country levies a (constant) emission tax. If the home country exports the dirty good and reduces its emissions to improve its terms of trade, the other country will shift capital from the clean sector into the dirty sector, as this production has become more profitable. This increases the marginal productivity of the use of the environment and, therefore, with a constant emission tax, increases emissions in the foreign country. With pollution being transboundary it is hence possible to obtain the perverse result that pollution is increased in the home country although it has reduced its own emissions. In that case, the second-best tax on the production of exportables must be lower than in the case of strictly local pollution.[10] Note, that this model still does not portray strategic interaction of the two countries since the policy of the foreign country remains unaltered.

Rauscher (1994) extends his previous results by introducing a third, non-tradable good. He assumes that the two countries are completely specialized in the production of one of the two tradables and produce in addition a non-tradable good. Both sectors use the environment and capital as factor inputs; the social welfare function is additively separable in each of the three consumption goods and in environmental quality; the model is of the Heckscher–Ohlin type. He shows that a small country will apply the same environmental

---

[9] Bommer and Schulze (1999) show that the result is turned upside down, if the political sector is taken into consideration. Their result is driven by the government's balancing of opposing interests; in Rauscher's setup – and this holds for all the papers surveyed in this section – heterogeneity of people and thus distributional conflicts are neglected. We will take up this issue in Chapter 4 when we discuss political-economic models.

[10] Also the first best policy (from a national perspective) has to account for transboundary pollution. In the case in which the foreign country sets emission taxes, the Pigouvian tax remains the same, but the optimal tariff is altered. If higher world prices increase foreign emissions, it is optimal to impose tariffs, which are higher than the standard optimal tariff on pollution-intensive imports. These taxes drive world prices even further down and reduce foreign emissions. Conversely, if the export good is environment-intensive, the export taxes should be lower than in the case of strictly local pollution. See Ulph (1994) on transboundary pollution; Ulph also analysis emission *standards*, which yield different results.

policy in both sectors. Moreover, Rauscher modifies in his model the general finding that a large economy should increase the price of the factor it is relatively well endowed with in order to improve its terms of trade. In the context of environmental policy this means for instance that the exporter of dirty goods should tighten its environmental standards beyond the Pigouvian level. In his model, a reduction of the environmental input throughout the economy diminishes the marginal product of capital in both sectors. It is however unclear in which sector this is more pronounced. If the return to capital is reduced more in the nontradables sector, capital will move out to the tradables sector, which might (but need not) increase the production of exportables and thereby deteriorate the terms of trade. This effect might call for sector-specific environmental standards. Under additional, reasonable assumptions, Rauscher shows that a reduction of emissions in the tradables sector reduces output of tradables and improves the terms of trade. Letting emissions increase in the nontradables sector yields an ambiguous effect: this policy raises the marginal product of capital, and thereby attracts capital from the tradables sector, as a result the terms of trade tend to improve. On the other hand, the increase of emissions boosts output and therefore reduces the price of nontradables. This price reduction tends to decrease the price of capital in this sector and thus drives capital out to the tradables sector, which deteriorates the terms of trade. Policy recommendations for this sector thus depend on the parameter values. Note, however, that these results are driven by (i) the assumed complete specialization and (ii) the fact that both sectors use the environment as an input.

Copeland (1994) extends the analysis to many sectors and many trade and pollution distortions and studies the welfare effects of piecemeal trade and environmental policy reforms. With many distortions present, the theory of the second best tells us that it is impossible to establish general propositions regarding welfare improving reforms. Copeland emphasizes the importance to consider the interdependence of the two policy fields and hence the need to coordinate policy reforms. For special cases he derives welfare improving reform schemes: if strongly polluting industries are also subject to substantial trade protection, an equiproportionate reduction in either trade barriers or pollution distortions will, in general (but not necessarily), improve welfare. Important for the analysis is the identification of spillover effects of policy reforms – the reduction of one distortion may exacerbate (or reduce) another distortion. In that respect, quotas may be preferable to taxes as they tend to annihilate (undesired) spillovers. In a sense, Copeland's paper represents a consequent generalization of the low dimension analyses reported above, and he incurs all the indeterminancies known, for example, from high dimensional issues in international trade theory; it is an application of general models of policy reforms in open economies to the case of trade and environmental policy.[11]

---

[11] For higher dimensional issues in international trade theory see e.g. Ethier (1984), for the analysis of piecemeal policy reforms see, *inter alia*, Vousden (1990) or Dixit (1985).

Schweinberger (1997) takes Copeland's analysis one step further. He analyzes a multihousehold economy that taxes emissions, provides public goods and opens up the economy to free trade in private goods. For this economy he derives necessary and sufficient conditions for gains from trade; these conditions are very general since he allows for (i) under or overproduction of public goods, (ii) individual evaluations of the environment, and (iii) different income levels. To derive the necessary and sufficient conditions he assumes that lump-sum transfers to households and lump-sum taxes are possible. He shows that a simple policy rule that keeps the level of production of public and polluting private goods fixed is sufficient to guarantee the optimality of free trade in an economy with multilateral nondepletable externalities.

2.3.4. North–South Models

In the Heckscher–Ohlin framework, trade patterns are determined by relative factor endowments which are assumed to be exogenous; the endowment with "environment" as a factor of production, however, is not exogenously given. To be sure, a country with huge wooded areas is better able to absorb CO_2 emissions than other countries; but the amount of environmental inputs available to the production process (i.e., the permitted quantity of emissions) is predominantly determined by environmental regulation. Regulation is of course the outcome of a political process, which needs to be studied from a political-economic perspective, but within the fictitious framework of a government maximizing a social welfare function this policy reflects the preferences of the representative individual. This is the policy perspective underlying the model by Copeland and Taylor (1994). They assume two regions, North and South, which are identical save for the level of the human capital. Both regions produce a continuum of goods with the help of labor, measured in efficiency units, and the environmental pollution is local. The difference in per capita income levels is the driving force of their model: because environmental quality is a superior good, the richer North imposes stricter pollution taxes than the South to offset the higher marginal damage. In other words, emission control is optimally provided for under autarky and free trade. Trade liberalization has three effects on environmental quality: the scale effect of increased economic activity raises pollution; the technique effect describes the switch to cleaner production processes in the course of rising income levels. Lastly, the composition effect reflects the pattern of international specialization. Because the relative price of pollution-intensive goods is higher in the North, the South specializes in the production of dirty goods. Under the model's assumptions, pollution in the North decreases, but increases in the South and world pollution rises in the course of trade liberalization. Since pollution taxes are set optimally, both regions gain from trade.

In a companion paper, Copeland and Taylor (1995) investigate the case of transboundary pollution. Under these circumstances uncoordinated national environmental policies cannot eliminate market failures and, consequently, gains from trade need not arise. As before, free trade and a sufficiently large difference in human-capital levels give rise to a concentration of pollution-intensive industries in the South which has laxer environmental standards, and increased global pollution. With transboundary pollution the population of the North is also exposed to the increase in pollution, but does not have the leverage to efficiently reduce pollution with the help of Pigouvian taxes. In the resulting noncooperative Nash equilibrium the North loses from trade while the South gains. This is because the South can commit to lower environmental standards (due to lower per capita incomes); this leaves the North with the option to accept higher pollution or to cut back on its own pollution. In Copeland and Taylor's (1995) model, the North exactly offsets the increase in emissions originating from the South if factor prices are equalized, so that global pollution remains constant.[12] The reduction of emissions is equivalent to a reduction in factor endowment and therefore shifts the budget constraint inwards. The result that the North unambiguously loses from trade depends on the specifics of the model; the basic switch in perspective however, remains valid in a much wider class of models. With only local pollution and non-Pigouvian pollution taxes, the exporter of the pollution-intensive good is potentially worse off (because pollution increases). In the case of trans-boundary pollution and different preferences across countries, the finding is reversed. The exporter of the cleaner goods suffers from increased global pollution. If this effect is strong enough to offset the gains from trade in terms of increased consumption, the country's welfare is diminished. The rationale behind this result is that a region cannot unilaterally internalize the global externality and that the lower preference for environmental quality gives the high-pollution country a strategic advantage since it can commit to higher emissions. That makes cooperation difficult. Copeland and Taylor demonstrate that international trade in pollution permits would curb global pollution since it equalizes the price of the environment and eliminates the pollution-haven effect. A globally efficient solution, of course, requires that each country sets the emission tax equal to the global marginal damage of their emissions. To obtain such a solution, side payments may be needed (see Markusen 1975b).

While Copeland and Taylor (1994, 1995) focus on income differences, and thus differences in environmental regulation, as the sole reason for trade, they also include differences in relative factor endowments as determinants of trade and pollution patterns in their 1997 paper. While the North is richer and therefore tends to have a more stringent regulation, the (richer) Northern

---

[12] If factor prices are not equalized, because income levels and therefore endowments differ too much, global pollution will rise. Copeland and Taylor (1995) produce a number of other interesting results. They show for instance that transfers from the North to the South reduce pollution in the South and raise pollution in the North. Transfers may increase North's welfare.

countries are also more capital abundant. If pollution rises positively with capital intensity of production, the resulting trade pattern depends on the relative strength of the two opposing effects. If income differences are small and thus – given equal national utility functions – environmental taxes are similar, trade patterns are determined by differences in relative factor endowments: the more capital abundant North exports the pollution intensive capital intensive good. Conversely, if relative endowment differences are small but differences in income levels are not, the North will specialize in the production of the clean good.[13]

Copeland and Taylor portray the interaction of the two effects in a specific factors model with labor and capital being the specific factors and the environment being the mobile factor; pollution is assumed to be confined within national boundaries. As in their previous papers, pollution is endogenously determined via maximization of the representative agent's utility. They show that if the endowment effect dominates the income effect, world pollution falls by shifting to free trade: the North specializes in the dirty good but applies stricter regulation than the South. If the income effect dominates, world pollution increases; these correspond to the result of their 1994 paper.

Chichilnisky (1994) focuses on a different reason for North–South trade in environment-intensive goods; she shows that ill defined property rights in the South create trade between otherwise identical regions. In a two region model, two tradable goods are produced by two factors, capital and environment. Both countries are identical in tastes, technologies and endowments, except for the fact that property rights of the global common, the environment, are poorly defined in the South, but well defined and protected in the North. This leads to an overuse of the environmental resources in the South since these resources are underpriced. International trade exacerbates the problem: trade equalizes factor and goods prices and thereby exports the misallocation to the North. The South overproduces the environment-intensive good, and the world overconsumes it. Apparent comparative advantages need no longer reflect actual comparative advantages.[14]

---

[13] There is a small inconsistency in the model: affluence should be defined in per capita income levels since the individual income of the representative agent determines his preference for environmental quality. With only two primary factors of production, capital and labor, however, differences in relative endowments are proportional to differences in per capita income levels. It is thus not possible that countries are similar in endowment ratios but unlike in per capita income levels. This inconsistency can easily be removed if labor is measured in efficiency units and labor exhibits different efficiency levels across countries. Under these circumstances different human capital levels contribute to income differences across countries as in Copeland and Taylor (1994, 1995).

[14] In this model of identical countries, no trade would occur if property rights did not establish an incentive for trade. In an extension of the Chichilnisky (1994) model, Chichilnisky and Di Matteo (1996) show that South–North migration triggered by wage differentials decreases pollution in the South and therefore increases welfare in the South, but may decrease welfare in the North.

2.3.5. Imperfect Competition

Up to now we have discussed only papers which analyze environmental policy in a perfectly competitive framework. Yet many industries are characterized by scale economies, which give rise to *imperfectly competitive markets* and international intra-industry trade. This observation raises the question as to how environmental policy is conducted when market power and thus profits are an issue. Under these circumstances governments have an incentive to behave strategically in order to shift rents from foreign to domestic producers; they do this by committing to a certain environmental policy prior to firms' production decisions of the firms, with the intention of giving the domestic firms a competitive edge over their foreign competitors. The literature on strategic environmental policy closely follows the strategic trade literature pioneered by Spencer and Brander (1983) and Brander and Spencer (1985). Examples are Barrett (1994b), Conrad (1993, 1996), Kennedy (1994), and Ulph (1996).[15] The basic setup is this: in a partial equilibrium model, two firms in two different countries produce exclusively for a third country market. The production of the good generates strictly local pollution, which the firm can abate at convex abatement costs. The remaining emissions produce a social damage, which is convex in the emissions. The government maximizes social welfare, that is, the difference of net profits of the domestic firm (total revenue minus production and abatement costs) and the social damage. In a two stage game, the two governments move first by setting environmental standards or emission taxes,[16] then the two firms seek to maximize profits in a standard Cournot-Nash game.

Strategic environmental policy is, of course, only second best; if the governments can use standard trade policy instruments, like production subsidies, there is no need to distort environmental policies (Barrett 1994b; Conrad 1993). Within the second best framework, the results are as follows (see Barrett 1994b). If producers compete in quantities (Cournot competition), the governments will set the emission standards too low as compared to the environmental optimum.[17] The lax domestic standard shifts the domestic producer's reaction curve outwards and *ceteris paribus* increases his output. If the foreign government did not react, this policy would shift profits to the domestic producer (the reason being that the government can commit the producer to a higher output which results in a reduction of foreign output). In

---

[15] Since strategic trade models are well known, we restrict the exposition to a necessary minimum. Although this branch of the literature is relatively new, there are more contributions than we can survey. For a survey see Ulph (1994).

[16] Most papers analyze either emission standards (Barrett 1994b) or emission taxes (Conrad 1993, Kennedy 1994); Ulph (1996) introduces a third stage where governments choose the instrument prior to choosing the level at which it is adopted.

[17] Environmentally emission standards are optimal if the ensuing marginal damage of pollution equals the marginal abatement costs.

the Nash equilibrium, however, both countries set respective standards too low, which increases emissions and total output, and hence, reduces both firms' profits as well as social welfare. This finding is in stark contrast to the case of a large country exporting pollution-intensive goods produced under perfect competition. In that framework, the large country has an incentive to tighten the environmental standards beyond the Pigouvian level, whereas here, the incentives work in the opposite direction.

The standard Cournot competition of the duopolists describes a scenario of "ecological dumping"; this result however is not robust. First, if the domestic industry is oligopolistic (instead of consisting of a single firm), the strategically optimal environmental policy might be tighter or laxer than the environmentally optimal policy. The reason for this ambiguity is that while a laxer emission standard shifts profits from the foreign producers to the domestic industry, a stricter standard reduces domestic output which is already too high – the domestic firms would be better off if they colluded and reduced their output. A strict environmental policy thus serves as an imperfect substitute for collusion. Second, if firms compete in prices and not in quantities (Bertrand competition) the results are reversed. Governments then have an incentive to set environmental standards which are stricter than the Pigouvian standards. This is not surprising since this kind of policy reversal is well known in the strategic trade literature (see Eaton and Grossman 1986).[18] Ulph (1994) and Conrad (1996) show that the results derived for emission standards carry over to the case of emission taxes. Moreover, Conrad (1999) shows that it makes a significant difference whether, in the usual two stage game, the government moves first by setting taxes and the firms react by determining their output, or whether the firms anticipate the environmental regulation and commit themselves to an output level before the government introduces the tax. All of these nonrobustness results, of course cast serious doubt on the usefulness of the policy recommendations derived by the literature on strategic environmental policy.[19]

---

[18] See also Conrad (2000) for a very insightful analysis on the lacking robustness of results in strategic trade models.

[19] The basic strategic environmental policy model has been extended in many other ways. Introducing domestic consumers, for example, tends to further lower the optimal standard since domestic output under imperfect competition is too small to begin with. In this version, social welfare also includes the domestic consumer rent. Of course, lax environmental standards are again only a second-best policy to increase production. Moreover, if pollution is transboundary, the rent shifting incentive is reinforced (Kennedy 1994). With local pollution the shifting of profits and thus production to the home country comes at the social cost of higher pollution. This is not (as much) the case with transboundary pollution since the increase of domestic pollution is (partly) offset by the reduction in pollution spillovers from abroad. Ulph (1997) studies the role of a supranational agency which is to prevent strategic ecodumping. The agency is assumed to be less well informed about national environmental damage costs than the respective national governments. He shows that neither a harmonization of standards nor minimum standards need to represent a Pareto-improvement if environmental damage costs differ across countries.

Haupt (1999) also assumes imperfectly competitive markets, but follows an entirely different approach pioneered by Krugman (1980):[20] he models two countries producing differentiated goods under increasing returns to scale with labor being the only factor of production. This setup results in monopolistic competition, and free entry guarantees zero profits. Profit shifting is therefore not an issue. Pollution occurs in *consumption* which implies that the decision whether to serve both markets or only the domestic market depends on the (difference in the) product standards set in the two countries. Environmental standards are assumed to affect the fixed costs of production, standards therefore have a direct impact on the number of varieties produced. The novelty of this approach is that only a fraction of firms in the low standard country might adopt the higher production standard of the foreign country and thus carry on international trade. The number and the size of the domestic firms thus depend on the domestic and foreign product standards. In this setup, environmental policies have uncommon consequences: stricter product standards abroad increase welfare if the home country has very low standards but reduce welfare if the home country's standards are high.

2.4. FACTOR MOBILITY

We have seen that in a regime of free trade countries specialize in the production of environmental-intensive or -extensive goods according to their respective relative endowments. Given optimal environmental policies, free trade will benefit all countries, including the exporter of the pollution-intensive good. We now introduce international factor mobility. This raises at least two issues: first, if commodity trade and factor mobility are substitutes, differences in relative factor endowments could give rise to factor movements instead of trade. This is particularly important if trade alone cannot bring about factor price equalization (FPE). Yet, even if factor endowments are not too different, factor prices will hardly be equalized through trade, because of the domestic "distortions" created by environmental policies. Optimal environmental policies (emission standards or taxes) typically differ between countries since pollution levels differ.[21] This is a direct consequence of free trade and the resulting specialization of the production structure. Unequal regulation causes differences in factor prices which would otherwise have been equalized through

---

[20] Gürtzgen and Rauscher (1999) also use a Dixit–Stiglitz type model of monopolistic competition to show how environmental policy affects the market structure at home and abroad and thereby emissions in both countries. These authors show that tighter environmental standards at home may reduce emissions abroad. This contradicts the conventional wisdom which presupposes a positive "leakage effect" (stricter environmental policy at home leads to more pollution abroad, and if pollution is partly transboundary, to a moderating or even perverse effect on pollution at home).

[21] Another reason for different levels of regulation is of course diverging national preferences with respect to environmental quality, see for instance Copeland and Taylor (1995).

trade (McGuire 1982). If however, optimal environmental policies trigger factor movements, is free trade still advantageous for all participating countries? Or, more fundamentally, how does environmental policy affect the allocation of capital and how does, as a consequence, increased factor mobility feed back to a (second-best) environmental policy?

There are basically three approaches to the analysis of international factor mobility and the environment. The first approach focuses on factor mobility and uses the traditional MacDougall–Kemp model. In this *single-good* general equilibrium model trade occurs only to pay interest on foreign investment (Oates and Schwab 1988, Long and Siebert 1991, Rauscher 1991b, 1992, 1993). The second approach is based on multicommodity models, which incorporate the interaction of trade patterns and international factor mobility (McGuire 1982, Merrifield 1988). While the first two approaches employ perfect competition general equilibrium models, a quite different third approach investigates location decisions in *imperfectly competitive* markets (Markusen, Morey, Olewiler 1993, 1995; Rauscher 1995, Pflüger 1996, Hoel 1997a, Markusen 1997). We start with the first approach.

2.4.1. Single Commodity Models

Oates and Schwab (1988) analyze in a traditional neoclassical setup how a small open economy with mobile capital and immobile workers should optimally tax capital and set emission standards. The environment serves as a third input in the production. Pollution is only local and they assume that each firm is allowed to emit pollutants in a fixed, administered proportion to its labor force, which makes the model *de facto* a two factor model. If tax revenues are redistributed to the identical residents, it is optimal not to tax capital[22] and to set emission standards such that the marginal rate of substitution between consumption and environmental quality equals the marginal product of the environment. Neither the capital allocation is distorted, nor is the environmental policy suboptimal, although jurisdictions compete for mobile capital. If however, jurisdictions must rely on capital taxation to finance public goods (or the government seeks to maximize revenue according to the Brennan and Buchanan (1980) hypothesis), environmental standards will be set at inefficiently low levels. A laxer environmental policy does not only increase domestic production and pollution, it also attracts capital, which generates additional revenue and thus finances additional public goods. The marginal social benefit from improving the environment (i.e. the marginal rate of substitution between consumption and environmental quality) now exceeds the marginal social costs of reduced production. The reason is that an improvement of the environment creates a positive externality for the other countries as

---

[22] This is a standard result in the tax competition literature. For a survey of this literature see Koch and Schulze (1998) and Wilson (1999).

capital exits the country and diminishes the tax revenue. Governments must take this into account. As a result, environmental standards are too low and the public good is underprovided. This result is an extension of a standard result in the tax competition literature, *viz.* that public goods are underprovided because competing governments are not able to internalize the positive externalities they create through taxation of mobile capital.

Long and Siebert (1991) corroborate the first result of Oates and Schwab (1988) that decentralized environmental policy formation is globally efficient if capital is mobile and pollution remains local. Oates and Schwab's finding hinges on the assumption that there are no other distortions such as capital taxes (as we have seen) and that countries are either too small to be able or unwilling to act strategically. Long and Siebert remove the second assumption and show in a standard single-good neoclassical model with two large countries how governments can use environmental taxes to manipulate international capital flows. A capital exporting country will set environmental standards too low, curb the capital outflow and thus improve its factor terms of trade. The resulting equilibrium is globally inefficient. Long and Siebert go on to show that if pollution spills over to the other country, decentralized environmental policy cannot produce efficient equilibria as in the case of pure commodity trade.

Rauscher's (1991b) setup is similar. He analyzes a neoclassical two-country model with mobile capital and emissions as factors of production. (Long and Siebert assume capital and labor as production factors and the use of capital produces emissions.) Within this framework he denies the factor terms of trade effect of environmental policies. An increase of emissions in the capital exporting country increases transboundary pollution and reduces capital exports, thereby increasing capital tax revenue; this policy thus hurts the capital importing country. Conversely, rising emissions in the capital importing country increase capital imports and the rental rate, but increase again transborder pollution – the total effect of the externality on the capital exporters' country is indeterminate. Increased capital market integration (modeled as lower mobility costs) causes the capital exporter to lower his emissions. The marginal productivity of emissions decreases due to the outflux of capital and so does the marginal utility of consumption since the income increases in the course of integration. In order to equalize the (reduced) marginal utility of emissions with the (constant) disutility of emissions, emissions have to be reduced. For the capital importing country the effect is ambiguous because capital market integration decreases the marginal utility of consumption, but increases the marginal productivity of emissions due to the inflow of capital. Also the welfare effects are ambiguous.[23] Rauscher (1993) studies trade

---

[23] Note that Rauscher's results hold only under the assumption of zero third derivatives of the production functions. This restricts the power of his results. Rauscher (1992) analyzes the effect of economic integration in a three-country model, which is otherwise similar to his (1991b) paper. Welfare results are again ambiguous.

patterns, trade-induced changes in emissions, factor terms of trade effects and welfare effects of increased capital mobility in a unified framework. The twist is that emissions, which again serve as a factor of production and generate pollution now, not only cause disutility of the eye-sore type (deterioration of environmental quality), but also reduce the overall productivity of the economy.[24] As a result, an increase in emissions has a factor-augmenting and a productivity-reducing effect, which work in opposite directions. This constitutes another source of ambiguity. While many of Rauscher's results are ambiguous, he manages to show that increased capital mobility is welfare-enhancing if the country is small and environmental policies are optimal. Large open economies may lose from increased capital mobility even if environmental policies are optimal. The reason is that the efficiency gain for the capital exporting country is thwarted by a negative terms of trade effect and higher *transborder* pollution; the impact working through altered *domestic* pollution on welfare is zero if optimal policies are in place.

2.4.2. Multi-Sector Models

The models reported so far bury some important aspects of the relationship between environmental policy on the one hand and international trade and factor movements on the other. Because they are essentially single commodity models, they cannot portray the effect of environmental policy on the production and trade *structure* when factors are mobile across countries. It will turn out for instance that environmental regulation affects sectors and factors quite differently and that the regulation impact depends on whether the factors have an escape option by moving abroad.

McGuire (1982) shows in a two sector Heckscher–Ohlin model with one polluting sector that country-specific regulation destroys the link between national factor prices established through trade. If one factor is mobile internationally, this gives rise to factor movements until factor rewards are equalized intersectorally and internationally. In the Heckscher–Ohlin model this implies that the country with the stricter regulation will lose the entire regulated industry if the country is small. If the country is large, the factors will flow out of the more heavily regulated industry until one country is completely specialized and factor prices are equalized again. Factor and goods prices are altered in the course of this adjustment.

---

[24] Rauscher's idea behind that formulation is that, industries that rely on clean air and water, e.g. agriculture and tourism, are negatively affected by environmental damage (Rauscher, 1993: 3). That would call for a multi-sector model, which would then allow for the explicit analysis of sectoral interests with respect to pollution. A political-economic model would be the appropriate method of analysis. Because Rauscher has chosen a single good model, he portrays the public input aspect of environmental quality as an efficiency parameter. Merrifield (1988) had a similar idea, namely that pollution damages the capital stock.

Rather than looking at the impact of environmental regulation on the production structure within a country, Merrifield (1988) focuses on the different effects of abatement requirements and production taxes when countries are completely specialized. In a two-country model with transborder pollution each economy produces a single composite commodity with the help of internationally mobile capital and immobile labor. Production generates emissions which can be abated through an abatement technology, which in turn is produced by capital and labor. Governments can levy production taxes to reduce output and hence emissions or impose a minimum ratio of abatement technology to output. Finally, pollution damages the capital stock so that abatement increases the efficiency of the existing capital stock used in the final good's production. The results are as follows: Only an increase in abatement requirements will unambiguously reduce pollution. A production tax in country A may or may not increase pollution in A. The tax on A's production increases consumer prices, reduces demand and diminishes factor prices in A; consequently, some capital moves to country B and increases output there. If B's production is more (less) pollution-intensive than production in A, total pollution increases (decreases) and the total capital stock in efficiency units is therefore reduced (augmented). This loss in (increase of) overall capital stock resulting from increased (decreased) pollution is disproportionally distributed: the bulk of it is taken from A (employed in B), as the tax in A drives capital from A to B. In contrast, a unilateral increase in abatement requirements reduces pollution and therefore is likely to increase the capital stock in efficiency units in both countries (unless the abatement technology is terribly capital-intensive), capital's remuneration thus declines. The country which imposes the abatement requirement will experience a reduction in its output (under normal demand conditions) due to the additional capital requirement for abatement; the increased scarcity of its output is however alleviated if the country absorbs most of the increase in capital (in efficiency units) that is brought about by a reduction in pollution.

Copeland and Taylor (1997) study capital mobility in a two-sector specific factors model with capital and labor being sector-specific and pollution being the mobile factor. The North is richer, values environmental quality more highly and imposes therefore a higher tax on environmental consumption. Because free trade equalizes commodity prices and the shadow price of pollution is higher in the North, the wage rate and return to capital must be lower in the North (there are no technology differences and perfect competition ensures zero profits). If capital becomes mobile, capital will migrate to the South either until all capital has exited the North or until the shadow price of pollution is equalized, which implies equal factor prices and incomes.[25]

[25] Because regulation is endogenous, McGuire's (1982) extreme result of complete specialization need not occur; in other words, if per capita levels are equalized before all capital has left the North, both regions continue to produce both commodities.

2.4.3. Imperfect Competition

Like most other models on factor mobility, the models that describe environmental policy under international capital mobility usually assume perfect
competition and perfect divisibility of capital. These assumptions ensure continuous relocation of capital as regulation changes. With imperfectly competitive markets, not amorphous capital but entire firms change location.
Consequently, marginal policy changes can give rise to discrete jumps in
national welfare. Markusen, Morey and Olewiler (1993) derive such discrete
changes in a two firms, two country, two stage model: In stage one the firms
decide whether to enter, where to produce and whether to export to the other
country if they set up only one plant. Stage two is a simple one-shot Cournot
game. The firms' decisions depend on the pollution tax set by the domestic
government, the foreign country remains passive. The authors show how the
resulting market structure feeds back to prices, pollution, tax revenue, and
firms' profits. These partial welfare effects move in opposite directions so that
results can be derived only for given parameter values.

In an extended version, Markusen, Morey and Olewiler (1995) incorporate
strategic interaction of the governments. They set up a two country, two sector
general equilibrium model, with one sector being perfectly competitive and
clean and the other sector consisting of one polluting company only. The dirty
monopolistic firm decides whether to produce in one country only (and to serve
the other country via exports), in both countries, or not to produce at all. The
decision depends on the tax rates in the two countries, plant-specific fixed costs,
and transportation costs. The two countries are identical and tax only the
production of the monopolistic, dirty good, but at potentially different rates for
exports and domestic consumption. A lower tax rate will increase effective
demand, thereby tending to make two plants profitable; at the same time it will
attract the single plant to the home country, which brings about a large discrete
welfare change. The governments will undercut each other's tax rates in order
to attract the single plant or, in the case of two plants, to induce the monopolist
to close down its plant in the other country. In the first case, the country moves
away from a no plant situation, in which it has to import at high costs and
receives no tax revenue; in the second case, the country experiences a welfare
gain as long as the export tax revenue exceeds the disutility from increased
pollution. Undercutting will come to a standstill when capturing the other
market does no longer make sense, that is, when the disutility from pollution
due to production for export purposes is just offset by the export tax revenues.
If undercutting does not make sense, the domestic tax is set as high as possible
without losing the domestic plant.[26] ·

---

[26] In a sense, Markusen, Morey and Olewiler's 1995 model is a refined version of a Bertrand
game; it yields the analog to the well known Bertrand result that in equilibrium prices equal
marginal costs.

The setup of Markusen *et al.* (1995) is very rich in that they explicitly analyze the interaction of production cost structures, transport costs and tax policies in both countries; this complexity entails that it is impossible to derive closed form solutions; as a consequence one needs to rely on numerical solutions instead. Rauscher (1995) and Hoel (1997a) extend Markusen *et al.*'s idea by simplifying the original setup and incorporating other features. Both papers are partial equilibrium analyses which assume that transport costs are zero. Thus, the firm will set up at most one plant and serve both countries from it; the location decision is reduced to the question in which country to locate. Rauscher (1995) models a monopolist which uses environmental resources as inputs and is subject to an emission tax. The monopoly's production generates a consumer surplus and causes an environmental damage which is convex in output; this damage can be local or transboundary. Coordinated first-best policies would require an emission tax that equals the combined marginal environmental damage and a price control to eliminate monopoly power. If the second instrument is not available, the optimal emission tax will be lower in order to increase production since the monopolist's output falls short of the socially optimal level. The tax would be set such that the marginal environmental damage equals the sum of marginal tax revenue and marginal consumer surplus in *all* countries. The location of the firm would depend on the pollution-assimilation capacities of the countries and the nature of the pollution (i.e. the extent to which it spills over to other countries). In the noncooperative case individual countries base their decisions only on tax revenues and environmental damage because their consumer surplus is realized independently of which country hosts the firm. Three cases can be distinguished in the case of strictly local pollution. For low environmental damage, tax revenue exceeds environmental damage over a wide range of tax rates and hence countries are better off hosting the polluting firm. They compete the tax rate down in order to attract the firm until the tax revenue equals environmental damage.[27] This tax rate is too low as compared with the optimal tax which is realized in the cooperative regime. The reason is that under strategic interaction maximization of total consumer surplus and tax revenue net of environmental damage is realized at different tax rates. For high environmental damages the resulting tax rate is higher than the optimal one under cooperation. The principle at work becomes apparent for extremely high environmental damages: if the environmental damage is larger than the tax revenue for all possible tax rates, no country will want to host the firm although the total benefit for one country, that is, tax revenue plus consumer surplus in that country, may exceed the environmental damage. This is a chicken game situation and it is not clear whether a pure strategy equilibrium will be reached. The environmental

---

[27] This presupposes, of course, that the emission tax is the only instrument. It would be more efficient to set the tax rate at the Pigouvian level of marginal environmental damage and to use subsidies to attract the foreign investment.

damage may even exceed individual countries' net benefit and thus the firm will not be established although it may be in the interest of the world as a whole to have the firm produce. The reason is that each individual country's consumer surplus is only a fraction of the world's consumer surplus and that total benefit may exceed environmental damage, but individual countries' benefits do not. This is the "not-in-my-backyard" (NIMBY) case. In case of global pollution, countries will experience the environmental damage regardless of whether they host the firm or not. Hence, if some countries are willing to host the firm at all (i.e., if consumer surplus plus tax revenue exceed environmental damage) they will compete the tax rate down toward zero as the opportunity costs of undercutting become infinitesimally small.

We see that in the noncooperative regime the tax rate can be either too high or too low as compared to the coordinated efficient solution, depending on the nature of pollution (transboundary or local) and its intensity. Hoel (1997b) employs a very similar setup and comes to the same conclusion. He extends the model to include domestic ownership of the firm, which implies that the respective share of the firm's profits enter the social welfare function. This tends to reduce the cooperative and the Nash equilibrium tax rates. The same holds true if the establishment of the firm generates higher employment.

Ulph and Valentini (1997) extend the strategic environmental policy analysis by allowing for intersectoral linkages which provide an incentive for agglomeration. There are two industries, upstream and downstream, with one firm of each type owned by the residents of each of the two countries. Transport costs for the intermediates establish an incentive for agglomeration, transport costs of the final good create an incentive for market proximity. Governments levy an emission tax which is equal to a production tax since the emission/output ratio is fixed.

The game comprises three stages: first the two governments set their environmental policy, then the firms locate their plants (none, one, or one in each country), finally, the firms choose their output. Like Markusen *et al.*, Ulph and Valentini find multiple equilibria; moreover marginal shifts in the tax rates may give rise to location reversals and discrete changes in welfare. The agglomeration effects reinforce the discontinuity. The authors also show that tax increases may lead to an attraction of firms because of changed agglomeration incentives. Lastly, hysteresis effects may emerge due to agglomeration: an increase of taxes may drive out firms which do not return in the course of a subsequent tax reduction. This paper is located at the intersection of the strategic trade (and environmental policy) literature and the geography and trade literature and derives a rich set of results. However, like other papers in the strategic trade literature, the analyzed setup is rather special and it is not entirely clear how robust the results are.

Pflüger (1996) studies the effects of environmental taxes on firm location in a general equilibrium, two region imperfect competition model. A competitive sector produces only with labor and its output is traded costlessly. This ties down the global wage rate. Then there is an imperfectly competitive sector of the Dixit-Stiglitz-Krugman type, producing with labor and emissions. Its output is traded at some positive cost. The fixed amount of start-up capital per firm determines the number of firms since the capital stock is given; profits are redistributed to the representative agents. Imperfectly competitive firms locate according to the demand conditions and the availability of emission certificates. An increase in the foreign emission tax rate attracts additional firms to the home country. This reduces the prices of the goods which are now domestically produced since transportation costs do not accrue anymore (home market effect), it also raises the tax revenue and pollution at home. Depending on the disutility of pollution, the net effect can have either sign; in the competitive equilibrium the tax rates thus are either too high or too low to internalize the negative externality. Because Pflüger only studies symmetric equilibria, he finds that both regions are worse off, having failed to attract additional firms and having suboptimal taxes in place.[28]

Markusen (1997) analyzes the impact of environmental regulation on the prevailing market structure and on welfare in a general equilibrium two country–two sector model.[29] One sector produces under constant returns to scale, the other sector is described by scale economies, caused by fixed costs on the firm level as well as on the plant level, and free entry. The imperfectly competitive sector uses only labor, which is mobile between the sectors, the other sector uses an additional sector-specific factor. While factors are not mobile internationally, plant location is. The number and type of firms in equilibrium depend on the (relative) marginal costs, the two types of fixed costs and the transportation costs. There are national firms that produce only in one country and may or may not export to the other market, and multinational firms which produce in both markets, but incur the firm-specific fixed costs only in the country of residence. Trade policy is reflected by the level of transport costs while environmental policy might fall on fixed or marginal costs. Markusen derives a set of interesting results in his simulation analysis. Increased protection tends to lead to the emergence of multinationals while the intensification of environmental control tends to shift the regime away from multinational firms towards national firms. Protection tends to decrease welfare as well as the regulation-induced reduction in output. Stated differently, under free trade plant location is more sensitive to differences in regulation. This increased regulatory competition is welfare superior to the introduction of

---

[28] Of course, it would be very interesting to study asymmetric equilibria to see whether some country would gain under these circumstances even though a welfare-inferior situation emerges for the whole world. This would parallel results from the tax competition literature, see for instance Bucovetsky (1991). Moreover, it would be interesting to include a second instrument such as a tax on production.

[29] For a similar setup see Markusen and Venables (1998).

trade barriers.[30] The multinational firms are shown to smooth the regulation induced cost increase in one country over both countries, and increases of the fixed costs lead to the exit of firms, but reduce welfare only mildly whereas rising variable costs reduce output per firm and thereby reduce scale economies, which decreases welfare considerably.

In a literal sense, Markusen's model is not about environmental policy as such, because the negative externality is left in the background. It rather demonstrates how (policy-induced) changes in the cost structure affect the market structure and give rise to regime shifts. However, environmental regulation and trade policy are a special application of his general model in that they affect relative costs and international arbitrage; the impact of the environmental policy on the externality would have to be modeled explicitly in order to derive "true" welfare results. One of the merits of this model is to show that environmental policy might have very different effects on allocation and welfare, depending on whether the policy affects the fixed or variable costs. This important difference, once noted, might be useful in order to rank specific environmental policy instruments.

2.5. SUMMARY

The normative literature on trade and environment is quite extensive, and we were only able to survey the most prominent contributions. What is most striking is the fact that a large part of the literature follows very closely the traditional literature on trade theory and public economics.

What have we learnt from this literature? We list below our own main conclusions:

(1) Without environmental policy in place and strictly local pollution, trade liberalization is beneficial for the country specializing in the clean goods, but may deteriorate welfare in the country exporting the pollution-intensive goods since pollution in this country increases. For a small economy the optimal policy is free trade and standards that fully internalize the environmental externality. If environmental policy is optimal, all countries benefit from free trade although world pollution rises.

(2) Large countries can use environmental policy to improve their terms of trade; this is, however, only a second-best policy when they are restricted in the use of tariffs. Through environmental control they can increase the relative scarcity of the factor they are relatively well endowed with, which increases the relative price of their exportables.

---

[30] This argument does not consider any welfare effects from changes in pollution, since Markusen does not model the environmental externality. Taking this effect into consideration would reinforce his result.

(3) A country's endowment with "environment" as a factor of production is largely determined by its environmental policy. The prevailing environmental policy in turn, is influenced by the representative individual's preferences, which are income-dependent since environmental quality is a superior good. To be sure, if property rights are more poorly established in some countries, they will have inferior environmental control and thus suffer from an overproduction of environment-intensive goods. This distortion will be exacerbated by free trade. The first best policy response under these circumstances is however to enforce property rights, not to restrict trade.

(4) In the case of transboundary pollution and different preferences for pollution across countries, trade liberalization may actually decrease the welfare of the country with a high preference for environmental quality. A policy that is optimal from a global perspective calls for international coordination, since a policy, which is optimal from a purely national point of view, will, for strategic reasons, not lead to a full internalization of the environmental damage.

(5) If markets are imperfectly competitive, say with one national firm only, strategically acting governments will set environmental policies which are too lenient when firms engage in Cournot competition. This incentive is reinforced if the pollution-intensive good is also consumed domestically and if pollution is transboundary. Yet, environmental policy may be too strict if the domestic industry is oligopolistic, and will be too strict if the firms compete in prices (Bertrand competition). In other words, the strategic environmental policy results are not robust.

(6) If capital is mobile and commodities are traded only to settle the balance of payments (one-good MacDougall Kemp model), the environmental policy induced effects on factor trade are analogous to the effects on commodity trade – environmental policy can be used to improve the terms of trade if the country is large. If optimal environmental policies are in place and the countries behave nonstrategically, capital mobility enhances welfare.

(7) In a pure Heckscher–Ohlin model with international trade, capital mobility leads to the abolition of the polluting sector in the country imposing the stricter regulation. This contrasts the earlier result (see point 1) that free trade is beneficial if all countries adopt optimal environmental policies. If capital is mobile, this nice result thus breaks down.

(8) If markets are imperfectly competitive, an environmental policy targeted at attracting capital can lead to discrete jumps in welfare; such a policy will increase tax revenues and pollution, on the other hand it will dispose of transportation costs and hence decrease domestic prices. Depending on the net effect on utility derived from increased home production, imperfect competition may bring about environmental taxes, which are

too high, or too low. An environmental policy, which increases the fixed costs, has less severe impacts than an environmental policy, which increases the variable costs, because scale economies are reduced.

Appendix. Gains from trade in the presence of a negative environmental externality

In this appendix[31] we show in a simple setup how the gains from trade depend on the direction of trade and on the source of the environmental externality (pollution via production versus pollution via consumption). Assume a two-good economy and that the (local) externality occurs only in connection with good X. We compare free trade with autarky. Pollution has a "shadow price" q_X. Superscript f denotes free trade, superscript a autarky. Subscript p denotes production, subscript c denotes consumption.

Pollution occurs in production

The value of free-trade production is maximized at free-trade prices.

$$p_X^f X_p^f + p_Y^f Y_p^f > p_X^f X_p^a + p_Y^f Y_p^a. \tag{1}$$

We use the balance-of-trade constraint on the left and the autarky market clearing conditions on the right hand side and obtain

$$p_X^f X_c^f + p_Y^f Y_c^f > p_X^f X_c^a + p_Y^f Y_c^a. \tag{2}$$

We then add and subtract the value of pollution in free trade and in autarky from the left and right-hand sides of (2) respectively. This manipulation results in

$$p_X^f X_c^f + p_Y^f Y_c^f + q_X X_p^f - q_X X_p^a > p_X^f X_c^a + p_Y^f Y_c^f + q_X X_p^f - q_X X_p^a. \tag{3}$$

Inequality (3) can be rearranged to yield

$$p_X^f X_c^f + p_Y^f Y_c^f - q_X X_p^f > p_X^f X_c^a + p_Y^f Y_c^f - q_X X_p^a + q_X(X_p^a - X_p^f). \tag{4}$$

Letting *RI* denote "real income" (utility), this can be written as:

$$RI^f > RI^a + (q_X(X_p^a - X_p^f)). \tag{5}$$

A sufficient condition for the existence of gains from trade is that the production of the polluting good decreases with trade. This will hold if X is the *import* good.

---

[31] We are indebted to Jim Markusen who suggested this appendix.

Pollution occurs in consumption

Equations (1) and (2) carry over to this case as well. Again, add the value of pollution in free trade and subtract the value of pollution in autarky on both sides of the inequality (2). Note that the externality occurs now in consumption.

$$p_X^f X_c^f + p_Y^f Y_c^f + q_X X_c^f - q_X X_c^a > p_X^f X_c^a + p_Y^f Y_c^f + q_X X_c^f - q_X X_c^a \tag{6}$$

Rearrange (6) to obtain

$$p_X^f X_c^f + p_Y^f Y_c^f - q_X X_c^f > p_X^f X_c^a + p_Y^f Y_c^f - q_X X_c^a + q_X (X_c^a - X_c^f). \tag{7}$$

In analogy to the first case we can rewrite eq. (7) as

$$RI^f > RI^a + (q_X(X_c^a - X_c^f)). \tag{8}$$

A sufficient condition for the existence of gains from trade is that the consumption of the polluting good decreases with trade. This *may hold* if X is the export good, but *might not* hold because income increases with trade; that is, consumption of both goods can rise in the course of trade liberalization.[32] The inequality (8) does not hold if X is the import good.

---

[32] This effect might occur due to the "traditional" gains from trade (which disregard environmental externalities), i.e. the enhanced consumption possibilities from trade liberalization.

3

The Empirical Evidence on Trade, Investment, and the Environment

GÜNTHER G. SCHULZE AND HEINRICH W. URSPRUNG

3.1. INTRODUCTION

As we have seen in the previous chapter, the theoretical literature on international trade, capital mobility and the environment is chiefly concerned with two issues: First, how does environmental policy influence international trade and investment flows and, second, how do trade and investment policies affect domestic environmental quality? These questions have been and continue to be of great concern to economists, policymakers and environmentalists alike.

If internationally diverging environmental policies are able to shape international trade and investment flows, they can serve as a substitute for classical trade and investment policy instruments in a situation of strategic interaction between countries. If countries want to improve competitiveness of their domestic products or compete for mobile capital they might be tempted to reduce environmental standards at home, which could lead to a competitive reduction of environmental standards to suboptimal levels ("ecodumping").[1] Alternatively, they might use environmental policies to shelter domestic industries from foreign competition ("ecoprotectionism"). If such policies are effective, they change national – sectoral – capital allocation; and to the extent that differences in environmental policies trigger international capital movements they affect also international capital allocation. Dirty industries migrate to countries with lax environmental control ("pollution haven hypothesis"). Conversely, trade policy alters the pattern of trade and the structure of domestic production and thereby overall domestic pollution as the pollution intensity of production differs across sectors. If this is so, does trade liberalization lead to a significant degradation of the environment? And if it does, which countries are affected and to what extent? These are important theoretical and empirical questions. Of course, all these effects feed back to the process of environmental and trade policy formation (see Chapter 4).

We are grateful to Dave Erwin for helpful comments. The usual disclaimer applies.

[1] A similar case can be made for the terms of trade argument advanced in the previous chapter.

Whatever view on government behavior one may adopt, the traditional social welfare-maximizing approach (Chapter 2) and the political-economic approach (Chapter 4) suggest that governments' environmental policies are effective. In particular, they assume that environmental control, and thus the availability of the environment as a factor of production, influences the trade pattern and the international allocation of mobile capital in a significant manner. This, however, is by no means obvious – whether regulation actually matters is, of course, an empirical question. Also, the influence of the trade regime on environmental quality hinges upon sectoral differences in pollution via production and the availability of cleaner technologies in open economies – the extent of both effects are again empirical issues.

In this chapter we review the empirical evidence on the relationship of trade, international factor allocation, and the environment.[2] We seek answers to three *empirical* questions:

1. To what extent do differences in environmental policies affect the pattern of international trade? The answer to this question allows us to asses whether "ecodumping" is at all an issue under the circumstances identified by the theoretical literature.
2. What is the quantitative impact of the trade regime on environmental quality? Can we identify channels through which trade liberalization affects environmental quality?
3. To what extent do international differences in environmental control lead to relocation of pollution-intensive industries to countries with low standards? In other words, is the "pollution-haven hypothesis" supported by the available empirical evidence?

The empirical findings presented below shed light on the relevance of the theoretical analyses surveyed in the previous chapter. The survey in this chapter is organized as follows. First we present in section 3.2 the two concepts for measuring the pollution intensity of (sectoral) production, environmental control costs and toxic release, and discuss their methodological problems. Section 3.3 is devoted to the empirical evidence concerning the influence of environmental policy on the pattern of trade. Section 3.4 presents evidence concerning the influence of the trade regime on environmental quality, and section 3.5 surveys the empirical analyses on the link between environmental policies and the relocation of pollution intensive industries. Section 3.6 summarizes and concludes.

3.2. "ENVIRONMETRICS"

How can one measure the quality and the availability of "the environment" as a factor of production, or, alternatively, the stringency of environmental

---

[2] Although we include most of the major contributions we cannot be complete; other surveys of the empirical literature include Dean (1992), Jaffe *et al.* (1995), and Bommer (1998).

regulation? Two approaches have been used – and both suffer from severe shortcomings. The Environmental Control Costs (ECC) approach measures the total costs of compliance with environmental regulation, which include capital costs of abatement equipment (including depreciation), operating costs of environmental management, and R&D expenditures for abatement technology. The ECC approach to measuring the stringency of environmental regulation has various shortcomings. First, it is debatable whether ECC cover all relevant costs caused by environmental regulation. Chapman (1991) argues that ECC have been highly underestimated since they do not comprise workplace health and safety costs. Chapman *et al.* (1999) note that ECC do not include any cost components related to environmental monitoring and planning, interest expenses for investment in protection equipment, or unreported labor intensive environmental activities that are part of the production process. Low (1992a) criticizes that ECC refer only to "end-of-pipe" expenditures and not to all capital expenditures related to environmental considerations. Moreover, outright prohibitions that do not lead to increased abatement, but to different production processes are not covered by this measure at all. Second, ECC represent (at best) only the costs of meeting a certain regulation, whereas total costs depend also on the firm's output and pollution decisions; ECC are not a direct measure of how strict a regulation is. For instance, a firm in a country with more lenient standards may incur higher ECC than a comparable firm in a more regulated economy, because it chooses a different factor input combination as a consequence of different factor prices, including the price for using the environment (pollution). On the other hand, ECC do not account for benefits of lower pollution for the firm itself. Inasmuch as pollution abatement reduces damages of the production equipment, the ECC overstate the true costs of abatement to the firm (Jaffe *et al.* 1995). All these shortcomings notwithstanding, the ECC provide an imperfect indication of how important the environment is as a cost factor compared to total production costs.

The second approach measures the actual discharge of pollutants per unit of output; it is based on data from the Toxic Release Inventory (TRI) as published by the US Environmental Protection Agency (EPA). These data on toxic release of 15,000 US companies are matched with their output data as published by the US Bureau of Census and aggregated over industries to yield sectoral pollution intensities. It is therefore a more direct measure of the real environmental consumption of the individual sectors than the ECC approach. However, adding up the physical amount of all 320 toxic releases – whether atmospheric, effluent, or solid – per dollar of output does not take into account the damage inflicted by the various pollutants. To address this obvious shortcoming, Lucas *et al.* (1992) weigh the different toxic releases, based on the EPA's Human Health and Ecotoxity Database which divides the emissions into four groups. They use linear weights (1–4) as well as exponentially formed weights (1, 10, 100, 1000). Clearly, this procedure remains somewhat arbitrary, and thus, toxic release data are only a very crude measure for the real environmental harm inflicted by sectoral production. Yet, since there are no better data available, we must rely on these measures if we want to shed some light on the empirical relevance of theoretical studies.

In the following, we turn to empirical studies on the relationship between environmental regulation and international trade patterns. More specifically, we are concerned with the following questions: How important is the environment as a factor of production? Do differences in environmental regulation and, consequently, in relative factor endowments explain the existing pattern of trade?

3.3. THE INFLUENCE OF ENVIRONMENTAL POLICY ON THE PATTERN OF TRADE

Walter (1973) looks at the importance of ECC for the US trade. He uses ECC data from the late 1960s and a 1966 input–output table to determine the effect of environmental regulation on trade. ECC comprise capital costs for pollution abatement equipment (including depreciation costs), operating and management costs for the equipment, and R&D expenditures in environmental technology. He finds that ECC amount on average to 1.75 percent of the total value of US exports and 1.52 percent of US imports assuming that foreign producers have the same ECC as import competing US producers.[3]

Less than 2 percent is a rather small fraction; ECC are therefore not liable to influence trade flows or output decisions. Some studies which directly investigate the output response have been summarized by Ugelow (1982): a study carried out by the OECD (1978), for example, investigates output effects of ECC in Japan, the Netherlands, Italy and the US; Yezer and Philipson (1974) do the same for selected US industrial sectors. They all reach the conclusion that the increase in prices and the reduction in output due to intensified environmental control are rather small, so that the effect of environmental policy on production (and hence trade) are clearly of second order. Richardson and Mutti (1976, 1977) estimate domestic and import market supply and demand functions for 81 industries in a general equilibrium approach. They find that output and price reactions to (existing) ECC are rather small, but that they might range up to 5 percent for selected industries if the polluter-pays-principle prevails. If general equilibrium effects through changes in income and exchange rates are taken into consideration, these effects are decreased even further.

Robison (1988) analyzes for the United States the effect of an overall price increase of 1 percent of the final goods (brought about by a rise in ECC) on

---

[3] Of course, this assumption is not innocent; it is mainly dictated by data availability – ECC are not compiled on a consistent international basis.

output of 78 sectors and the balance of trade. He uses a partial equilibrium model that accounts for inter-industry trade in intermediate goods, and bases his results on input–output tables and estimated trade elasticities. The neglect of moderating general equilibrium effects implies that his results establish an upper bound for the actual reaction of production and the balance of trade. The 1 percent price increase seems incremental in terms of total costs; compared to the existing level of ECC this is however quite substantial. Robison estimates the average ECC as a share of total export to amount to 0.37 percent in 1973 and 0.72 percent in 1982. The 1973–82 period is characterized by increasing environmental regulation in the US: ECC's share in total imports have risen from 0.48 percent in 1973 to 0.99 percent in 1982. This implies that the ratio of ECC content of imports to exports has risen from 1.15 in 1973 to 1.39 in 1982, implying that the US have specialized in cleaner products. Like Walter (1973), he is forced to assume (because of data availability) that the ECC of the imports are equal to the ECC of the import-competing US sectors.[4] Robinson's results show that the total (and individual sectors') output reaction is negligible; the sectors' trade values decline between 0.12 percent and 7.08 percent, averaging 2.69 percent. The trade balance reduces net by 0.67 percent. In other words, the doubling of ECC has only very moderate effects on aggregated output and trade.

Tobey (1990) measures the impact of environmental policy on the pattern of trade by testing whether the Heckscher–Ohlin model of international trade better explains the existing pattern of trade when variations in environmental policies are accounted for. Out of 64 3-digit SITC industries he selects the 24 most pollution-intensive industries, that is those that have direct and indirect pollution abatement costs exceeding 1.85 percent of total costs. They are aggregated into five groups (mining, paper, chemicals, steel, and nonferrous metals) and, subsequently, their trade vectors are regressed on a set of factor endowments (various kinds of labor, land, and natural resources). A dummy variable for the stringency of environmental regulation is then introduced to capture the effect of environmental policy on the trade pattern. It turns out to be insignificant. An omitted variable test corroborates the result. If ECC influenced the net exports, but are omitted from the regression, the error term should have a negative sign more often for countries adopting more stringent regulations (industrialized countries) than for less developed countries. Tobey however is unable to reject the null hypothesis postulating an equal distribution of errors in both groups of countries. His result is corroborated by Murrell and Ryterman (1991) who also use an omitted variable test for 1975 data in a

---

[4] This shortcoming is not specific to these two studies. In fact it is shared by all other empirical studies. The empirical analyses focus on composition effects, but do not measure international differences in regulation. Therefore, Robison's study is not a test of the Heckscher–Ohlin model, since the trade partners' environmental regulation and thus their endowment with "environment" as a production factor cannot be measured. The Heckscher–Ohlin model explains international trade with relative endowment differences.

similar framework and cannot find any impact of environmental policy differences on trade flows. Kalt (1988) also applies a Heckscher–Ohlin model and regresses changes in net exports on changes in ECC, among other variables. He finds that the relationship between abatement expenditures and net trade performance is insignificant for the whole population of 78 sectors, but becomes significantly negative if applied only to manufacturing sectors. His study covers the 1967–77 period.

Low (1992a) investigates the pollution content of Mexican exports to the US in 1988, again using for Mexican industries pollution abatement data which were compiled for the US. He shows that Mexico is not overly dependent on pollution-intensive industries, only 18 out of 123 3-digit SIC industries incur ECC exceeding 1 percent of total output. The weighted average of all industries is only 0.54 percent. Mexico's export of dirty goods (those goods, where the ECC are greater than 0.5 percent of total output) accounts only for a little over 10 percent of total exports. However, those dirty exports to the US have increased three times as fast (9 percent annually) as all commodities together (3 percent p.a.). The traditionally more lenient environmental policy of Mexico has not led to a specialization in production of dirty goods, although it might have had some impact on sectoral growth. The growth effect, however, could just as well be the result of structural shifts occurring in the course of development, as we will see below.

Ferrantino (1997) shows that abatement costs did not have any significant impact over time on the "revealed comparative advantage" of the US, that is, its sectoral trade pattern. The trade patterns of the dirty industries rather followed closely the overall trend. He concludes that the overall level of abatement costs is too small to constitute a major (dis-)advantage.[5] US pollution abatement operating costs have steadily risen from 0.3 percent of output value in the early 1970s to 0.8 percent in 1992 (averages for manufactures). Although abatement costs are concentrated in only a few sectors, pollution abatement operating costs and capital expenditures amount to around only 2 or 3 percent of total costs for some of the dirtiest industries (see Ferrantino 1997: 52).

Van Beers and van den Bergh (1997) have challenged the common finding that environmental regulation does not affect the trade pattern. Using a gravity model they regress bilateral trade in 1992 for 21 Organisation Economic Cooperation and Development countries on Gross Domestic Product, population, and land area of both countries, distance, dummies for customs unions and neighbors, and two different indices for the stringency of environmental regulation, both for the exporting and the importing country. The index which turns out significant is constructed as the unweighted average of the ranks that a country has in the group of 21 countries with regard to (i) energy intensity per unit GDP, and (ii) the change therein, normalized to lie in the unit interval.

---

[5] He draws on the data from *Current Industrial Reports* and *Pollution Abatement Costs and Expenditure*, published by the US Department of Commerce (various years).

They find this index to have a significant negative impact on total bilateral trade flows and on trade in goods of nonresource based pollution-intensive industries, both for the exporting and the importing country. This is sensible because resource-based dirty industries typically cannot escape regulation via international reallocation since the resource is immobile. Note, however, that the study does not analyze *sectoral* but only aggregated cross-country differences in trade performance so that possible structural changes due to environmental regulation are blurred. Moreover, it is not clear that the index actually reflects differences in regulation. First, it is constructed from a ranking that does not take the actual differences in regulation into account but the mere order of stringency; second, energy intensity of production is at best an indirect measure of overall environmental regulation – it may as well be influenced by different factor prices for other factors which influence the factor combination, or by a different sectoral composition of output (with sectorally different energy intensities) over which the authors aggregate, to name just two influences. Lastly, subsidized energy may go hand in hand with a stringent regulation of emissions. Still, their measure tries to capture possible indirect effects of government regulation on productivity through higher input prices (energy), which the ECC do not cover.

Porter (1991) and Porter and van der Linde (1995) postulate the opposite relationship: flexibly designed environmental regulation which does not prescribe the method of pollution reduction but only the environmental goal (the amount by which pollution has to be reduced) may lead to a productivity increase and thus benefit firms. Managers will be induced by the regulation to reconsider the entire production process which will eventually lead to a better combination of production factors. Empirical evidence has so far been largely anecdotal and controversial (see Porter and Linde 1995; Jaffe and Palmer 1997). Measurement problems make this hypothesis hard to test (Jaffe and Palmer 1997); yet it is hard to conceive that regulations will lead to efficiency gains that cannot be realized through competition in the market place. Still, the so-called "Porter Hypothesis" has led to a small literature on environmental regulation and innovation. Jaffe and Palmer find in a panel regression (US, 1975–91) at the two and three digit industry level that increases in compliance costs are associated with higher R&D expenditures in the immediate future when controlling for industry-specific fixed effects. The authors find little evidence, however, that the industries' process innovations as measured by successful patent applications is related to compliance costs.[6] Schneeweis and Schulze (2001) use microeconomic data from the IFO Institute for Economic Research enterprise panel to address the environmental regulation–international competitiveness relationship. Their rich data set comprises 2405 German companies and covers the time span 1982–90, which is a major advantage since

---

[6] See Albrecht (1998) for further evidence on the Porter Hypothesis.

previous studies used only aggregated data or only very few firms. They find that investment in environmental protection has risen considerably in absolute levels and that environmental protection has also gained in relative importance as an investment goal as compared to other investment motives. More importantly, companies which state that environmental improvement is a major investment goal tend to be companies with higher employment, more outward oriented firms, and firms with higher rate of innovation expenditures as a share of turnover.

To summarize these findings, it seems fair to conclude that environmental *policy* overall has but a very limited effect on the trade pattern. This result may in part be due to the insufficient data quality and the conceptual difficulties with ECC as a measure for the stringency of environmental regulation (Jaffe *et al.* 1995). The most important reason, however, is obviously the very small fraction the ECC contribute to total production costs. This renders the comparative advantage created by more lenient environmental regulations relatively unimportant compared to other sources of comparative advantages such as differences in technologies, and in endowments with natural resources, human and physical capital, labor, and other production factors (see also section 3.5). At most, environmental regulation may have an effect on selected high-polluting industries or firms at the margin.[7] But even if environmental policy has no significant effect on the trade pattern, we can still pose the opposite question: how does the trade regime affect the pollution pattern? Does pollution change as a byproduct of trade liberalization? We present the empirical evidence on this issue below.

3.4. THE INFLUENCE OF THE TRADE REGIME ON THE ENVIRONMENT

Trade liberalization influences the environment mainly through four channels: First, the countries restructure their production according to their comparative advantage, which implies a change in overall pollution because pollution varies across sectors (composition effect). Second, liberalization yields a better factor allocation and thus increases production and pollution (gains from trade or income effect). The income effect is moderated through a higher preference for the environment and political considerations (see Bommer and Schulze 1999) which lead to stricter environmental control (regulation effect). Last, liberalization is often accompanied by an increased technology transfer, which implies cleaner production processes and a more efficient abatement equipment

---

[7] See also Ulph (1993) for a calibrated imperfect competition model for the fertilizer industry. He finds a significant impact of environmental policy. Things may be different for environmental policies explicitly targeted at reducing the energy content of production in order to reduce global warming. This might have an impact on energy prices and thus on terms of trade, see Clarke *et al.* (1996).

(technology effect). These effects could be discerned if data for toxic releases or ECC were available for a variety of countries; since this is not the case, only the income and composition effects can be analyzed and even this analysis pre-supposes that the ranking of industries according to their pollution content does not change from country to country.[8]

Lucas, Wheeler, and Hettige (1992) use 1987 toxic release data for 15,000 US manufacturing firms as published by the EPA (see section 3.2), match the toxic release data with individual output data and aggregate them to sectoral data to obtain toxic release per dollar of output. They assume that these pollution intensities are equal for some 80 countries and over the time span 1960–88. They confirm the result of an inverse U-shaped relationship between GDP per capita and total toxic release from manufacturing output per unit of GDP.[9] The decline in pollution intensity (not in levels!) for high income countries is due to a declining share of manufacturing output in GDP, not to a less toxic mix of industries within manufacturing. In fact, the relationship for pollution per unit of manufacturing output goes the other way. In devel-oping countries, however, the growth in toxic intensity has been far more rapid. Lucas *et al.* (1992) rank the countries according to a price distortion index and find that the change in pollution intensity depends crucially on the trade regime and the growth rate. Open economies had a toxic-neutral change in the 1970s and a strong shift towards less-toxic structures in the 1980s, whereas closed economies had a rapid change towards more polluting struc-tures. The trend was much more pronounced for fast growing economies. These results are reinforced by a similar study by Birdsall and Wheeler (1992) for Latin American countries.

Mani and Wheeler (1999) investigate production of pollution-intensive products for a number of countries. They find that pollution-intensive pro-duction as a share of total manufacturing has fallen consistently in OECD countries, but has risen in the developing world. This is, however, not due to a regulation-induced comparative advantage of developing countries in polluting production which has led firms from the rest of the world to relocate their respective production facilities in order to serve the world market ("pollution haven hypothesis"). Rather it is driven by a higher domestic demand for pol-lution-intensive products as the average income rose from low to middle levels; with further rising incomes, environmental regulation will become more stringent and production will shift to cleaner sectors.

---

[8] Case studies are possible however. Wheeler and Martin (1992) in a case study on the wood pulp industry show that open economies tend to adopt cleaner technologies more rapidly. The technology transfer effect could, in principle, also go the other way and include new and dirtier products. Likewise, Repplin-Hill (1999) shows in a study of 30 steel producing countries in 1970–94 that the cleaner electric arc furnace technology is diffused faster in countries that have more open trade policies.

[9] See Radetzki (1992) for a discussion of the link between growth and environmental quality.

The link between income and environmental quality is considered by Grossman and Krueger (1995). They investigate urban air quality and water pollution in different countries and regress pollution on per capita income. The big advantage of their study is that they have fairly reliable data, compiled by the Global Environmental Monitoring System (GEMS), from various cities and river basins in many developed and developing countries. They find pollution to be an inverted U-shaped function of per capita income for most pollutants. The turning point is at an income level of $8000 per capita (see also Seldon and Song (1994) for a related analysis). To survey the literature on the relationship between economic growth and environmental quality (the so called "environmental Kuznets curve") lies beyond the scope of this chapter.[10] The inverted U-shape relationship between environmental quality and per capita income level has, in general, been corroborated, although the relationship is by no means uniform across all dimensions of environmental quality. Note also that there is no automatic reversal of environmental degradation as income rises. A major problem with this relationship is that it largely depends on a change in environmental *policy* as per capita income rises (see Ervin 1997a). Moreover, policy reversals are likely to depend on the degree of publicness of the environmental problem at hand.[11] In any case, policy reversals may come too late since there are ecological irreversibilities (e.g., extinct species).

The changes that have taken place in the international trade in pollution-intensive goods are analyzed in the paper by Low and Yeats (1992). These authors identify 40 three-digit SITC industries with ECC of at least 1 percent of total sales and analyze the changing composition of countries with a revealed comparative advantage in dirty industries. The latter is measured by the quotient of the share of country i in world exports of commodity j and the share of country i in total world exports. They find that in the period 1965–88 the share of dirty products in world trade has declined from almost 19 percent to less than 16 percent, and that the industrialized countries account for the lion's share (77 percent and 74 percent of world trade, respectively). The comparative advantage in dirty industries is dispersing, and the dispersion is greatest in direction of the LDCs. The deconcentration of comparative advantage occurred also in nonpolluting industries, but at a much lower rate.

Ray (1995) studies the structure of US protection in the seventies and eighties and points out that trade policy was tilted towards protection of heavily polluting domestic industries, thereby making market access for their competitors from other industrial countries like Japan or Europe more difficult.

---

[10] See Chapter 9, and de Bruyn and Heintz (1999). Dean (1999) incorporates the income effect on emissions in a model that portrays the composition effect and applies the model to China. She finds that increased openness leads to more emissions which indicates that China may have a comparative advantage in pollution-intensive goods. At the same time income growth reduces pollution via a technique effect.

[11] Another problem is the need for international cooperation if the environmental problem is transboundary, or even global; see Chapter 10.

The generalized system of preferences however favored dirty industries' exports to the US more than average.

Grossman and Krueger (1993) analyze the environmental consequences of the North American Free Trade Agreement (NAFTA). First, they test whether the US–Mexican trade pattern is influenced by environmental regulation. They regress the ratio of US imports from Mexico to total US shipments per industry on the factor shares of labor, human capital, and physical capital, the injury rate as a proxy for labor standards, and the share of ECC in total output. They find the influence of ECC (and of the injury rate) to be small and insignificant. They repeat the exercise for the maquiladoras industries, which are primarily low-skilled labor-intensive assembly activities in Mexico along the US–Mexican border with an 80 percent reexport requirement to the US. But even in this special case, which suggests itself as an escape from environmental regulation, they do not find any significant relationship between pollution abatement costs and sectoral activity in Mexico. In a second step, the authors calculate the expected changes in pollution due to the NAFTA-induced trade liberalization. They combine the predictions of the Brown, Deardorff and Stern (1992) computable general equilibrium model for sectoral post-NAFTA changes with toxic release data for the sectors and find that toxic releases decrease somewhat in Mexico (mainly due to a decline of chemicals) and increase in the US and Canada.

Bommer and Schulze (1999) analyze the pollution content of trade before and after NAFTA. Like Grossman and Krueger (1993), they combine estimated trade changes, generated from computable general equilibrium models, with sector-specific toxic release. They find that US export production is more pollution-intensive than US imports (by 37 percent), but that the toxic release content of expected trade *changes* is much more dramatic – the export changes are two and a half times dirtier than the import changes.

These results on NAFTA liberalization are in line with the more general observation by Anderson (1993) who argues that developing countries tend to protect their import-competing dirty heavy industries, while disadvantaging their primary and labor intensive sectors. On the other hand, industrial countries often protect labor intensive sectors such as textiles and agriculture. Trade liberalization will shift production to the sectors with comparative advantage: poor countries most likely will get relief from environmental pressure by specializing in labor-intensive and relatively clean production, whereas industrial countries will tend to specialize in dirtier industries. Middle income countries, the Asian tigers for example, are, however, likely to experience a shift to pollution-intensive industries such as steel and shipbuilding. This would be in accordance with the inverted U-shape relationship between pollution and development. There are more sectoral studies which reinforce these findings. A survey is to be found in Bommer (1998).

Two computable general equilibrium (CGE) models try to estimate the effects of the Uruguay round on the environment. Perroni and Wigle (1994) construct a multisector, multiregion model with suboptimal abatement levels and show that trade liberalization *per se* has modest negative impacts on the environment in all three regions – North America, other developed countries, and low and middle income countries. In contrast, a move to full internalization, both coordinated and unilaterally, has much more considerable positive effects on environmental quality. However, on top of the usual substantial difficulties with CGE models the choice of the existing internalization rates is necessarily speculative. The model should therefore be viewed as an appropriate structure to think about this problem, and the results to represent a rough educated guess rather than a definitive estimate. This applies to the model of Cole *et al.* (1998) as well. These authors adopt the Grossman and Krueger (1993) research design to analyze the impact of the Uruguay round liberalization on five air pollutants for a number of countries and regions. By combining a CGE model of sectoral changes induced by the Uruguay round (Francois *et al.* 1995) and a data set on local air pollution (Hettige *et al.* 1994) they calculate the composition effect resulting from a changed production pattern. Moreover, they calculate the scaling and technology effect by using the environmental Kuznets curve (EKC) which relates pollution intensities to GDP per capita. Expected gains from trade liberalization give rise to a movement on the EKC, whereas the composition effect shifts this curve. The authors find that the composition effect harms the environment in the developed countries and alleviates the environmental situation in the developing world as the textile industry moves to the developing countries while heavy industries concentrate in the developed world. The combined scale and technology effect (which includes the income-induced increased preference for environmental control) affect the developed world positively and the developing world negatively. The aggregated effect predicts an increase in all five emissions; in the developed countries the emissions of three local air pollutants will fall, while nitrogen dioxide and carbon dioxide emissions will generally rise.

Rather than simulating, Antweiler, Copeland and Taylor (1998) estimate the impact of the trade regime on pollution intensities with the help of a panel regression that measures the sulphur dioxide concentration at various sites. Like Grossman and Krueger, they employ the data compiled by the Global Environmental Monitoring System (GEMS); their data set consists of 2621 observations from 293 sites located in 109 cities of 44 countries, spanning the years 1971–96. The challenge they take on (unlike most of the other empirical papers) is to estimate composition, scale and technology effects separately. In the theoretical section that guides their empirical approach, they present a two sector trade model with one polluting sector to derive the determinants of these three effects. Firms choose gross output, inputs and abatement levels to maximize profits; optimal abatement activity is shown to be a function of the goods price and the pollution tax which the government sets in order to maximize the representative agents' utility. Trade policy, say trade liberalization, influences on pollution through three channels: (i) the scale effect captures the increase in pollution that stems from increased economic activity after liberalizing trade (holding composition and abatement technology constant), (ii) the composition effect captures the shift in the production structure towards the pattern of comparative advantage. This comparative advantage is determined in the model by the relative capital–labor endowment ratio (Heckscher–Ohlin approach) and the level of abatement costs which is determined by the pollution tax (possible pollution–haven effect). Relative capital abundance will increase pollution and therefore trade liberalization, *ceteris paribus*, increase pollution in the capital abundant countries (the exporters of the polluting good) and lower it in relatively labor abundant economies, (iii) the technology effect captures tax-induced higher abatement activities; the tax is determined by the level of income (environmental quality is a normal good) and by the disutility of pollution which follows a time trend. Thus, high income countries will have a policy-created comparative disadvantage in pollution intensive goods, other things equal.

In their empirical approach, Antweiler *et al.* regress SO_2 concentrations (emissions per square kilometer) on GDP per km^2 to capture the scale effect, on the capital–labor ratio and its square to capture the composition effect, and on income per capita and its square and population exposed to pollution to measure the technology effect. They add site-specific fixed effect and city-specific weather variables to control for different natural cleansing abilities and seasonal effects. A time trend is to capture trends in public awareness of environmental problems, in abatement technology and in the world price of the polluting good. A dummy for communist countries captures the smaller public opposition against pollution in these countries. Their results confirm the model's predictions – pollution increases significantly with GDP/km^2 and with capital abundance, and decreases with income per capita (for the latter two components at a diminishing rate); communist countries pollute more and the time trend for pollution is significantly negative. No measure of openness to trade reaches standard significance levels. This is not surprising since openness affects countries differently and the aggregate effect need, therefore, not be significant. If openness (measured as exports plus imports as a share of GDP) is interacted with the employed country characteristics, it becomes highly significant. The direct effect now becomes significantly negative! The interaction effect with income is positive: openness makes richer countries cleaner and poorer countries dirtier which the authors interpret to support the pollution-haven hypothesis to be discussed further in the next section. The interaction with capital abundance is negative which reflects the environmental consequences of the factor-endowment explanation of international trade patterns. Trade thus influences pollution intensities through all three channels. The authors estimate scale and technology elasticities and conclude that if trade increases both output and income by 1 percent, pollution falls by approximately 1 percent. Trade is thus good for the environment. The Antweiler *et al.* analysis is a very thorough piece of work which produces very insightful results.

While it covers many different sites, it is limited to a single pollutant and includes only very few observations for many sites. Further research is highly desirable along the lines pioneered by Grossman and Krueger, and Antweiler, Copeland, and Taylor.

To sum up, while the influence of environmental policies on trade patterns is rather weak, there is considerable empirical evidence that trade policy has a strong impact on environmental quality. It must be noted though, that with a few notable exceptions described above most empirical studies apply US data on toxic release or environmental control costs also to foreign countries; this lowers the value of the estimates of actual changes in pollution caused by trade liberalization. Thus, these analyses do not account for international differences in regulation and technology, but reflect at best the environmental consequences of changing sectoral activities in the course of trade liberalization. Obviously, this is restrictive; the results must be interpreted with caution.

3.5. ENVIRONMENTAL CONTROL AND THE RELOCATION OF INDUSTRIES

Theoretical studies on environmental policy in an open economy in the international context (see Chapter 2) suggest that dirty industries have an incentive to relocate to countries with lower environmental standards ("industrial flight hypothesis"). Similar fears have been voiced in the political debate. What is more, less developed countries (LDCs) have an incentive to strategically lower their standards to attract capital from industrialized countries ("pollution haven hypothesis"). Do these hypotheses stand up to closer empirical investigation? Apparently, such a regulation induced capital flight has not taken place on a large scale, otherwise the ECC approach would have significantly contributed to the explanation of the existing trade pattern. If compliance cost differences had established a significant comparative advantage, this would have created incentives for restructuring of resources within the LDCs and would have changed foreign direct investment (FDI) flows. But the findings presented in sections 3.3 and 3.4 do not support this conception. Still, for few selected sectors environmental regulation may have played a considerable role, especially if the low level of ECC does not reflect the "true costs" of environmental regulation (see section 2.2).

Walter (1982) studies the pattern of FDI flows from Western Europe, the USA and Japan and finds little evidence that it has been influenced by environmental regulation. Duerksen and Leonard (1980) points out that most of the FDI in dirty industries has gone to other industrialized countries with similar standards. This pattern has been stable over time. Bartik (1988) showed that the location decision of the Fortune-500 firms within the USA was not influenced by state-imposed pollution control expenditures, and that it had only a very small effect on the startup rate for new businesses (Bartik 1989). Leonard (1988) presents case studies of FDI in Ireland, Mexico, Romania, and

Spain; he cannot find any systematic evidence that FDI in these countries were influenced by US environmental regulations. The reason is that investment decisions are very complex and that ECC differences play only a very minor part in it since they are so small compared to other cost differences. Friedman *et al.* (1992) study the location decision of foreign capital within the United States. They show that environmental control, measured by the ratio of state-specific abatement expenditures and state GDP, is not significant in explaining the choice of state. It is not clear, though, that their indicator accurately measures the stringency of regulation. There are several more, mostly sectoral studies, surveyed by Pearson (1987) and Jaffe *et al.* (1995) which confirm the overall picture.

Levinson (1996) uses a conditional logit model of plant location choice to analyze the impact of different environmental regulations across US states on the geographical distribution of pollution-intensive industries. Using microdata from the census of manufactures and various measures of regulation stringency he finds very little evidence that regulation differences affect the interstate location of newly established plants. The surveys by Levinson (1995, 1996) on interstate plant location and environmental regulation corroborate this result. Levinson conjectures that it might simply be too costly to design different production processes for each location/regulation. Alternatively, the pollution-intensive industries may happen to be the least footloose. Mani *et al.* (1996) apply a conditional logit model of location choice to all new investment in India in 1994. After controlling for factor price differentials, infrastructure, and agglomeration they find no evidence for an impact of different enforcement levels of environmental regulation at the state level on the location choice. If differences in environmental regulation do not affect the allocation on the federal level, it is not very surprising that they do not do so on the international level, where mobility costs are much higher. On the other hand, regulation differences may be more pronounced on the international level. In spite of that, Henderson (1996) shows that US firms take air quality regulation into account when deciding where to locate within the US. Those localities that are not in compliance with the national ambient air quality standards (NAAQS) are subject to stricter enforcement of federal regulations; for instance, newly established plants will be subject to stricter regulations concerning equipment specifications. Henderson shows that pollution-intensive firms have relocated to areas that comply with these standards and therefore have more lenient regulations.

The study by Gray and Walter (1983) represents a further exception to the overall picture. This study investigates the relocation of copper-smelting, refineries, and factories in the chemical industry in response to regulation in Western Europe. The evidence is not incompatible with regulation-induced relocation to developing countries which allow a more polluting technology. Leonard (1988) also provides some evidence that US environmental regula-tion has led to some overseas investment for narrowly defined, highly pollution-intensive industries.[12] Diwan and Shafik (1992, table 15.1–2) report that developing countries account for only 22 percent of world output, but 43 percent of world carbondioxide emissions. This is due to lower emissions per unit of capital in the developed countries, which the authors attribute to stricter environmental policies. Ferrantino (1997) shows that US foreign investment undertaken by the pollution-intensive chemical industry is no different from all other manufactures; in particular, FDI does not flow increasingly into developing countries which can be assumed to have a more lenient environmental regulation. Ferrantino's research is motivated by the observation that although operating abatement costs are quite low, capital expenditures for abatement are significant at 10 percent of total capital expenditures for US manufacturing.

Why then have not more industries moved to pollution havens as McGuire (1982) suggested?[13] The first reason is the relatively low overall pollution abatement costs incurred in the north.[14] The benefits of relocation are simply far too low to justify a revision of the location decision which is determined by many other factors. The very small fraction of heavily polluting manufacturing may be an exception to this rule. In a 1990 survey carried out by the United Nations Conference on Trade and Development (UNCTAD) program on transnational corporations, most multinational firms claimed that environmental practices overseas are determined by environmental regulations in their home countries (UNCTAD 1993). Birdsall and Wheeler (1992) and Wheeler and Martin (1992) offer some explanations for this observation: even if multinational firms locate some of their production in developing countries with lower standards, these firms might have strong incentives to apply higher "Western" standards, because of fears that they might lose reputation and, as a result, consumers in their markets in developed countries ("green consumerism") or that their products may not meet the standards in the export markets.[15] Moreover, in case of an environmental accident, law-abiding multinational firms might nevertheless be subject to liability claims. Firms, finally, invest in their equipment as a package – they install the latest, most efficient and cleanest equipment. To unbundle the equipment in favor of dirtier processes might not be profitable or feasible. If multinational firms duplicate the cleaner technology used in advanced countries with more stringent regulation, they might even have an incentive to lobby for higher environmental

---

[12] Some minor relocation took place when some wood-furniture firms from Los Angeles moved to Mexico as a reaction to intensified environmental control by the city of Los Angeles; this is documented by the US General Accounting Office (Levinson 1995: 434).

[13] Theoretical models portraying the pollution-haven hypothesis are surveyed in Chapter 3.

[14] ECC might, however, constitute a nontrivial cost component in the south, because the relative scarcity of capital makes capital-intensive pollution abatement relatively more expensive than in the north.

[15] Paper produced with chlorine, for example, has traces of dioxin and can therefore not be exported to Germany (Birdsall and Wheeler 1992).

standards in order to raise domestic rivals' costs (Ferrantino 1997). The cleaner technology of multinational firms provides another argument for liberalizing (sectoral) capital controls and provides an explanation for the rather small shift of comparative advantage in dirty industries towards the developing countries, as noted by Low and Yeats (1992).

3.6. SUMMARY

The empirical analyses provide a somewhat mixed picture. They are hampered by conceptual difficulties as well as insufficient data quality and availability, the latter of which makes cross-country comparisons problematic. Given these caveats, the following results emerge: First, trade flows seem hardly to be influenced by differences in environmental standards. Environmental control costs are too small to establish a substantial comparative advantage. Second, the trade regime has a strong influence on environmental quality. While there seems to exist an inverted U-shape relationship between pollution intensity and the stage of development, it is remarkable that closed developing countries seem to be much more polluting than open developing economies. This difference is the more pronounced, the faster the LDCs are growing. Trade liberalization thus gives rise to a reduction of pollution in developing countries, but may increase pollution in middle income and industrialized countries. Lastly, there is no evidence that firms from industrialized countries have relocated to a large extent to "pollution havens" to escape stricter regulation.

4

The Political Economy of International Trade and the Environment

GÜNTHER G. SCHULZE AND HEINRICH W. URSPRUNG

In this chapter we pick up the thread from Chapter 2 and turn to the political-economy aspects of the interrelationship between economic integration and environmental policymaking. As in Chapter 2, we first survey models which analyze trade and environmental policies in a regime of low international capital mobility (sections 4.1–4.4), then turn in section 4.5 to a regime of high international capital mobility and address the so-called industrial flight hypothesis, that is, the issue of industry relocation induced by environmental regulation (production standards). Since the literature on the political-economy aspects of economic integration and environmental policy is still in its infancy, we will discuss the few existing contributions in somewhat more depth than we did in Chapter 2. We conclude by presenting and discussing the sparse empirical evidence relating to the political economic view in section 4.6.

4.1. THE POLITICAL-ECONOMIC INTERRELATION BETWEEN TRADE AND ENVIRONMENTAL POLICY

Even though it appears expedient to formulate trade and environmental policies separately, these policy domains have never been neatly kept apart in the political sphere. Kym Anderson and Richard Blackhurst, in their influential 1992 collection of studies on the *Greening of World Trade Issues*, introduce the political-economy view into the scholarly debate by pointing out that "the trade/environment area has an above-average risk of being exploited by special-interests groups to their own benefit and at the expense of the general interest." In their capacity – at the time of writing – as members of the Economic Research Division of the GATT Secretariat, they emphasized in particular "the risk that traditional protectionist groups will manipulate environmental concerns in order to reduce competition from imports" (p. 20). While the influence of environmental interests on endogenous trade policy (ecological protectionism) is, of course, very conspicuous, it should be noted that the reverse relation, that is, the influence of international trade on the determination of environmental policy (ecological dumping), plays a similarly notorious role in endogenous policy formation. In any case, it is obvious that both trade and environmental policy are determined simultaneously in the political process; regarding one of the two policy regimes as exogenous is thus clearly not a matter of portraying political reality, but rather a matter of analytical convenience. Using the exogenous or endogenous treatment of the two policy regimes as the main distinguishing feature of the models covered in this chapter, we arrive at the taxonomy presented in table 4.1. The upper right hand cell refers to models in which neither of the two policies is formed in a way that could lay claim to a realistic portrait of the political process. One can thus subsume in this cell the normative models surveyed in Chapter 2. We begin our survey in section 4.2 with the upper left hand cell, that is, with a model in which environmental policy formation is not analyzed. The focus is thus on the nexus singled out by Kym Anderson and Richard Blackhurst in the above quote: to what extent can

Table 4.1. *Political economy of trade and environmental policy: the theoretical literature*

	Trade policy endogenous	Trade policy exogenous
Environmental policy exogenous	Section 4.2 *Hillman and Ursprung (1992, 1994)* • LOPEC • imperfect competition • partial equilibrium	Chapter 2
Environmental policy endogenous	Section 4.3 *Schleich (1999)* • SMOPEC • perfect competition • general equilibrium[a] *Schleich and Orden (1999)[b]* • LOPEC • perfect competition • general equilibrium[a] *Rauscher (1997)* • SMOPEC/LOPEC • perfect competition • partial equilibrium	Section 4.4 *Bommer and Schulze (1999)* • SMOPEC and LOPEC • perfect competition • general equilibrium *Fredriksson (1997a, 1999a)* *(Aidt 1998; Fredriksson 1997b)* • SMOPEC • perfect competition • general equilibrium[a] *Bommer (1996)[b]* • LOPEC • imperfect competition • partial equilibrium

[a] We will qualify this characterization below.

[b] These models portray also transboundary pollution, the other ones only local pollution.

Notes: SMOPEC: small open economy; LOPEC: large open economy.

environmental concerns be exploited by protectionist interests? We then turn in section 4.3 to models in which both trade and environmental policies are determined endogenously (lower left hand cell) and conclude in section 4.4 by presenting models addressing the question as to whether a downward competition of environmental standards ("race to the bottom") is inevitable in the course of global economic integration which is then taken as given (lower right hand cell).[1]

4.2. THE INFLUENCE OF ENVIRONMENTAL CONCERNS ON ENDOGENOUS TRADE POLICY FORMATION

The political influence of environmental groups on the determination of international trade policy has been formally portrayed in a model based on electoral competition in Hillman and Ursprung (1992) and, more formally, in Hillman and Ursprung (1994). The focus on trade policy implies that first best environmental policy intervention is ruled out. This framework lends itself to answering two kinds of interrelated questions. The first question is, will environmental groups use their political influence to support free trade or protectionist policies? In other words, are environmental groups agents of free trade or protectionism? A further aspect of this question concerns the potential for consensus or conflict of interest among environmental groups in different countries: do environmental groups in different political jurisdictions have common trade policy objectives? If environmental groups in two trading economies both seek mutual free trade or mutual protectionism, there is consensus; if one group seeks protectionism and the other free trade, there is a conflict of interest. Strategic elements may also, in principle, affect environmentalists' decisions regarding support for alternative trade policies: a Prisoners' Dilemma arises if the environmental interest groups in two trading economies confront a mutually beneficial policy choice that maximizes the aggregate environmental gains (or minimizes losses), but it is nevertheless individually optimal for country's environmentalists to defect and choose the alternative policy. To state a major finding in advance, the environmentalists' position depends on whether the externality occurs in *production* or in *consumption* and whether they also care about environmental quality *abroad*. After having established the trade-policy stance of the environmentalists, the second question arises as to how the environmentalists influence trade policymaking. To answer this second question, the political process needs to be portrayed in detail.

To establish the interest of environmentalists in influencing trade policy, Arye Hillman and Heinrich Ursprung, following their 1988 article, use a partial equilibrium model of a representative import-competing oligopolistic industry.

---

[1] Related to this literature are the studies investigating the influence of administered (contingent) protection on endogenous environmental-policy formation (see Leidy and Hoekman 1994, 1996).

The trade policy choices are restricted to the policy pronouncements of two competing candidates or political parties. The losers from protectionist trade policies are foreign producers seeking market access, and the gainers are domestic import-competing producers. Domestic consumers are of course also losers from protectionist policies, but, on grounds of free-rider problems and small stakes in the outcome of policy determination, consumers can be presumed to be "rationally ignorant" (as they in general are) with respect to the trade policy that is to be adopted for any one industry. The protagonists with sufficient stakes in the outcome are producers who gain or lose as domestic market access to imports is denied.[2]

In this well established framework, Arye Hillman and Heinrich Ursprung now introduce environmental interests as a third group seeking to influence the political determination of trade policy. The issue is, whether the environmentalists side with the free trade or the protectionistic candidate and, consequently, whose prospects are improved through the entry of the environmentalists into the political arena.

The answer to this first set of questions – are environmentalists agents of free trade or protectionism, do national environmental groups have common cause with the trade policies of their comrades abroad, or do they confront strategic considerations as in a Prisoners' Dilemma? – depends upon the source of the adverse environmental impact (which lies in the consumption or production of a good) and on whether the environmentalists' concerns with the adverse environmental impact transcends their national boundaries. One can define a "green" as an environmentalist who is concerned with the adverse environmental impact in his home country only, and a "supergreen" as an environmentalist who is concerned with an adverse environmental impact in both his home country and in the country of his foreign trading partner.[3]

When the adverse environmental impact is associated with *domestic consumption*, the environmentalists described as greens oppose imports because this increases the domestic price and therefore curbs domestic consumption and environmental degradation. The greens are thus allies of protectionist domestic interests. The less competitive the domestic industry is, the greater the benefits of this political alliance are. The greens would ideally wish production to be undertaken by a protected monopolist – for the higher the concentration in the domestic industry, the greater the decline in domestic production, and therefore in autarky the greater the decline in domestic consumption (the cause of the adverse environmental impact). The protectionist policy not only reduces

---

[2] An alternative portrait in which the domestic export industry supports the liberal trade policy stance at home is drawn by Gould and Woodbridge (1998).

[3] Some environmentalists are especially concerned about spillovers, that is, domestic actions which cause pollution abroad. "Greens" do not have these concerns. "Supergreens" do, but notice that their concerns are much more encompassing. A true supergreen cares about the state of the environment independent of the source of pollution. Of course, if the pollution is global, this distinction becomes meaningless.

consumption by curbing imports, but also allows domestic producers to take advantage of reduced import competition to exploit their domestic market power, thus further decreasing consumption. Supergreens as well as greens are protectionist agents in these circumstances: supergreens wish to minimize consumption everywhere, which again is achieved by protectionist policies in each country.

However, when the source of the adverse environmental impact is *domestic production*, the situation is more complex and environmentalists confront strategic problems in their choice of which trade policy to support. Environmentalists who are greens wish to minimize production at home, and benefit if imports replace domestically produced output in domestic consumption. However, imports produced abroad disadvantage the foreign environmentalists, who are concerned with pollution in their own country. Therefore, a potential conflict arises between environmental interest groups in the two trading economies, and the environmental interest groups potentially confront a Prisoners' Dilemma. The potential for the Prisoners' Dilemma is present because of the incentive confronting environmentalists in each country to "free ride" off one another by supporting free trade policies at home that increase pollution abroad via foreign production for export. The best outcome for environmentalists in both countries is mutual protectionism where no country's environmentalists free ride off the other. Whether or not the Prisoners' Dilemma arises to confront the environmentalists in two trading economies depends in a somewhat complex way upon market structures in the industries producing the offending good in the two economies, and on the degree of substitutability in consumption between the domestically produced good and imports in each economy. The prospects for defection (free riding behavior) depend upon the foreign supply response in producing exports for domestic consumption. Asymmetries in market structures in the two trading economies therefore give rise to asymmetries in free riding incentives. When market structures are similar, there are shared benefits from a mutual end to free riding, that is, from protectionist policies in each economy. Substitutability in consumption affects the likelihood of a Prisoners' Dilemma in support for trade policies since, with low substitutability, protection evokes a lower domestic output response – in the limiting case of zero substitutability, protection would not affect domestic output (and hence domestic pollution) at all. Lower levels of substitutability therefore expand the range of combinations of market structures in the two trading economies for which there is no Prisoners' Dilemma and environmentalists in both economies support protectionist policies.

When those conditions which give rise to the Prisoners' Dilemma are present, the environmentalists in different countries confront a problem of international coordination if they are to be effective and not work at cross-purposes to one another. However, such international coordination is difficult because of the gains to any one country's environment from defection. On the other hand, environmentalists who are supergreens do not confront the potential for

Prisoners' Dilemma situations when deciding which trade policy to support. The supergreens internalize the potential conflicts since they seek to minimize the total adverse environmental impact associated with international trade in both countries. Hence, the supergreens have no incentive to free ride off each others' imports, and they support protectionist candidates.

Overall, then, only a sufficiently asymmetric market structure or the potential for the Prisoners' Dilemma stand in the way of the solution that environmentalists are agents of protectionist interests or at least supportive of these interests. Under such circumstances, environmental interests can be "captured" by protectionist producer interests. Moreover, it may be more effective for protectionist producer interests to channel their financial political contributions – that are the expression of political support – via environmentalist interest groups: it is politically more rewarding to plead protectionism via the environmental cause than via the self-interest of producer rents in an import-competing industry.

In order to assess the *influence* of environmental trade-policy preferences, Arye Hillman and Heinrich Ursprung proceed to endogenize the competing parties' trade policy pronouncements by assuming that the parties' constituencies, on the basis of the publicized trade policy platforms, provide them with financial contributions which, in turn, determine the probability of electoral success. The policy pronouncements are assumed to be chosen with the intention to maximize the probability of election. This portrait of the political process in which two games take place (the policy pronouncement game played by the political parties and the subsequent campaign contribution game played by the interest groups) represents an application of the interest-group cum electoral-competition approach which Rauscher (1997: 221) regards to be the most realistic approach to modeling the political process.

When environmentalists are absent from the political arena it has been demonstrated (see Hillman and Ursprung 1988) that the outcome of political competition is the following:[4] if policy pronouncements are made in terms of tariffs, each candidate takes an extreme position on the spectrum of policies in that one assumes a policy of free trade and the other a policy of prohibitive protection. If, however, policy pronouncements are made in terms of voluntary export restraints (VERs), the political equilibrium is characterized by the announcement of a common policy which, in general, implies incomplete protection. These results, however, do not survive if environmental interests enter the political arena. In the tariff regime, environmental interests can, for example, give rise to a partial convergence of the trade policy stances in a manner which is detrimental to the environmental interests. In the VER regime, the result of complete convergence at an intermediate level of protection continues to hold only as long as environmental concerns do not exceed a lower bound, that is, for sufficiently strong environmental concerns the political

---

[4] See Hillman and Ursprung (1994b) for a discussion of the employed equilibrium concept.

process either generates complete polarization or an extreme policy outcome (autarky or free trade).

4.3. SIMULTANEOUS DETERMINATION OF TRADE AND ENVIRONMENTAL POLICIES

The corruption approach to explaining endogenous policy formation has especially been designed to explain the *structure* of the implemented policies. In its simplest version (see Grossman and Helpman 1994), lobbies offer the incumbent government contributions which are contingent on the policy decisions taken. Since contributions are aimed at influencing the government's policy and not – as was the case in the electoral-competition model discussed in the previous section – at affecting the election outcome, the portrayed inter-action between organized interest groups and the government meets the cir-cumstances of corruption. The objective of the government is to maximize political support which is specified as a weighted sum of social welfare and total financial contributions received.

Grossman and Helpman (1994) use this modeling framework to investigate the structure of trade policy across industries in a specific-factor model of a small open economy. Their model falls just a little short of a full-fledged gen-eral-equilibrium model since it is based on quasilinear preferences and a numeraire sector; income and cross-price effects are thus suppressed. Also, the lobbies' bribes do not enter the circular flow of the economy except from decreasing the income of the lobby members. Most importantly, in the original model by Gene Grossman and Elhanan Helpman, no pollution externalities are supposed to exist and the focus is on trade-policy. Dixit (1996) has introduced more than one policy instrument in the Grossman–Helpman model, that is, production and consumption taxes/subsidies, still assuming no pollution. In a recent paper, Schleich (1999) extended the Dixit model to develop a positive theory of trade *and* environmental policy. To that end he assumes that either consumption or production of some industry output generates pollution. Two kinds of policies are analyzed: (i) domestic policies, that is, consumption and production taxes/subsidies which, of course, efficiently internalize production and consumption externalities, and (ii) trade policy measures, that is, tariffs/subsidies on imports and exports. The government cares about the environ-ment since pollution is assumed to directly affect social welfare and thereby political support. Environmental interest groups, however, are not supposed to enter the picture to compete with the producer interests in the bribery contest. Environmental concerns nevertheless do appear on the demand side of the political market since the producer lobby groups are supposed to internalize the environmental damage that their members suffer from pollution.

The qualitative features of the political-economic equilibrium of Joachim Schleich's model are quite intuitive. In the case of *pollution via consumption*, the equilibrium policy is simply the Pigou consumption tax when only domestic policies are available. In this case, the special interests of the industry lobbies cannot be accommodated since, in a small open economy, producer prices cannot be influenced with the help of consumption policies. When, however, domestic *and* trade policies are available, import tariffs or export taxes will be applied to benefit the organized producer interests and to harm the unorganized ones. Also consumption policy is used to adjust the domestic consumer prices to reflect world prices plus the Pigou tax. Political-support maximization and social-welfare maximization thus give rise to identical consumer prices and hence to the same (optimal) level of pollution. The inefficiency of the political process makes itself be felt solely in the protection of organized producer interests: organized industries produce an inefficiently high output. When only trade policies are available, pollution via consumption calls for an import tariff or an export subsidy since either reduces domestic consumption. The political support and the environmental motive of trade-policy intervention thus reinforce each other for organized industries. Whether trade policies alone give rise to a level of environmental protection which exceeds or falls short of the efficient level cannot be determined a priori.

In the case of *pollution via production*, it is not optimal for the government to apply trade policies when production policies are available since both policies change producer prices. However, trade policies additionally distort consumer prices. When production policies *are* available, organized clean industries obtain a production subsidy whereas organized polluting industries receive either a subsidy or a tax, depending on the political support impact of pollution as compared to the potential financial contribution of the industry. Environmental quality in the political equilibrium, as a rule, turns out to be suboptimal; only if marginal damage costs are decreasing, then lobbying may improve environmental quality too much. When solely trade policies are available, organized clean industries obtain import tariffs or export subsidies. Polluting industries receive, *in addition*, an import subsidy or an export tax – the joint effect being again ambiguous. Interestingly, it is also ambiguous whether production policies or trade policies alone will generate more pollution. In contrast to standard theory it is thus possible that the inefficient instrument (trade policy) gives rise to a better environmental quality than the efficient instrument (production policy).

In a follow-up paper to Schleich (1999), Schleich and Orden (1999) generalize the model from a small-country to a large-country economy. By doing so, three new aspects can be considered: market power, the nature of the externality which is either purely local or global, and the reactions of the other countries' governments. The last aspect allows for a comparison of a regime in which the governments do *not* cooperate on policies and a regime in which they do cooperate.[5]

---

[5] In the trade policy context this distinction corresponds to the "trade wars" and "trade talks" introduced by Grossman and Helpman (1995).

The results can be summarized as follows. In the noncooperative regime, the policies resemble the small-country results, plus an effect reflecting terms-of-trade considerations, that is, an optimal tariff argument, and an effect referring to global externalities. In the cooperative regime, the optimal tariff effect disappears and the global externalities are internalized. More specifically, if pollution occurs via consumption and consumption policies are available, environmental quality will no longer be optimal, as was the case in a small economy because the terms-of-trade are manipulated to accommodate the producer lobbies. This holds for the noncooperative as well as for the cooperative regime. If pollution occurs via production, just as in a small economy, the socially desirable level of environmental quality will not be reached. This result holds even if the governments cooperate. Moreover, it can be shown that the more efficient regime (cooperation) may give rise to a lower level of environmental quality than the less efficient regime (noncooperation).[6]

An intriguing aspect of these models is that they provide a rationale for the political choice of inefficient policy instruments. While *governments* always prefer efficient instruments in the setup portrayed by the corruption approach, the involved *interest groups* may not do so because efficient instruments elicit higher bribes (see Grossman and Helpman 1994: 847–8). In small economies, the industry lobbies might thus prefer trade policies over the more efficient domestic policies (see Schleich 1999: section 2.4) whereas, in large economies, the industry lobbies in both countries might prefer the less efficient non-cooperative over the more efficient cooperative regime. Restrictions of the set of policy instruments available to the government can thus, in principle, be endogenized by modeling the interaction between the political stakeholders at a prior constitutional stage of the political-economic game (see Schleich and Orden 1999). On the liability side of the corruption approach to modeling the political process, two aspects are in the forefront, namely (i) the unrealistic informational requirements with respect to computing and signaling the optimal contribution schedules and (ii) the kind of portrait drawn of the political process by the assumed sequencing of the moves. Portraying the political process as an explicit exchange of gifts or a "sale" of policies does, after all, seem to restrict the applicability of the approach to a subset of political systems and cultures which is not congruent with the political settings usually investigated.

In his monograph, Rauscher (1997) employs a less cynical approach to modeling the political process. He attempts, however, to answer a similar set of questions: can *environmental production standards* which are too lax be explained by the particularities of the political process in open economies? Are *environmental consumption standards* liable to be abused as an instrument of disguised protection? In contrast to the paper by Joachim Schleich, Michael

---

[6] International environmental cooperation is analyzed in Chapter 10.

Rauscher employs a partial equilibrium model to analyze these issues. To justify this choice, he convincingly argues that neglecting general-equilibrium repercussions of policy measures is not unrealistic in a positive portrait of the behavior of political agents who are usually unaware of the general equilibrium effects of their activities. The second main difference between the two models consists, as mentioned above, in the modeling of the political sector. Michael Rauscher uses the political support function approach which maintains that the policymaker's objective is a weighted average of the interests directly affected by the respective policy measures. This modeling approach has its roots in the pioneering work by Stigler (1971) and Peltzman (1976) on regulatory capture. It can be interpreted as a reduced form of a more complicated model of political-economic interactions, such as, a voting game under incomplete information (see Yang 1995) or, indeed, a Grossman–Helpman type model of corruptible politicians.

Michael Rauscher considers a single import-competing industry. The imported good and the domestically produced good, which is not exported, are incompletely substitutable in consumption. The domestic firms are price takers. The production technology exhibits constant returns to scale and uses the environment (pollution) as well as a specific factor, which is fully employed. The government can use three instruments to protect the environment from pollution and/or the domestic industry from import competition. The first instrument is a consumption tax which, in principle, can be used to internalize consumption externalities. If, however, the consumption tax on the imported good contains an element of discrimination against imports, the surtax is nothing but an import tariff. Consumption taxes (tariffs) directly increase the consumer price of the respective good and thus have a positive influence (via reduced demand) on the environment. Second, the government may specify product standards for domestically produced and for imported goods in order to put a cap on pollution via consumption. Product standards increase the unit-costs of production and decrease pollution via consumption at home. They also have a great potential for being abused as hidden protectionist measures, maybe even more so than consumption taxes. The third policy measure is an emission tax on the production process. Emission taxes also increase the unit-costs of production and reduce pollution; in Rauscher's model they have effects which are qualitatively similar to those of product standards.

Social welfare maximization in this framework gives rise to the standard optimality conditions: the optimal tax rates equal the marginal environmental damage and the optimal product standards are chosen such that the marginal cost of increasing product quality equals the marginal improvement in environmental quality. Michael Rauscher compares this first-best policy with the results which emerge if one or more of the policy instruments are captured by an interest group. To portray the capture of policy instruments by specific-factor or environmentalist interests, he assumes that the government maximizes political support which, in addition to traditional social welfare, encompasses (i) the income accruing to the industry-specific factor, (iia) the quality of the domestic environment (greens), and (iib) environmental quality abroad (supergreens). Environmental quality and the income accruing to the owners of the specific factor are weighted with a parameter measuring the political influence of the respective interests. Assuming that environmental quality is a linear function of pollution, Michael Rauscher arrives at the following conclusion for the capture of a single instrument. As a consequence of the political influence of the specific-factor interests, quality requirements which imported goods have to meet and the tax on consumption of the imported good is too high and the tax on consumption of domestic goods is too low. The influence of the environmental interests (for example with respect to emission taxes) is less straightforward. Since policies that reduce domestic emissions and consumption of the domestic good (direct effect) tend to raise foreign emissions and the consumption of the foreign good (indirect leakage effect), their position always depends on the parameters of the model. If the direct effects of environmental regulation, however, dominate the indirect effects, then industrial and green interests have a common interest in strict quality standards for the foreign good and high consumption taxes (tariffs) on the foreign good. Since the protectionist content of quality standards can more easily be obfuscated than that of tariffs, such standards are likely to be the first choice if environmental and trade protection are to be combined. Somewhat surprising is the result that environmental *and* industry interests may both advocate and thus implement high emission taxes and strict quality standards for the domestically produced good. This result, however, is an artifact of the model in which the sector-specific factor is used for production *and* pollution abatement.

If *all* available policy instruments are subject to special interest capture, the political-support maximizing policy palette becomes somewhat counterintuitive and perhaps also counterfactual. A large weight of the *industry interests* in political support brings about low consumption taxes levied on the domestic good, *high* emission taxes, and *high* environmental quality standards of domestic goods. A large weight of the *green interests* yields the standard results (high consumption and emission taxes and tight product standards) and a large weight of the *supergreen interests* leads to higher tariffs and to less restrictive quality standards for the imported good. As in the model by Joachim Schleich surveyed above, the mechanism which generates these striking results is the feature that the government always uses the most efficient instrument to secure political support from the respective interests. Owners of specific production capital are offered low consumption taxes, owners of pollution-abatement capital are offered high product standards and emission taxes, and supergreens are offered tariffs in exchange for political support.

Michael Rauscher concludes that his model obviously still lacks some of the aspects which are important in the real policymaking process. He also singles out one feature which he suspects to represent the most prominent omission, namely obfuscation (see Magee *et al.* 1989: chapter 18). To properly model obfuscation, one would need to introduce incomplete information and the strategic use thereof. We turn to a model using such a setup in section 4.5. Michael Rauscher brings his analysis to an end by adding some remarks on how the derived results may change in a large economy framework.

4.4. THE INFLUENCE OF ECONOMIC INTEGRATION ON ENDOGENOUS ENVIRONMENTAL POLICY FORMATION

The question as to how a given round of trade liberalization is likely to influence environmental policymaking is investigated in a recent paper by Bommer and Schulze (1999). The model used by these authors is basically a standard perfect-competition, two-sector trade-model with sector-specific factors. The standard model is augmented by the assumptions that, first, the environment is used as an additional factor of production (pollution) and second, environmental regulation is endogenous via a political-support maximizing government. The focus of the paper is thus on the general equilibrium effects of inter-industry trade on factor remuneration which, in turn, influences environmental regulation.

In the model by Rolf Bommer and Günther Schulze, the income-redistribution effect of environmental policy is a consequence of the assumption that one sector relies (more) on the environment as an additional input to its production process. This could, in principle, be either sector. Since, however, industrial countries' export sectors tend to be more polluting than the import-competing sectors, it is assumed that only the export sector uses the environment in the production process. Pollution via production can be restricted with the help of environmental standards set by the regulator. This policy brings about the crucial redistribution effects: tight environmental standards reduce the productivity of capital and labor in the "dirty" export industry thereby driving labor from the dirty export industry to the competing "clean" import industry where the relative productivity of capital is enhanced. The two sectors compete with each other for the mobile factor, that is, labor. The dirty export sector is directly affected by environmental regulation, whereas the import-competing sector is indirectly affected via the ensuing inter-industry labor movements. The resulting conflict of interests with respect to environmental policy is readily identified: environmentalists and the owners of capital in the import-competing sector favor tight environmental regulation, whereas the owners of capital in the export sector and workers favor a more lenient regulation. The regulator maximizes political support (which derives from the capitalists' residual claims in the two sectors, the wage rate and the level of pollution) by balancing the various sources of political support at the margin. The result depends, of course, on the overall state of the economy which, in turn, is determined by the ruling constraints. Starting out from a political-economic equilibrium of a protectionist trade-policy regime, an exogenous policy shift towards a more liberal trade policy will change the conditions on which environmental policymaking is based. Trade liberalization increases profits earned in the export sector, reduces profits earned in the import-competing sector, and increases the wage rate in terms of the import good and thus presumably also benefits labor interests. Since environmental regulation is portrayed as a restriction on total pollution, trade policy does not affect the interest of the environmentalists.[7]

Trade liberalization, by benefiting the export sector and labor interests, and hurting the import-competing sector thus disturbs the economic basis underlying the political-support maximization calculus. In order to reestablish a political-economic equilibrium, the government will transfer some of the liberalization gains from the export industry and from labor to the losers of economic integration by tightening the environmental standards. Environmental policy and trade policy are substitutive tools for income redistribution, but they do have a different substitution relationship for environmental interests. These interests stand to gain from a policy substitution which works according to the rule "share the gain and share the pain". The overall conclusion of the paper by Rolf Bommer and Günther Schulze is thus that trade liberalization will give rise to an endogenous shift in environmental regulation towards tighter standards in industrial countries in which the export sector is more polluting than the import-competing sector with the consequence that environmental quality improves in the course of increasing trade integration.

Fredriksson (1997a) also deals with the determination of environmental policies in small open economies. Even though the focus is not on trade policy, some important insights with respect to trade integration can be gleaned from Fredriksson's model. The political sector is portrayed with the help of the corruption approach popularized by Grossman and Helpman (1994). However, unlike the model by Schleich (1999) (see section 4.3 above), which also employs this approach, trade policy is assumed to be given – a regime of free trade prevails. Moreover, Fredriksson restricts his analysis to only two sectors: a clean numeraire sector which only uses labor in the production process, and a dirty sector which uses labor and capital as inputs. A second difference as compared to the model by Joachim Schleich consists in the specification of the agents' preferences. There are three types of agents: workers and capitalists, who only care for their respective market incomes, and environmentalists, who also care for the environment. Workers are assumed not to form a lobby group whereas environmentalist and industrialists do.

Explicitly taking into account lobbying by environmentalists, Frederiksson (1997a) arrives at the following result: the equilibrium pollution tax varies

---

[7] In an alternative setting, Rolf Bommer and Günther Schulze analyze the effect of environmental standards per unit of output. In this context environmentalists oppose trade liberalization, other things being equal, since it increases the export sector's output and thus pollution.

negatively and total pollution varies positively with the given producer price. The producer price can be interpreted to be composed of the world market price and a component reflecting protectionist trade policy intervention (e.g., a tariff). Assuming that the polluting sector is import competing, we can thus conclude that trade integration, in this model, gives rise to a more restrictive environmental policy and to a cleaner environment. The intuition behind this result is straightforward. First of all, environmental policy deviates from the welfare maximizing policy because not all interests form lobbies. Second, the unambiguous effect of trade integration is due to the fact that the industrialists' *marginal* profits decrease as the producer prices decrease whereas marginal pollution remains constant. This gives the lobby of the environmentalists an advantage *vis-à-vis* the lobby of the industrialists, whose marginal benefit of lobbying is continuously reduced in the course of economic integration. Pollution itself decreases in the course of trade integration because the direct effect of the producer price on output is augmented by the indirect political effect which also reduces production via a more restrictive environmental policy.[8]

Interpreting decreasing world market prices as a reduction in trade protection implies the assumption that redistribution effects via tariff *revenues* do not influence the political-economic equilibrium. This appears to be a reasonable simplification. Nevertheless, in his 1999 follow-up paper, Per Fredriksson provides a more general analysis of the influence of trade liberalization on environmental policy which includes the tariff-revenue effect. There is a second difference between the two models. The 1999 paper allows firms to engage in (costly) pollution abatement in order to reduce pollution-tax payments. It is not surprising that the more general model incorporating tariff-revenue and pollution-abatement effects as well as convex (instead of linear) supply functions produces results which are less sharp than those arrived at in the more special model. Under these more general circumstances it can be shown that the pollution tax rate may decrease and total pollution may increase with a more liberal trade policy.

Two further variations on the theme of endogenous environmental policy formation in open economies are to be found in Aidt (1998) and Fredriksson (1997b). Both of these articles, however, focus on the political choice of the environmental policy instrument and not on the influence of trade integration on the environment. The model analyzed by Fredriksson (1997b) is very similar to the one presented in the companion paper of the same year. The environmental policy choice is between a pollution tax and a pollution abatement subsidy. It is shown that the inefficient pollution abatement subsidies are nevertheless used in the political-economic equilibrium to redistribute income

---

[8] To be sure, the assumptions underlying this result are quite restrictive. The assumed marginal disutility of pollution is at odds with the typical notion of increasing marginal disutility of pollution. Moreover, it is unclear why the clean sector should not form a lobby of its own.

by governments seeking political contributions. Moreover, it transpires that total pollution may well be increasing in the pollution abatement subsidy rate if the subsidy rate is treated as an exogenous policy variable (see Fredriksson 1997a: section 4). Toke Aidt's model differs from the models designed by Per Fredriksson on three accounts. First, it consists of *many* industries each of them producing an internationally traded good. Second, production in all non-numeraire industries makes use of raw materials, and the processing of raw materials (energy) is assumed to have a negative external effect on the environment. All individuals, finally, care about market income *and* the state of the environment. The interest groups thus have "multiple goals", that is, they advocate industry-specific *and* environmental interests. The two environmental policy instruments which are considered are production and input (raw material) taxes/subsidies. Input taxes are of course in this context equivalent to pollution taxes. Again, one can conclude that inefficient instruments (the production tax) will be used in the political-economic equilibrium due to the government's income redistribution objectives. If not all industries are organized in lobby groups, the policy outcome will not completely internalize the economic externality. In the general setup employed by Toke Aidt the overall level of pollution can go either way as compared to the efficient level. Unambiguous results only emerge in very special cases.

Trade integration, by its very nature, does have international repercussions. Merely on account of this, restricting the analysis to the framework of small open economies endowed with industries operating under perfect competition cannot really do justice to the investigated phenomenon. Moreover, it is often observed that national policymakers are very hesitant to forge ahead with restrictive environmental policies in the form of production standards if their main trading partners are not following suit. Because policymakers show consideration for the competitive standing of domestic producers, international harmonization is often seen as a prerequisite for strict environmental policies. Bommer (1996) acknowledges this fact and investigates the scope for environmental policy harmonization in the course of trade integration. He uses a two-country model akin to the industrial-organization models developed by Ludema and Wooton (1994) and Barrett (1994) to portray the economy, and the political-support-function approach to portray the polity. Rolf Bommer's model encompasses two asymmetries. First, the political influence of the environmental interests need not be the same in the two countries and, second, trade integration may have different consequences for the producers located at home and abroad. Trade liberalization in such a context thus changes the political impact of the involved interests across countries. Rolf Bommer shows that for reasonably calibrated models it is quite possible that trade liberalization gives rise to a harmonization of environmental policies in cases where harmonization has not been feasible in the preintegration environment. This result is robust with respect to the existence of international environmental spillovers.

4.5. FACTOR MOBILITY

The integration of the world economy has recently gained additional momentum by way of increased international capital mobility. Some scholars even single out a high degree of capital mobility as the dividing line between international integration and "globalization". Whatever the significance of capital mobility may be, it is clear that international capital mobility adds a new dimension to the political-economic interaction in the environmental policy game. Crucial in this context is not so much the fact that multinational firms in some cases actually choose to withdraw from a jurisdiction, either in order to avoid high domestic production standards or to take advantage of soft ones abroad. It is rather the potential for respective political threats that emerge if the exit option becomes viable. The scope for strategic relocation is exacerbated in the presence of informational asymmetries since, by transmitting deceptive information, firms may well be able to avoid unwelcome production standards even if these standards are absolutely compatible with production at the ancestral location.

The political-economic literature on environmental policymaking in a "globalized" world characterized by a high degree of international capital mobility is very small indeed. Oates and Schwab (1988) provide a natural starting point for the subject matter at hand. Their model of interjurisdictional competition is built around a constant returns-to-scale production function with the inputs capital, labor, and environment (local polluting emissions). The capital stock is fixed, and perfect capital mobility gives rise to an equalization of net returns on capital across the jurisdictions. Labor is immobile and allowable emissions are directly proportional to the labor input. The government determines the allowable emissions–labor ratio and the capital tax rate. Capital tax revenues are distributed uniformly among the residents and the agents' utility derives from consumption and the quality of the environment. In order to portray the local public decision process, Wallace Oates and Robert Schwab apply two different modeling approaches: the Leviathan and the median voter approach.

The base-line result has already been presented in Chapter 2: if all residents (voters) within a jurisdiction are identical, the median voter model and social welfare maximization are equivalent. One, therefore, arrives at the result that interjurisdictional competition neither distorts capital allocation (the capital tax rate is zero) nor does it bring about an inefficient level of environmental protection. This optimistic result hinges, of course, on the assumption that no conflict of interest exists within the individual jurisdictions. If one assumes some degree of heterogeneity among the voters, the efficiency result breaks down. Wallace Oates and Robert Schwab investigate a scenario with two types of residents: workers who profit from an inflow of capital because capital increases the marginal productivity of labor and thus the workers' income, and nonworkers whose income is fixed. Nonworkers are thus more concerned about the environment than the workers. In such a dichotomous body of voters, the median voter is either a worker or a nonworker, depending on which group enjoys a majority.[9] If the workers outnumber the nonworkers, the political-economic equilibrium implies a subsidy to capital, otherwise a positive tax on capital is levied. The outcome is thus not socially optimal in either case. The identity of the majority also determines the level of pollution. It is not a priori clear whether a worker or a nonworker majority will generate a higher level of environmental protection.

If the political process is portrayed with the help of the Leviathan approach, the objective of the government is no longer tied to the preferences of the political stake holders. It is, therefore, not surprising that such a political system is liable to generate inefficient outcomes. Oates and Schwab (1988) assume that the Leviathan government maximizes its own utility which encompasses capital tax revenues and the utility of the representative voter. Maximizing the Leviathan's utility subject to the budget constraint in this setup obviously results in a positive capital tax rate and an inefficiently high level of pollution.

The analysis by Wallace Oates and Robert Schwab represents a first step towards an indepth investigation of the influence of global capital market integration on environmental policy. Among the obvious shortcomings of the sketched portrait are the following: (i) the implications of political integration, that is, policy harmonization, are not elaborated upon, (ii) interest group activities do not play any role in the political process, and (iii) incomplete information on the part of the policymaker with respect to the firms' incentives to relocate are not considered. Two recent papers address some of these issues: Fredriksson and Gaston (2000) focuses on the first two issues, and Bommer (1999) on the last one.

The paper by Fredriksson and Gaston (2000) is based on the model by Oates and Schwab (1988). Per Fredriksson and Noel Gaston do not, however, employ either the Leviathan or the median voter approach but rather the corruption approach to modeling the political process. Moreover, they distinguish three types of agents: workers, capitalists, and environmentalists. Only the environmentalists, whose income is exogenously determined, care about pollution. The investigation squarely addresses the following question which is arguably the most important one in the policy discussion on international environmental economics: does the increasing integration of capital markets and the ensuing debilitation of the coercive power of the nation states call for uniform environmental rules on a supranational level in order to avoid an

---

[9] Notice that heterogeneity with respect to capital ownership is inconsequential for policy formation since the capitalists do not have any stake in the policy outcome. The net return on capital is independent of an individual jursidiction's capital tax policy because of the exit option.

environmental-policy race to the bottom?[10] The answer given by Per Fredriksson and Noel Gaston is in the negative: environmental regulation is independent of the institutional design, that is, centralized and decentralized endogenous policy determination yield identical results.

The explanation for this striking assertion lies in the substitutability of the lobbying efforts of workers and capital owners. In a *decentralized regime*, capital has an exit-option and faces a given net return. Capital owners, therefore, have no incentive to engage in costly political activities. Labor, however, unambiguously gains from lax environmental standards, whereas the environmentalists' welfare is negatively affected. These two interest groups therefore have an incentive to be politically active. If both or neither of these are organized, the employed corruption approach (or rather the employed refinement criterion) yields the familiar result that the policy outcome is efficient. Inefficient policies are a consequence of asymmetric interest organization. This, of course, contrasts with the median voter model employed by Oates and Schwab (1988) which predicts inefficient policy outcomes in heterogeneous communities. In a *centralized regime* it is assumed that, for all practical purposes, capital is fixed at the relevant supranational level which implies that the return on capital is affected by centralized regulation. As a consequence, the capital owners now have an incentive to join labor in lobbying for less stringent standards. It turns out that capital-owner lobbying in the centralized regime has the same effect on environmental regulation as labor lobbying induced by capital competition in the decentralized regime. The emission standard is thus independent of the constitutional arrangement as long as capital and labor interests are either *both* organized or both not organized.

Bommer (1998b) analyzes to what extent *partial* industry relocation can take place for purely strategic reasons, that is, in order to signal an alleged loss of competitiveness to the home government. Such costly signals are individually rational if they convince the receiver to back down from a proposed policy with the consequence of larger domestic profits in the future. Bommer combines the political support function approach with a simple model portraying the interaction between a domestic monopolist and an uninformed regulator. Foreign direct investments (FDI) undertaken by the monopolist represent the only way of signaling the producer's inability to adapt to a proposed production standard. In this framework, Rolf Bommer shows that partial relocation which is not profitable from an economic viewpoint may nevertheless be carried out (i.e., such a behavior may be part of a perfect Bayesian equilibrium of the analyzed signaling game) in order to deter the regulator from imposing further regulatory constraints. Moreover, it is shown that import competing firms are more likely to take refuge in strategic relocation the more

---

[10] Per Fredriksson and Noel Gaston couch their argument in terms of policymaking in a federal system. They realize, however, that the idea of their model also extends to multicountry systems such as the European Union.

international trade is liberalized. For export firms, trade liberalization has the opposite effect.

4.6. THE EMPIRICAL EVIDENCE

The empirical literature on the political-economic linkage between trade and the environment is rather slim. The most basic issue in this context is, of course, whether (i) environmental and trade-policy decisions are really influenced by political support considerations, and if so, whether (ii) political pressure exerted by environmentalist interests spills over to trade-policy making and vice versa.

Cross-section analyses of representatives' voting behavior, undertaken with the intention of identifying the political influence of interest groups, have a long tradition, in particular, in the political economy of trade policy (see Marks and McArthur 1990). With respect to the first issue raised above, the evidence is quite clear: political-support considerations do play a significant role in trade-policy determination and the available evidence for environmental policy making points to the same direction (see e.g. Coates 1996). Whether, however, spillover effects between the two policy areas are likely to influence the policymakers' trade and/or environmental policy stance to a significant extent, is still an under-researched question. An analysis of congressional voting patterns on NAFTA (see Kahane 1996) indicates that such spillover effects may indeed be important.

Closely related to the spillover issue is the "Baptist-and-Bootlegger" coalition issue: to what extent are the interests of environmentalists ("Baptists") captured and abused by domestic producers ("Bootleggers") who, of course, seek less august objectives than their coalition partners? Whereas the theoretical studies by Hillman and Ursprung (1991, 1994) delineated the scope for "eco-protectionist" coalitions by identifying circumstances of common interest, the studies by van Grasstek (1992) and Körber (1998) provide empirical evidence for the virulence of political alliances between environmentalists and industries who stand to gain from impeding domestic market access to foreign competitors. Van Grasstek (1992) addresses the question by asking whether the voting records of US senators support the claim that protectionist bills are more likely to be supported if they can be presented in an environmental guise. Craig van Grasstek regresses the senators' votes cast on issues linking trade and environment in a probit analysis on the strength of the protectionist, anti-protectionist and green interests in the senators' respective constituencies (states). He uncovers strong evidence that linking environmental concerns with trade policy issues can increase congressional support for protectionist policies, that is, "pure" protectionist bills are less attractive to legislators than protectionist bills wrapped in a neat environmentalist cover. Given the protectionist stance of environmentalists, the "green connection" might also make trade liberalization less attractive than it might otherwise be – the respective evidence is, however, not conclusive.

Whether investigations of representatives' voting records are meaningful or not hinges crucially on how well the political pressure exerted by interest groups can be measured at the single constituency level. Since there is little agreement on how to measure interest group influences at the constituency level, various proxy variables have been constructed using a wide variety of data. As a consequence, the results of theses studies will never be undisputed. An alternative approach would be to analyze the development of the legislative process over time (for meaningful time-series studies, the track record of eco-protectionism, however, is still too short) or to focus on specific exemplary cases. Some cases of extraterritorial application of domestic environmental regulations have indeed attracted a great deal of attention. The most notorious instance of this was the tuna-dolphin case with the attendant GATT ruling of 1991 which maintained that the United States violated Mexico's trade rights when it banned imports of tuna which were not dolphin safe.

The case study by Körber (1998) analyzes the political-economic reasons behind the policy change in the US dolphin safe legislation which resulted in the controversial embargo against Mexican tuna. The fact that the US canneries had to reverse their harvesting policy has traditionally been interpreted as being indicative of the rising strength of the US environmentalist movement which was able to force a billion dollar industry into submission. The study by Achim Körber, however, convincingly argues that the large US companies derived benefits from this policy reversal, too. The new policy provided them with the leverage to reinstate the embargo against Mexico and to raise their smaller domestic rivals' cost while they themselves had lost their interest in the Eastern Pacific Ocean as a source of raw tuna.

To political economists, the conclusion drawn by van Grasstek (1992) and Körber (1998) will come as no surprise. Producers seeking protection from foreign competition find themselves more and more on the defensive. These interest groups would not be likely to let the opportunity slip to capture a new ally with a seemingly immaculate political aura. Ecoprotectionism, on the other hand, also offers environmental lobbies an important advantage: the producers harmed by environmental trade restrictions are foreign companies who have limited opportunity to participate in the domestic political process (see David Vogel's comment on van Grasstek (1992)).

More specific than the spillover issue is the question surrounding the issue of compensation of losers from policy shifts. The studies by Fredriksson (1997, 1999a) and Bommer and Schulze (1999), for example, suggest that trade lib-eralization, by benefiting the polluting export industry, may give rise to an endogenous adjustment of environmental policy which compensates the losers from liberalization, the clean import competing industry, via stricter environ-mental standards. Conversely, stricter environmental standards may give rise to compensations by way of trade liberalization. The study by Eliste and

Fredriksson (1999) attempts to disentangle the empirical relationship between environmental regulations and transfer policies which, in particular, encompass trade policies, by analyzing cross-country data collected for the agricultural sectors of 49 countries. In the presented regressions producer subsidies do not appear to have any influence on the stringency of environmental regulations, whereas the farm lobby group appears to become relatively more powerful as the stringency of environmental regulations increases. Taking into account the limited coverage and quality of the data, the empirical evidence at least does not seem to be at variance with the compensation hypothesis.

If one cannot but describe the empirical literature on the relationship between trade and environmental policy as slim, the literature on the relationship between capital mobility and environmental policy is almost nonexistent. Considering the substantial empirical literature which investigates how the ongoing globalization of the economy affects the traditional nation states' ability to conduct independent fiscal policies (for a survey of this literature see Schulze and Ursprung 1999), it is probably not too daring to predict that globalization-induced changes in national regulation policies will become the subject of intensive research in the near future. A first empirical investigation of the effects of international capital market liberalization on environmental regulation is to be found in the second part of the paper by Fredriksson and Gaston (2000). Two empirical tests are presented which corroborate the authors hypothesis that capital mobility or, alternatively, institutional design are not likely to have any significant effect on the stringency of environmental standards. The first test shows that, in a cross-section regression covering the industrial sectors of 31 countries, the standard capital-control measures do not have a statistically significant effect on an index of environmental regulations and enforcement. The second test analyzes roll call votes on environmental bills in state legislatures and in the US Senate and House. No discernible differences in the support for environmental policy at the different levels of policymaking can be identified. These results seem to indicate that the apprehended downward-competition of environmental standards is not an inevitable consequence of economic globalization.[11]

4.7. SUMMARY

Looking at the political-economic literature on environmental policymaking in open economies from a bird's-eye perspective, three interrelated impressions arise. First, this literature convincingly demonstrates that the extent and scope of the deformities of the policyformation process are such that by neglecting

---

[11] Notice that these results are in line with the empirical findings on the effects of environmental regulation on the relocation of industries (see section 3.5). Because these effects have been shown to be rather small, environmental policy is not an effective instrument to acquire mobile capital and therefore a competitive downward pressure on standards is not to be expected.

these aspects, the casual observer is led to a completely inadequate picture of the state of affairs. Only by calling attention to and by analyzing the political-economic forces underlying the political process in depth can one ever hope to be in a position "to design laws and institutions that will properly constrain interest group behavior or harness it to the public interest" (Hoekman and Leidy 1992: 241). Because of the complexity of the political-economic relationships – this is the second observation – one cannot expect to arrive at general results which, in a straightforward manner, then suggest recommendations for constitutional design. Normative statements which are worth their salt need to be based on the surrounding institutional and economic setting. The third insight refers to the often heard argument that the increasing integration of the world economy will give rise to environmental plight. Even though the surveyed literature clearly shows that economic integration does have strong consequences with respect to distribution, a careful analysis of the interaction of the attendant political forces does not support this unequivocal judgment. It rather transpires that economic integration can have positive or negative effects on environmental policy and the environment. Given this general result, the notorious critics of economic integration, including environmental organizations and "green" political parties, would be well advised to rethink their position.

5

Trade, Agriculture, and Environment

DAVID E. ERVIN

5.1. INTRODUCTION

Most developed nations are pursuing liberalized and expanded agricultural trade as well as improved environmental quality. Therein lie several potential conflicts. Will the trade expansion be hindered by stronger environmental programs for agriculture? Or, will environmental quality be sacrificed, especially in developing countries, in the rush to capture growing global food markets? Will nontariff trade barriers appear disguised as environmental protection efforts? Will the opening of new trade pathways spread harmful nonindigenous pests? How can transboundary environmental resources affected by food production and marketing be protected without raising trade tensions? These diverse concerns form a central question: How can trade expansion and environmental protection be managed such that the pursuit of one objective complements the other, or has negligible adverse effects?

Despite the popular fears of conflict, trade liberalization and the reduction of significant environmental problems (i.e., those for which net benefits exist) are complementary in principle. Both objectives must be satisfied to achieve maximum economic welfare for a country (Anderson 1992). The pursuit of each seeks to improve economic efficiency by ensuring full consideration of benefits and costs in private and public decisions. However, in practice, trade and environmental programs conflict because of ill-designed government policy and missing markets (Ervin 1997a).

Trade in food and fiber products is a large and fast-growing enterprise. Agricultural trade was valued at approximately $465 billion (US) in 1997, comprising approximately 11 percent of total world trade (FAO 1998; WTO 1998). From 1990 to 1997, the value of global agricultural trade rose 43 percent. Liberalized trading rules approved in the Uruguay Round Agreement (URA) Agreement on Agriculture (AoA) and regional trade liberalization pacts,

The author gratefully acknowledges the editorial assistance of Suzanne DeMuth.

such as the North American Free Trade Agreement (NAFTA), have accelerated the expansion. The AoA established schedules to reduce government subsidies for farming and to convert nontariff trade restrictions to declining tariffs for signatory countries. In so doing, agricultural supplies from many developed countries that have been subsidized will gradually decrease in share, world prices will rise, and other countries, many in developing phases, will expand their production. The new trading rules should increase the volume of international agricultural commerce, create new trade pathways that heretofore have been unexplored, and shift regional production patterns.

Concern about the risks to the environment from growth in agricultural production and trade rose over the last decade. The shifts will affect domestic and transboundary environmental resources in complex ways. More transport from trade expansion will release more emissions into air and water resources that migrate over country borders. Intentional and accidental introductions of invasive plant and animal species through trade can affect environmental conditions significantly. Countries that have not subsidized their farmers will experience pressure to expand their agricultural land base (i.e., extensification), and to intensify production, for example, higher fertilizer and pesticide use per hectare, in response to higher prices. Countries that have subsidized their farm sectors will experience decreased production pressure at the extensive and intensive margins. The contractions may diminish or enhance environmental services.

The public demand to improve environmental quality affected by farming remains robust in most developed nations. The demand, driven by rising income levels, has led to more and stronger agricultural environmental (agri-environmental) programs. In contrast, there are weaker agri-environmental programs in many developing countries. Hence, for different reasons, fears of environmental risks from expanded agricultural trade have surfaced in developed and developing countries. The URA acknowledged such risks and created a Committee for Trade and Environment (CTE) within the World Trade Organization (WTO). Similarly, the NAFTA was accompanied by an environmental side-agreement that created a Commission for Environmental Cooperation (CEC). Yet, the structure of these trade-related environmental initiatives and their sluggish progress have drawn criticism (Charnovitz 1997; Esty 1994). In short, there has been weak integration of environmental management into trade institutions and vice versa to date.

Widespread losses to agricultural trade or to the environment have not occurred to date. Current problems tend to concentrate in specific areas, such as along border zones, or on specific crops, often creating "pockets of stress" (OTA 1995a). Nonetheless, serious conflicts are likely to grow because trade and environmental policies have not evolved in complementary fashion. Many countries do not have comprehensive policies for agri-environmental management or for trade liberalization. Moreover, periodic episodes of weak agricultural prices, as always seems to occur, may precipitate a return to state interventions in agriculture and protectionist environmental policies.

This chapter analyzes the theory and evidence on two basic linkages between trade, agriculture, and the environment. The first covers the complex array of environmental effects from liberalized agricultural trade. The second assesses the existing and emerging impacts of agri-environmental programs on trade flows. In the closing section, some principles to move trade and environmental management in agriculture to more complementary and less conflicting postures are offered.

The scope of issues covered by "environment" requires brief comment. The term refers to the natural environment composed of physical and biological resources, such as water, soil, air, and biological diversity. These resources serve production needs, satisfy consumption, or hold intrinsic value by their very existence. They may be owned and managed for commercial production, such as cropland and orchards, or held in common, such as rivers, streams, and pest gene pools. This analysis excludes effects that are not directly linked to a natural environment medium, such as food safety issues arising from pesticide contamination. This is an arbitrary definition since humans are part of the natural ecosystem, but the management of human risks and benefits from agricultural trade extends beyond the scope of this chapter.

5.2. IMPACTS OF AGRICULTURAL TRADE LIBERALIZATION ON ENVIRONMENTAL QUALITY

A growing body of evidence links a diversity of negative and positive environmental conditions to agriculture. A recent US assessment revealed widespread remaining problems and opportunities for improvement despite significant progress over the last decade (OTA 1995b). Water quality topped the list of problems, while large gains have been made in wildlife habitat and soil erosion control. In the Netherlands, livestock manure causes serious ammonia emissions and heavy metals pollution. Intensive dairies in New Zealand cause poor stream conditions. Australia confronts land degradation from salinization due to irrigation. Traditional haymaking and extensive grazing help conserve much of Sweden's biodiversity in plants, birds, and insects. The Nordic countries generally perceive traditional diverse agricultural landscapes as providing important environmental amenities. Japan considers rice farming to provide valuable rural landscape and other benefits.

How will the URA and related national agricultural policy reforms affect this broad range of environmental conditions? Both positive and negative environmental effects should be expected. The net effect depends on how the relative changes in output and input prices, such as land values, induce shifts in production and trade, and how those shifts map onto the environmental resource base of each country, and onto transboundary resources. The ultimate impacts are not set just by bio-physical processes, but also depend on the efficacy of agri-environmental programs. For this reason, some of the most serious risks may unfold in developing countries with immature environmental management institutions for agriculture.

It is important to stress that liberalized trade generally is not the root cause of environmental problems related to agriculture (see Chapter 1). The problems stem from the nonrival or nonexclusive nature of environmental goods and services, which hampers the formation and functioning of private markets. A related perspective is that incomplete property rights and missing markets hinder effective environmental resource management (Bromley 1996). Hence, we witness a variety of public or collective actions to protect and enhance environmental values affected by agriculture in developed countries. Although not a root cause, trade liberalization and accompanying agricultural policy reform can exacerbate or ameliorate the environmental problems.

Several typologies of potential environmental effects due to changes in trade or trade policy have been offered. Grossman and Krueger (1993) sort the effects into scale, product composition, and technology changes. Runge (1995) expands that list to include effects from improvements in resource efficiency (e.g. less nutrient waste) from reforming government policies that stimulate excessive input use and production. For this section's analysis, attention will be focused on questions related to: (i) transport, (ii) invasive species, and (iii) production extensification and intensification (Ervin 1997a). The conditioning role of agri-environmental policy and the risks from production expansion in developing countries close the section.

5.2.1. Will Increased Transport for Agricultural Trade Degrade Environmental Resources?

Perhaps the most obvious environmental effect to the public from increased agricultural trade comes from pollution caused by more international transport. Most transboundary environmental resources affected by transport are not subject to effective multilateral pollution control policies. Thus, the risk and likelihood of environmental degradation rises with more international commerce. For example, burning more fossil fuel to ship traded goods could exacerbate global air pollution. Some of those problems may be partially offset by reduced domestic transportation and air pollution where trade liberalization shifts production patterns to become more proximate to markets. For example, Canada has reformed its transport subsidies that encouraged eastward shipments of grain from its Prairie provinces as part of trade liberalization (Ervin and Fox 1999). Therefore an assessment of the total environmental effect from more trade requires consideration of shifts in national and international shipments.

The findings of a recent study for OECD bear on the section's title question:

It seems safe to conclude that international trade liberalization will necessitate some increase in the output of the transport sector of the OECD countries. Furthermore, it may lead to a shift in the mode of transport away from the relatively benign sea and rail transportation towards the more deleterious road and air transport. For these reasons, the implications of trade liberalization for the transport sector and the environment need careful attention and offsetting environmental protection measures may need to accompany trade liberalization measures. (Gabel 1994: 170)

Gabel also explains that if transport services and energy were properly priced to include their social and environmental costs, trade liberalization could lead to more efficient and environmentally beneficial transport patterns. But pricing inefficiencies in transport from market and government failures are pervasive, thus negating this potential gain. The author concludes that trade liberalization generally will exacerbate environmental damages from transport.

Correcting the transport pricing problems is a nontrivial task. Many nations have increased controls on air pollutants associated with transport, such as lead in gasoline, but likely have stopped short of full-cost pricing. Public programs that favor one or more transport sectors continue to distort shipment modes and likely have environmental consequences. Correcting the pricing inefficiencies for international environmental resources may be impossible in the foreseeable future. Most transboundary and global environmental resources, including the atmosphere and oceans, are not subject to effective multilateral pollution control policies (see Chapters 10 and 11).

Studies of the possible environmental effects of agricultural trade liberalization generally have not examined the effects of the transport pricing inefficiencies. Whalley (1996) analyzed the effects of a potential carbon tax. His analysis attempts to show the possible effects of instituting a global carbon tax to reduce greenhouse gas accumulations. If his large carbon tax scenario accurately approximates pricing corrections for global environmental damages, then the model predicts large trade changes. Because exports and imports of some agricultural products, such as fruits and vegetables, are in fresh, perishable form, it's likely that transportation plays a relatively large role in their trade compared to other crops.

5.2.2. Do Harmful Nonindigenous Species Pose Serious Environmental Risks?

A second set of environmental effects can be traced to the accidental or intentional introductions of harmful nonindigenous species (HNIS) of plants and animals. Without effective policies to screen the introductions, HNIS can damage environmental resources. Country actions, such as bans, quarantines, and inspections, are authorized by WTO rules to avoid serious damages to human, animal, and/or plant health from the pests. Most countries have institutions to minimize the risks to commercial resources, with New Zealand providing a model approach (OTA 1993).

Opening new pathways under trade liberalization will increase risks of invasions. Programs to control HNIS damages generally focus on human and livestock health rather than the natural environment. That focus is understandable if the environmental resources suffer externality or public good problems because those responsible for importing the destructive species do not pay their full cost.

A study has documented evidence of widespread HNIS problems in the US (OTA 1993). For example, the invasion of leafy spurge has extensively degraded western US rangelands. That assessment concluded that significant commercial and environmental damages resulted from past invasions, and that trade was a vector for the introductions. Although it did not conduct a formal cost-benefit analysis, the report findings suggest that the marginal costs of additional control fall well below the avoided damages for many cases. The study suggested that domestic control programs were often more cost-effective than trade-related restrictions. The analysis also noted that significantly less attention is given to environmental damages from HNIS because of missing market values for many environmental effects and their public good nature.

In a subsequent report, the nature of HNIS risks related to agricultural trade were discussed (OTA 1995a). That analysis concluded that the risks were significant and that existing institutions provided inadequate control over the invasions, in particular those threatening environmental resources. The original OTA (1993) study notes that about 40 percent of the pests affecting agriculture have been accidentally or deliberately introduced into the country, sometimes through trade. Notable cases of relevance to agriculture, including knapweed and cheatgrass/medusahead affecting western rangelands, and Russian wheat aphids. The difficulty in all of these cases is that the pathway and vehicle causing the damages are very difficult to pinpoint. Hence, trade protectionist measures may be disguised as legitimate actions. The control of HNIS damages also could fall short of the social optimum for the same reasons.

A huge HNIS issue looms on the horizon with broad potential consequence for agricultural trade. The transfer of genetically modified organisms (GMOs) through trade has caused serious concern in Europe and developing countries about their implications for indigenous flora and fauna.

5.2.3. How Will Agricultural Trade Liberalization Affect Extensification and Intensification?

The third set of environmental effects stems from production expansions and contractions fostered by trade liberalization. The shifts occur through extensification, that is, increases or decreases in farm land, and through the application of more or fewer inputs to a given land base, that is, intensification. By virtue of its large land and water requirements, agriculture affects a greater share of a nation's natural resources than most other industries. Its production processes inevitably alter vegetative cover, apply fertilizers and pesticides, use irrigation water, and generate animal wastes. Risks may not be immediately perceived because of the diffuse nature and time lags involved in nonpoint pollution, such as nutrients and pesticides accumulating in drinking waters. Moreover, the degradation of some natural resources may reach critical zones beyond which the effects cannot be remediated, as with the extinction of flora and fauna species.

The production shifts reflect farmer responses conditioned by agricultural policies and current technology. Hence, the baseline from which changes are measured reflects any negative or positive environmental effects stemming from current policies. In the longer-term, dynamic adjustments are more complicated as markets and technology evolve in response to new trade opportunities and environmental policy. Most empirical analyses of agricultural trade liberalization adopt a short-run, static view because of the uncertain policy, market, and technology paths. Moreover, available estimates of global production changes are mostly for countries or regions. Large-scale modeling analyses do not capture the diversity of adjustments necessary to forecast environmental changes. For example, analyses that project the production impacts of URA implementation in the European Union (EU) are of limited use. The heterogeneity of production adjustments and natural resource conditions across individual countries requires more disaggregation to anticipate environmental effects. For instance, a country may move to a more homogeneous agricultural system focusing on a few major crops, thus diminishing regional biodiversity derived from a range of crop and livestock enterprises.

Researchers increasingly argue that reliable estimates of environmental effects from production shifts require site-specific natural resource and production detail, as opposed to using aggregate models comprised of large resource units of assumed homogeneous character (Antle and Just 1991; Antle *et al.* 1996). In concept, the use of site-specific conditions gives more precision, but such analyses require extensive data that are unavailable for many regions. Using local data to inform and check the larger models seems a prudent step until more detailed data are at hand.

Global studies can identify regions that are likely to experience large adjustments for intensive study. For example, the effect on world food output of a complete removal of agricultural support policies has been estimated as "…negligible and the relocation of production is minor, e.g., grain and meat production would have been 5 to 6 percent lower in industrialized countries and 3 to 8 percent higher in developing countries" (Anderson and Strutt 1996: 155). These estimates portray the maximum regional shifts from fully liberalized trade, not those from the URA. Such extreme policy reform was not required by the URA. The AoA exempted key types of agricultural support from the total support ceiling and reduction schedule, and will phase in constraints over the next decade. Still, regions estimated to have larger adjustments offer clues to significant environmental changes. Japan and Western Europe would experience the largest decreases, from 15 to 50 percent of their baseline production. In contrast, Africa, Latin America, North America, and Australasia comprise increases ranging from 5 to 20 percent. Because the countries estimated to lose production use more chemicals and practice intensive livestock production compared to the bulk of those increasing production, Anderson and Strutt hypothesize that global environmental pressure from agriculture will fall. However, the net shift in a country's production will likely be composed of heterogeneous responses over its regions. Complicating matters, increased or decreased production may cause damages depending on the specific natural resource situation and public preferences. For example, land abandonment in many EU areas causes environmental loss from degraded landscapes, although in other nations it may restore valuable wildlife habitat or other environmental services.

Analyses of national agricultural program reforms, either required or stimulated by URA provisions, could offer clues to environmental effects within some countries. Several studies of the US provide consistent results. Generally, the reduction, decoupling, or elimination of national production subsidies reduces incentives for fertilizer and pesticide use, pressures to convert environmentally vulnerable lands to arable production, and other stresses such as irrigation water withdrawals (Ervin *et al.* 1998; Howitt 1991; Just *et al.* 1991; Miranowski *et al.* 1991; Ribaudo and Shoemaker 1995; Tobey and Reinert 1991). However, a survey of these studies concluded that reductions in US environmental stresses will be uneven and modest (Kuch and Reichelderfer 1992). Only one study has estimated that bilateral elimination of US and EU agricultural support programs would result in a significant increase in US production and chemical use (Abler and Shortle 1992). Another analysis of bilateral US–Mexico trade reform indicated modest regional production increases (Burfisher *et al.* 1992).

Evidence from unilateral policy reform in New Zealand, combined with a downturn in prices, suggests more pronounced effects in practice than from the simulations (Reynolds *et al.* 1993). Phosphate fertilizer, pesticide use, and conversion of marginal lands in pastoral agriculture appear to have fallen significantly. Moreover, the reduction in sheep numbers and the diversification of pastoral farming has reduced grazing pressure in some hill country areas. Pesticide use in other sectors, such as horticulture, has risen; land quality declined on certain farms due to income pressure during the short term transition period; and some weed and other pest problems rose, due to a decline in pest control assistance, among other factors. The mixed outcomes are consistent with the uneven effects predicted by modelling analyses of US policy reform. The New Zealand government has judged the overall net effects to be definitely positive. Perhaps most importantly, the government felt that it could not have begun to address the full range of agri-environmental issues while a complex web of agricultural subsidies was in place. Reform of agricultural support policies was a necessary, although not sufficient, first step to addressing agri-environmental issues. Sweden has also experienced environmental gains under policy reform from converting land in grain production to grazing.

Conservative simulation estimates are plausible because short-term economic analyses often underestimate the responsiveness of agricultural systems. The simulation analyses for the most part do not account for long-term input substitution and technological innovation that can further reduce environmental stress. This may account for more optimistic estimates of environmental gains from policy reform by studies projecting technological change (Faeth 1995). One study that factored in normal technological progress (without domestic policy reform) concluded that added environmental stress from projected US export increases under the NAFTA and URA pacts would be negligible (McCarl *et al.* 1994). Indeed, the major loading measures, such as erosion and fertilizer use, did not show increases and some even declined over baseline levels.

Most estimates of environmental effects from policy reform are broad inferences based on changes in inputs or loadings rather than shifts in ambient conditions because of inadequate science and data. Reliable estimates of changes in ambient environmental conditions require knowledge of the spatial and temporal distributions of production pressure on the local natural resource base (Antle and Just 1991). Those requirements raise the degree of analytical difficulty by an order of magnitude, due to the wide diversity of natural resource conditions within and across countries. Thus, it is not surprising to find either aggregate studies with broad-brush inferences about environmental loadings, or localized studies of environmental responses without the ability to generalize broadly (Antle *et al.* 1996). The uncertainty argues for adopting agri-environmental policies that allow flexible management responses to heterogeneous natural resource and production conditions.

5.2.4. The Role of Agri-Environmental Policy

Figure 5.1 portrays conventional theory about the lines of causation that translate trade shifts to environmental effects. Note that environmental policies provide a filter through which the product, technology, scale, and structural effects pass to determine environmental pressures, states, and responses. Seen from this perspective, the environmental effects of trade are not predetermined based on changes in chemical use or other production practices, but are endogenous and depend upon how the management policies shape the pressures and responses. Thus, any attempts to forecast the environmental effects of agricultural trade liberalization must account for agri-environmental policies. The coverage and efficacy of these (endogenous) policies become key determinants in the eventual effects on environmental quality (see Chapter 4). For example, the US Department of Agriculture has concluded that federal programs, such as the Conservation Reserve Program (CRP) and conservation compliance provisions, have resulted in significant improvements in soil erosion control, water quality, and wildlife habitat (USDA 1994).

Trade Liberalization (and domestic agricultural policy reform)

⇓

Improved Resource Allocative Efficiency (+)

⇓

Growth in Income/GDP per capita (+)

⇓

Scale (Pollution) Effects (-)

⇓

Demand for Environmental Quality (+)

⇓

Change in Environmental Policy (+)

⇓

Change in Product Composition   and   Change in Production Technology
(less polluting or more amenities)   (less polluting or more amenities)

Figure 5.1. *Conceptual impacts of trade liberalization and economic growth on the environment. Adapted from Runge (1995).*

Will agri-environmental programs in most countries provide adequate coverage to control significant problems that might come from trade liberalization? Most developed countries have improved the functioning of conservation and environmental programs for agriculture over the past two decades (OECD 1993; OTA 1995a). However, most countries still have incomplete incentives to treat the positive and negative environmental effects from agriculture. As one example, the effluent from irrigation discharges is not controlled in the US. Also, the US programs may not effectively induce public or private R&D processes to develop low cost technologies that meet competitiveness and environmental objectives simultaneously (Ervin and Schmitz 1996). These gaps are more pronounced for transboundary and international environmental effects, given the high cost of negotiating and implementing multilateral efforts (see Chapters 10 and 11).

5.2.5. What are the Risks in Developing Countries?

As noted above, studies of the global reallocation of agricultural production under trade liberalization estimate that some of the largest increases will occur in Latin America, Africa, and Asia (Anderson and Strutt 1996). Some of the countries that will experience production pressure are improving their environmental programs. However, most do not have adequate environmental policies to protect against the likely damages (Lutz 1992). Developing countries that respond to growing market opportunities from trade liberalization will affect their natural resources through extensification (land conversion), and intensification, such as more frequent tillage and increased fertilizer and pesticide applications (Lee and Roland-Holst 1993). Anticipating the environmental outcomes from these processes requires analysis of site-specific production and natural resource characteristics, in tandem with agricultural and environmental policies. This type of analysis for developed countries is very demanding because of information and science requirements. It is even more imposing for many developing countries where the study of environmental processes related to agriculture has just begun.

Antle *et al.* (1996) provide an example of this analysis, focusing on pesticide use in Ecuadorian potato production. Using detailed models of field production and natural resource conditions, they simulate the environmental effects of trade liberalization for Ecuador. In this particular case, trade liberalization would raise the effective price of pesticides and also increase the potato price relative to dairy, the chief competitor for agricultural resources in the area. They illustrate that policy liberalization that increases pesticide prices by 30 percent, coupled with potato price increases of up to 90 percent, will not necessarily have an adverse impact on the environment (water quality) or on production. This finding from combined extensive and intensive margin changes illustrates the importance of accounting for the specifics of the agricultural situation.

The simulation procedure also has the capability of estimating a shift in the production-environmental quality frontier caused by policy liberalization. The data intensive nature of analysis makes it an expensive undertaking.

Why should developed countries be concerned with environmental effects in developing nations? Because some of those effects have global significance, such as the clearing of rainforests for new production (Giordano 1994). Also, there are few institutions that permit developing countries to engage with developed countries in resolving these problems (Chichilnisky 1996).

5.2.6. Summary

The weight of the evidence does not suggest that production changes from trade liberalization and expansion will cause broad environmental damage in developed countries. Some improvements can be expected, but "pockets of stress" from concentrated production, land abandonment, and HNIS invasions will also occur. Larger risks lie in the developing countries when trade liberalization and a rapidly growing world population will push up food demand and prices. Fostering national and transboundary environmental policies for these uncertain problems with minimal trade effects presents a major challenge.

5.3. IMPACTS OF AGRI-ENVIRONMENTAL PROGRAMS ON TRADE

Two types of environmental programs apply to agricultural production and trade. The first attempts to reduce damages from production practices, such as soil erosion and water pollution, or increase the positive effects, such as countryside amenities. Examples include the regulation of pesticides and land set aside for wildlife habitat. The second applies restrictions or conditions to trade flows to guard against environmental degradation, such as phytosanitary measures.

5.3.1. Programs to Manage the Environmental Effects of Agricultural Production

Countries use a wide array of programs to manage the environment related to agriculture (OECD 1993b). The US has at least 35 programs to control negative and increase positive environmental services from farming (OTA 1995a). Most countries pursue a similar set of environmental objectives for agriculture, including reduced water pollution, soil erosion control, wildlife habitat protection, and landscape preservation, albeit with differing approaches. Whether the myriad programs affect food and fiber trade becomes a natural question.

Few studies of the effects of environmental programs on agricultural trade have been conducted, perhaps for two reasons. First, most countries use a variety of voluntary-payment programs that do not impose unwanted costs on farmers. For example, the US government spends about $3.5 billion per year on its programs, the vast majority of which goes for cost-sharing for practice adoption, land rental payments, and education/technical assistance (USDA 1997). Those programs should exert little or no competitiveness drag on individual farms and ranches. However, the effect on industry supply as a whole may indeed be significant, especially for land set-aside. Second, data on pollution control costs are not collected and reported for agriculture, unlike other industries. This omission likely traces to the difficulty of surveying the large number of diverse farms in most countries and the diffuse, nonpoint nature of much pollution.

The findings for other industrial sectors that have had extensive environmental regulation offer insight into potential competitiveness effects in agriculture. Diverse studies of the effects on trade and foreign investment flows have come to remarkably consistent findings. Schulze and Upsrung in Chapter 3 conduct an extensive review of this literature. They find little evidence of trade or industry relocation effects from environmental regulation. The lack of significant effects may reflect management or technology innovations that lessen compliance costs over time, and similar environmental programs across competing exporters.

Two qualifications to the conclusion of insignificant trade and industrial location effects by environmental regulation should be kept in mind. First, some industries spend very different amounts on pollution control and face considerably different degrees of world competition. Analyses of aggregate trade flows may miss the effects on specific sectors. Case analyses show some sectors with high compliance costs have been disadvantaged in trade (OTA 1992). These sectors may suffer from unfavorable pairwise differences with their competing exporters. Even small amounts may be important in increasingly competitive international markets under trade liberalization. This situation may apply to the impacts of the planned phase-out of methyl bromide on the US fruit and vegetable industry. Second, the studies are backward looking by necessity, and subsectors that anticipate strengthened environmental requirements require careful monitoring. For example, a trend to impose more direct controls in farming would warrant attention.

5.3.1.1. *Voluntary-Payment Approaches*
As noted, developed countries predominantly use voluntary-payment programs to reduce pollution and increase environmental services from agriculture (OECD 1993; OTA 1995a). The programs offer educational services to help identify problems, technical assistance to install and maintain recommended practices, cost-sharing to defray a portion of the expense of implementing a practice, and rental payments for land set-aside schemes that temporarily or permanently retire land from production. No other industry is offered such a wide array of voluntary-payment programs for environmental management.

The subsidies reflect that farmers generally hold the rights to dispose of wastes into environmental media under current legal rules, and hence require compensation to alter their behavior (Bromley 1996). Defining cost responsibility in favor of producers likely stems from the special political status given agriculture in most developed nations. It also reflects the technical difficulty of implementing regulations to control diffuse sources of pollution that cannot be readily traced to their sources over such a large land base and from millions of diverse production units, that is, nonpoint pollution.

Subsidies for conservation and environmental management in agriculture were sanctioned in the URA under two conditions: (1) that they be applied as part of a clearly defined government program to fulfill specific conditions; and (2) that the payment amount not exceed the cost of the management practice or the loss in income associated with program compliance. These conditions are part but not all of the requirements to ensure that subsidies do not cause trade distortions. In addition to providing minimum necessary compensation, the subsidies should be provided only in cases where the expected benefits of environmental improvement outweigh the anticipated costs. Another condition is that the subsidies stimulate producer and R&D innovations that minimize long-term compliance costs. In practice, if the subsidy exceeds the necessary compensation and does not encourage such cost savings, they can attract capital to the industry, enlarge supplies, and thereby aggravate negative environmental problems or oversupply positive amenities. Some US subsidies do not meet the minimum compensation and incentive-compatibility requirements (OTA 1995a).

5.3.1.2. *Compliance and Regulatory Approaches*
Some environmental compliance programs also apply to agriculture. For example, the US implemented soil conservation and wetlands compliance programs in its 1986 omnibus farm legislation. The compliance schemes require farmers participating in other agricultural programs to meet minimum conservation standards to remain eligible for program subsidies. The EU is considering compliance programs as well.

Three types of direct agri-environmental control programs exist. Virtually all countries regulate the introduction and use of pesticides. Human and environmental risks are controlled by registering only those compounds deemed to be without excessive risk encountered in application or through food, water, or air exposure. Many developed countries also regulate the alteration of certain lands that would cause environmental loss. For example, conversion or drainage of certain wetlands in the US is regulated. Finally, an increasing number of countries regulate the effluent from confined animal and other concentrated production operations into streams, rivers, lakes, and estuaries.

What has been the effect of this mix of voluntary-payment and regulatory programs on costs, supplies, and trade competitiveness? The general hypothesis is quite clear – with relatively little cost responsibility imposed on farmers, the supply and trade effects should be negligible. OECD policy supports the application of the polluter-pays principle (PPP) over subsidies, because full cost internalization stimulates incentives to correct significant damages and encourages innovation in pollution treatment. Unlike other industries, the PPP has been applied sparingly in agriculture (Tobey and Smets 1996). Some countries, for example, New Zealand and Australia, have imposed charges on inputs causing pollution and the charges generally have not affected use.

Few studies have attempted empirical estimates of the effects of agri-environmental programs on trade flows. Tight statistical analysis has proven difficult because of missing data on the net costs of the programs and their environmental benefits. Some analysts have attempted estimates, but their data are deficient (e.g. Gardner 1996). Looking at payments and costs is important because past programs have made substantial transfers into agriculture in many countries, such as for drainage (Paden 1994; Sutton 1989). Such subsidies may have boosted production and trade, especially when coupled with past production and export subsidies. As just explained, such subsidies can distort trade flows if their environmental benefits do not exceed their costs.

Tobey (1991) estimated the potential for different crops to generate pollution and correlated the estimates with the revealed comparative advantage performance of crops in the world market. He found that the crops that perform well in world markets also have the largest pollution potential. Therefore, stringent programs to control that pollution could affect their trade performance advantage. However, he concluded that the magnitude of trade competitiveness losses is likely to be quite modest for three reasons. First, most competing exporters have introduced similar agri-environmental programs, which implies that the relative trade competitiveness effects have not likely changed significantly. Second, less developed countries (LDCs) do not hold large market shares in most of the commodities. That is, the production patterns of non-OECD countries do not suggest that uneven agri-environmental requirements will greatly affect trade by OECD nations. Also, competitiveness effects of agri-environmental programs are likely to be swamped by larger forces such as labor costs and exchange rate fluctuations.

5.3.1.3. *Trends in Agri-Environmental Approaches*

Although agri-environmental programs do not appear to exert broad trade effects, some trends raise concern that production and trade may be increasingly affected (OTA 1995a). There is a distinct chance that conservation subsidies may grow in the form of green payments, as production and export subsidies are diminished under the URA. Such a rise in subsidies for environmental protection does not necessarily jeopardize trade, as explained above. However, in practice, subsidy programs often depart from the conditions required to avoid trade distortions. Moreover, the voluntary-payment approach may be losing strength. More countries are implementing policies with more cost responsibility for farms and ranches. New Zealand has moved in this direction by charging producers for the expense of operating a number of programs (Sinner *et al.* 1995). Australia is following a similar course. In the US, states are using more direct controls (Ervin and Schmitz 1996). About one half of the states now have laws that grant the authority to impose penalties on farmers for water pollution (Ribaudo 1997).

Some analysts speculate that the trends in agri-environmental programs could be a major factor in distorting agricultural trade, by raising production costs and providing justification for restrictive policies abroad (Gardner 1996). That assessment is not shared by others (Whalley 1996), but they admit that major future global initiatives, such as a carbon tax, could exert significant trade effects. Considering just the changes in one country's programs is not sufficient to judge competitiveness effects. That depends upon the actions taken by competing exporters, many of which are increasing the breadth of agri-environmental programs (OTA 1995a).

Two US agri-environmental programs illustrate the potential to affect trade. The first is land set aside for environmental purposes that diverts cropland from production. The second is controls on pesticides to avoid environmental damage. The phase-down and eventual banning of methyl bromide used to fumigate soils and preserve perishable exports of fruit, with no apparent substitute, is a good example. The effects of these programs are also influenced by the path of agricultural policy reform. For example, the 1996 US Federal Agricultural Improvement and Reform (FAIR) Act decoupled agricultural program payments from particular commodities, removed planting restrictions on program acreage, and reauthorized conservation set-aside.

Land set-aside. US commodity and conservation programs have used temporary land retirement for over half a century. Total diversions in the US exceeded 20 million hectares in approximately one-third of the years that land has been idled. It climbed to a peak of just over 30 million hectares in the late 1980s (USDA 1994).

Enrollment in the CRP reached 14.75 million hectares in 1994, and was reauthorized in the 1996 FAIR Act. The original CRP had dual objectives – to reduce soil erosion and to control crop supplies. However, the principal objective of the renewed CRP is to maximize environmental benefits per dollar of expenditure. It is unclear how much land will be retired under the new version and what regions and crops will be most affected. About 6.7 million hectares were signed up in the first enrollment in 1997 and the regional pattern was similar to the first, except for relative declines in the Southern Plains (Texas, Kansas, etc.) and the Pacific Northwest states (Oregon and Washington). The US Department of Agriculture implemented an improved environmental benefits index and rules to avoid paying rents in excess of market rates to help ensure net benefits, thus moving the program toward satisfying the URA conditions.

What effect will the new CRP have on agricultural production and trade? In theory, the CRP reduces the availability of cropland and raises its price as a factor of production. If the enrollment rules guard against excessive retirement, then the primary effects are improved environmental quality, higher land prices, the substitution of nonland inputs for more expensive land, and reduced crop supplies and higher prices. In practice, the effects are difficult to estimate because of uncertain total enrollment. An early assessment projected net benefits for the program, and concluded that it did not inappropriately reduce supplies and raise prices (Young and Osborn 1990). By implication, these findings imply that trade was not distorted, but the study did not address trade effects. A recent study estimated significant trade effects on corn, sorghum, and wheat, but did not include the environmental gains to assess its overall economic performance (Leetmaa and Smith 1996). Another study found significant trade gains from downsizing the old CRP and few environmental losses (Abel *et al.* 1994). The increased trade was not judged a major threat to competing exporters because the increased US trade would be mostly from sharing in global market expansion.

The EU's Common Agricultural Policy (CAP) has a set-aside program as well. The program serves the dual goals of supply control and environmental enhancement. Given that the CAP has partially decoupled agricultural payments, the set-asides would eventually constrict trading opportunities once existing stocks are dissipated. Prior to the time when stocks have reached normal carryover levels, set-aside rates may reach 25 percent of arable land. Although CAP reform has partially decoupled agricultural payments, supply-enhancing measures are still implemented and annual set-aside continues to play a role of supply control with few environmental benefits. For the long-term programs, the benefit-cost calculus of CAP set-aside beyond that point becomes the same as for the CRP.

Controls on Pesticides   Countries generally control the risks from pesticides by registering only those compounds deemed to be without excessive human and environmental risks and by issuing use restrictions. Pressure to reduce pesticide use in many countries has grown, and programs to foster reductions are becoming more common. For example, Sweden and the Netherlands have achieved reductions of 65 and 35 percent, respectively, over the last decade through a variety of efforts including integrated pest management. These reductions have come from relatively high base levels, especially for the Netherlands, and have apparently not caused severe production decreases. Understandable concern emerges about maintaining trade competitiveness under tighter controls.

Assessments have not shown that the US farm sector has suffered economically from pesticide regulation. Instead, consumer prices appear to have risen slightly (Osteen and Szmedra 1989). Regulation of pesticides is not likely to exert significant effects when good substitutes are available or can be developed in timely fashion. Also, the cancellation of a compound spurs adaptation and innovation of new pesticide products. These conclusions may not apply to the regulation of pesticides on "minor crops," which cover crops such as fruits and vegetables. The minor crops often provide small or narrow markets for specific pesticides with the effect that firms have insufficient incentive to register replacement compounds for those whose registration is cancelled. The US Environmental Protection Agency instituted a set of reforms to improve the timely supply of minor use compounds (OTA 1995a).

The 1996 Food Quality Protection Act (FQPA) which will trigger a massive review of existing chemical registrations under new criteria and standards. It is unclear at this point what pattern of pesticide changes will ensue under the FQPA. However, the potential for affecting trade is significant, but not for US exports as might be expected. Countries that desire to export foods into the lucrative US market must ensure that their products satisfy the FQPA standards, which will be imposed at border inspections in the form of minimum residue levels. Hence, the FQPA standards will become *de facto* world standards for exporters wishing to enter the US market.

A major global pesticide issue related to trade is the phase out of methyl bromide. This compound is used as a soil fumigant in the production of crops in certain regions, and as a treatment for fruit and vegetable exports and imports. Methyl bromide depletes ozone in the upper atmosphere, which provides a protective cover for harmful radiation from sunlight. Under provisions of the Montreal Protocol to reduce ozone-depleting substances, signatory countries have scheduled complete phase-outs or significant reductions of the chemical. The US originally scheduled an early phase-out of its use by 2001 to comply with the Montreal Protocol and the US Clean Air Act, but has recently relaxed the schedule to 2005. Although substitutes for methyl bromide (e.g., radiation, heat treatments, and use of carbon dioxide and diatomaceous earth) are being researched and developed, economical substitutes for the pesticide do not yet exist.

US fruit and vegetable growers have been estimated to risk losing export markets of $1.1 billion because of regulations by importers that require incoming products to be treated with methyl bromide to avoid unwanted pests (Forsythe and Evangelou 1994; Yarkin *et al.* 1994). Producers of fruits, vegetables, and other crops in the southeastern and western states face estimated losses of $1 billion per year if methyl bromide is banned as a soil fumigant, unless new economical substitutes emerge (Ferguson and Padula 1994). Despite these losses, the human health and environmental benefits from the phase-out have been estimated to far exceed the costs (US EPA 1988). Still, the regional distribution of costs are uneven with costs concentrating in specific production regions, such as Florida (Taylor 1997) and California. How to moderate these severe trade costs to specific regions and maintain an otherwise profitable environmental initiative becomes a major issue. The strategic use of R&D to help the fruit and vegetables sector adapt to such environmental regulations may be appropriate.

5.3.1.4. *Summary*

Two major findings emerge from this section. First, current agri-environmental programs likely exert negligible trade effects (Ervin and Fox 1999). Most of the programs employ voluntary payment approaches and most developed countries pursue similar objectives. Second, growth in those programs, driven by robust public sentiment for environmental protection, will enlarge any trade effects. Land set-aside programs hold the largest potential to affect agricultural trade and will become more restricting after agricultural policy reform. Restrictions on pesticide introduction and use could exert the largest effects, in particular the reregistration of compounds under new US law. Any trade distortion from these restrictions depends critically on the availability of substitutes, including targeted R&D efforts to offset the restricted compounds.

5.3.2. Environmental Measures Related to Trade

Most concern about the environmental effects from liberalized agricultural trade has centered on the shifts in production and resource use. But the pursuit of environmental protection may draw in food trade flows as well.

Environmental trade measures (ETMs) must avoid violation of the General Agreement on Tariffs and Trade (GATT) rules. Because it was drafted over 40 years ago, the original GATT did not refer to environmental measures. Nonetheless, Article XX provides for two exceptions from a country's GATT obligations that permit trade measures related to environmental management:

1. Article XX(b) provides an exception for measures "necessary to protect human, animal or plant life or health."
2. Article XX(g) provides an exception for measures "relating to the conservation of exhaustible natural resources if such measures are made effective in conjunction with restrictions on domestic production or consumption."

These exceptions provide the general basis upon which countries can take action to protect natural resources from trade-related risks. But the application of these exceptions is tightly circumscribed by restrictions on any measures to avoid unnecessary trade interference. In satisfying the restrictions, the net effect has been to severely limit their application to environmental problems (Esty 1994). Some restrictions understandably guard against obvious trade risks. For example, any measure should not be applied in a manner that would constitute arbitrary or unjustifiable discrimination between countries where the same conditions prevail, or constitute a disguised restriction on international trade. However, others may constrain what otherwise may be viewed as legitimate environmental protection issues related to trade. For example, the exceptions mention natural resources but not "environmental" objectives, and thus may not pertain to the atmosphere, oceans, and other resources outside the "exhaustible natural resources" category. Similarly, the "necessary" test has been defined to justify environmental policies only if no "less GATT-inconsistent" policy tool is "available" to achieve the goal, a very restrictive interpretation that may exclude least cost solutions (Esty 1994: 48). Other restrictions, such as no extraterritorial or extrajurisdictional measures or limitations on unilateral actions, may not fit with the science and dimensions of environmental problems.

The application of the Article XX exceptions has not resulted in a clear and consistent set of rulings by panels (Esty 1994). Thus, countries that consider the application of ETMs may not be assured that their actions will be held GATT-legal if challenged. Implementation of the URA should resolve some of these issues because its language recognizes trade and environment linkages for the first time in the GATT. In addition, it established the Committee on Trade and Environment to clarify the application of ETMs and other issues.

The application of either exception generally comes in the form of a product standard on such characteristics as the amount of pesticide residue or the presence of unwanted insects on imported fruits or vegetables. However, countries may also pursue product-related process standards, such as the satisfaction of sanitary processing conditions. Technically, the regulation is on the product, but it relates to the process of production. Like product standards, the same rules must apply to domestic and foreign production processes.

The URA added two agreements related to environmental management, the Agreement on the Application of Sanitary and Phytosanitary (SPS) Measures, and the Technical Barriers to Trade (TBT) Agreement. SPS measures pertain to achieving specific objectives, while the TBT code deals with product standards, technical regulations, and conformity assessment procedures.

5.3.2.1. *The SPS Agreement*

The SPS Agreement defines the conditions for countries to use trade measures to protect human, animal, or plant health from sanitary and phytosanitary risks and remain GATT-legal. Possible measures include laws, decrees, regulations, requirements, and procedures. More specific examples of measures include inspection; certification and approval procedures; quarantine treatments, including relevant requirements associated with the transport of animals and plants; food processes and production methods; meat slaughter and inspection rules; procedures for the approval of food additives; and the establishment of pesticide tolerances. This list makes clear that SPS measures do not stress natural environmental issues. Rather, the primary driving forces for passing the new SPS Agreement were food safety concerns and avoidance of unscientific measures that restrict food exports. Nonetheless, the SPS measures include animal and plant life and health and therefore pertain to the natural environment.

Several key provisions portray the character of SPS actions. First, to minimize the introduction of disguised trade barriers, countries must base SPS measures on scientific principles. This requirement does not restrict a country's choice of the level of food or environmental safety, but only the measures used to attain them. For example, an importing country may establish the permissible risk of introducing damaging pests at 5 percent, but must defend its choice of control measures based on the scientific evidence of their efficacy. The country must also conduct a risk assessment to assure that the purposes for the pest control measures are valid.

Second, countries must use international standards as minimums where they exist. This provision will help move counties toward at least partial harmonization and thereby remove potential barriers to the free flow of trade (Krissoff *et al.* 1996). Applicable standards include the Codex Alimentarius Commission for food safety, the International Office of Epizootics for animal health, the International Plant Protection Convention coupled with regional organizations for plant health, and other international organizations that the Committee on Sanitary and Phytosanitary Measures may use.

Third, the SPS Agreement requires the application of least trade restrictive practices. To judge that a measure is inconsistent with the Agreement, the following need to exist:

1. An alternative measure exists that would achieve the level of protection demanded by the importer.
2. The alternative measure would need to be reasonably available, and technically and economically feasible.
3. The alternative would need to be significantly less restrictive.

Under the SPS Agreement, if an exporting party "objectively demonstrates" that its measures achieve the importing party's proper level of SPS protection, then the importing party must accept the measures as equivalent, even when they differ from the importer's preferred methods. This issue has relevance, for example, to apple trade between the US and Japan. Other provisions include the avoidance of disguised restrictions on trade, and the provision of opportunities for governments to demonstrate equivalency of protection from different measures, for example, chemical and nonchemical treatments.

Agricultural SPS illustration. Because fruit and vegetable trade involves the shipment of fresh or raw goods, the introduction of unwanted pests is often a serious risk. In the absence of effective policy, the introduction of HNIS through new trade routes can damage environmental resources. Some notable environmental cases include possums in New Zealand, brown snakes in the South Pacific islands, and zebra mussels in the US. The cumulative economic damage from 1906 to 1991 caused by 79 HNIS organisms in the US, less than 14 percent of the total, was estimated at $97 billion (US) (OTA 1993). Future damages from just 15 of the very harmful animal and plant diseases were projected between $66 and $134 billion. It is noteworthy that these damage estimates do not include a full accounting of environmental effects due to missing market prices and observable quantities.

The difficulty of applying an SPS-based ETM can be illustrated with the Mexico–US avocado trade dispute (Bredahl and Holleran 1997; Roberts and Orden 1997). This case primarily concerns commercial rather than "natural environment" damages, but shows the challenges and tradeoffs of applying an SPS measure. US officials first established a quarantine that prohibited the importation of Mexican avocados in 1914, fearing such imports would carry avocado seed weevils that would damage domestic production. Mexican authorities petitioned for approval to export avocados from the states of Michoacan and Siniloa in the 1970s. US officials denied their application, citing the apparent ease with which the seed weevils were recovered in Michoacan, and that seed weevils and Mexican fruit flies were frequently intercepted in fruit contraband at the border. After expanding its groves and improving its production practices in the 1980s, Mexico issued requests for approval to export Hass avocados to the US in the early 1990s. The US Animal and Plant Health Inspection System (APHIS) approved a request in July 1993 to allow Mexican avocados grown in Michoacan to be shipped to Alaska. In February

1997, APHIS published a science-based rule that permits Mexican Hass avocados exports to 19 northeastern states and the District of Columbia during November through February.

Why did the US government cancel its full ban on the importation of Mexican avocados? Does the US–Mexican avocado rule herald a new era in which science plays a stronger role as envisioned under the new SPS Agreement? Clearly, a key element in reversing the quarantine for the selected states was the development of a "systems approach" to reduce the risk of pest infestation to an insignificant level. The systems approach includes nine components: host resistance; field surveys; trapping and field bait treatments; field sanitation practices; postharvest safeguards; winter shipping only; packing house inspection and fruit cutting; port-of-arrival inspection; and limited distribution. This comprehensive pest management approach sought to assure US interests that Mexican avocados could be imported with negligible risk to domestic avocado production. Moreover, the satisfaction of all requirements could impose new costs on Mexican growers that may diminish their considerable competitive advantage. Should pests in the avocados be detected at any stage, the imports can be suspended. Still, 85 percent of the over 2000 comments submitted on the proposed rule opposed its adoption. The high percentage is not surprising in that over one-half of the comments came from the US avocado industry. Another force at work to adopt the rule may have been the potential for Mexican technical barriers on other US farm exports. Adoption of the systems-based approach might also encourage Mexico to adopt similar approaches for apple, peach, cherry, and other US exports that could be disrupted with SPS actions. The benefits of taking this modest action outweighed the risks from pest infestation and other potential trade disruption in the eyes of US officials.

What does this action portend for other SPS environmental issues? One major difference stands out. Imports that threaten commercial losses stimulate private and political support for SPS restrictions according to the political theory of regulation (Roberts and Orden 1997). Basically, producer groups that fear losses from opening up trade put political pressure on regulatory agencies to protect their domestic markets from foreign competition. Small industries with few large members, such as the US avocado industry, are more proficient at this activity because they incur lower transaction costs than large diverse groups. The perceived threat may be justifiable or unjustifiable based on science. In contrast to commercial losses, natural environmental risks (or benefits from avoidance) are often diffusely spread with small individual effects, therefore mobilizing political support for implementing SPS-based environmental actions will likely be more difficult Moreover, the benefits of action may be public goods, either nonrival or nonexclusive, which means that private parties have insufficient incentives to remedy problems that offer overall net benefits. Consider leafy spurge that has caused large overall damages to public and private rangelands, but with small damages to individual ranchers and recreationists.

The US–Japanese apple case is another phytosanitary example. The Japanese government banned the import of US apples until 1995 fearing the introduction of pests such as coddling moth and fire blight. Like the avocado case, the primary concern centers on damages to commercial orchards. The recent change in policy by the Japanese to allow imports under systems-based pest control regimes for growers in the Pacific Northwest adds to the SPS literature. Ironically, after the import ban was lifted, the Japanese demand for US Red Delicious and Golden Delicious varieties fell to virtually zero despite their lower prices. Japanese consumers prefer different apple varieties, such as Fuji and Gala. US growers have responded to this market signal and are attempting to ship those varieties. However, the Japanese have requested new pest control protocols for the additional varieties, which differ from those for the Red and Golden Delicious. The process of designing and implementing new systems-based pest control measures is time consuming and expensive. The US government has taken the position that the previous protocol offers adequate protection, while the Japanese prefer new pest control protocols for individual varieties. Negotiations are underway to resolve this issue.

The apple case illustrates the central role that science plays in SPS disputes, which has special implication for environmental cases. Environmental science relevant to these issues may well be less developed than science for production purposes because of the public good problems noted above. Furthermore, GATT dispute panels have generally not involved environmental scientists compared to the roles for production-related scientists (Esty 1994). Unless specific attention is given to them, the shortcomings will limit the efficacy of SPS actions for environmental issues.

5.3.2.2. *The Agreement on Technical Barriers to Trade*
As part of the URA, GATT contracting parties also approved new provisions that define the application of technical measures to trade in goods and services (also referred to as the Standards Code). The Agreement on Technical Barriers to Trade (TBT) essentially defines the process for distinguishing legitimate uses of product standards, technical regulations, and conformity assessment procedures from efforts to use them as disguised barriers to trade. The TBT addresses the development, adoption, and implementation of mandatory and voluntary product standards that affect trade, and the procedures used to determine whether a particular product meets a standard. The TBT includes two specific mechanisms for implementation – technical assistance and notification of proposed standards.

Technical and labeling requirements, as well as packaging, product content, and processing methods that affect the characteristics of the product are covered in the Agreement. Unlike the SPS measures, which have as their objective the protection of human, animal, or plant life and health, the technical measures pertain to the type of approach taken (Stanton 1997). Thus, the SPS and TBT codes are guided by different purposes but can intersect, such as for the establishment of pesticide limits on fruit and vegetable products.

The TBT Agreement ensures a URA signatory country's rights to protect human health or safety, animal or plant life or health, and the environment as legitimate objectives. Like the SPS Agreement, the principal thrust of the TBT is to reduce the unjustified restriction of trade by product standards. Thus, they apply directly to agricultural trade, but mostly to food safety issues, not to environmental protection. One well known example is the EU's decision to ban imports of beef raised with the aid of growth hormones, which was ruled illegal by a GATT panel. Outside of agriculture, a well-known technical environmental measure is required air pollution emissions equipment on imported automobiles, just as for domestically produced cars.

Only environmental measures related to product standards, however, are covered by the TBT. It does not, therefore, cover most measures related to a nation's water, air, and land pollution legislation, which largely concern emissions from production processes. Key provisions of the agreement include nondiscrimination against imports, measures that do not restrict trade more than necessary, and measures that are established in transparent fashion. The TBT also promotes the use of international standards where they exist, but preserves the rights of countries to enforce more stringent standards at the national, state, province, or local level if they so choose. This latter provision eased fears of some signatory countries that the use of international standards might encourage downward harmonization of standards.

Advocates of liberalized agricultural trade worry about the rise of new technical barriers, including ETMs, in the wake of the URA's actions to reduce old barriers (Runge 1990). There is little empirical evidence to document the scope, degree, and effects of existing technical barriers. Anecdotal stories regularly surface, but usually without scientific documentation. Some conceptual studies are underway (Orden and Roberts 1997) to fill this void. Also, Thornsbury *et al.* (1997) developed a careful definition of technical barriers, distinguishing between those that have prima facie objectives to correct market inefficiencies and those that do not. They use this definition to review data collected by Roberts and DeRemer (1997) on "questionable" technical barriers affecting US agricultural exports, which is reported below. The technical measures were judged "questionable" if they appeared to violate one or more of the principles of the URA. Bredahl and Holleran (1997) build a conceptually consistent framework for analyzing technical regulations related to food safety. The focus of most work has been on clarifying the conditions under which technical regulations are legitimate exercises to protect human, animal, and plant health and safety.

Empirical measurement of TBTs is in its infancy. One survey analyzed the impact of environmental standards on the exports of southern US commodities (Marchant and Ballenger 1994). The authors interviewed experts for the region's major export crops to assess the extent to which domestic or foreign product and process standards affected trade. In general, their findings did not reveal extensive and significant effects on trade from either current domestic or foreign environmental actions. The phase-out of methyl bromide was the exception.

A comprehensive assessment of technical barriers to US agricultural exports is available (Roberts and DeRemer 1997). Technical barriers in this analysis encompass all product or product-related process standards that impede US exports regardless of their legality *vis-à-vis* GATT rules. Therefore they could include transparent violations of existing SPS and TBT codes, legitimate applications of the codes as judged by GATT rules, or applications of product and process standards that have questionable legitimacy. Using survey findings from US Department of Agriculture field staff and representatives of producer groups who identified approximately 300 "questionable" measures in 63 foreign markets, the authors estimated that the technical barriers threatened, constrained, or blocked nearly $5 billion in 1996 US exports.

The questionable barriers were unevenly distributed by value of impact, by region, and by purpose, and for SPS or TBT cases. The estimated trade impact (loss in producer gross sales revenue) was under $10 million for 70 percent of the barriers. On the other hand, just 20 barriers accounted for over 60 percent of the total impact. East Asian countries and the Americas led other regions in barriers. About 60 percent of the impact was attributable to measures that affected market expansion, followed by market retention, and then market access. Over 90 percent of the issues are SPS applications, and the remainder are other technical barriers. The barriers generally mirrored the broad pattern of trade flows for US agricultural products.

Relevance to agriculture and the environment. There is a dearth of scientific and case information on TBTs and the environment. Almost all cases pertain to SPS issues. Exports of processed and fruit and vegetable products were reported to suffer most from technical barriers, about 10 percent of the total, by Roberts and DeReemer (1997). Barriers in the fruit and vegetable commodity groups were concentrated in the East Asian markets. In the authors' judgement, only a very small proportion of this extensive list of technical barriers was exclusively directed to environmental issues. The largest categories on an economic basis were for plant health and food safety, and for processed and horticulture products. Plant health likely includes actions taken for joint commercial and natural environment purposes.

Another study examined the effect of a developed country environmental standard on the potential for exports by developing countries (Verbruggen *et al.* 1995). The issue was requirements for eco-labeling to enter the Dutch cut flower industry. To respond to public environmental concerns about domestic flower production, an industry association developed an eco-labeling scheme that creates segmented flower markets to allow consumers to reward positive environmental performance. These eco-labeling requirements based on national circumstances and preferences have the potential to disadvantage flower imports from developing countries. This case illustrates the difficulty of protecting domestic environmental values without unnecessarily restricting trade opportunities.

The few assessments reviewed in this section do not suggest that technical barriers related to environmental protection broadly restrict agricultural trade. The potential for new barriers to emerge after other nontariff barriers are phased out under the URA cannot be discounted. Conflicts over genetically engineered plants and animals likely will appear in the near future.

5.4. COMPLEMENTARY TRADE AND ENVIRONMENTAL INSTITUTIONS FOR AGRICULTURE[1]

Despite all of the WTO provisions pertaining to agri-environmental measures, including Article XX exceptions, the "Green Box," etc., the criteria to judge whether country schemes conform to trade liberalization precepts are unclear. Uncertainty pervades the design and implementation of effective environmental and conservation programs for agriculture that will not be challenged as trade restrictive. Still, countries are implementing agri-environmental programs to respond to their citizens' desires for landscape protection, improved water quality, wildlife habitat, and other environmental services affected by farming.

A set of principles, or a "Code of Good Process," for designing agri-environmental programs consistent with WTO rules and guidelines can be envisioned. The SPS agreement is instructive in this process. However, unlike SPS issues that rely heavily on biological science to judge validity of applications, environmental initiatives may be pursued for strictly social purposes. Concepts from economics can be rigorously applied to guide the determinations of GATT-legality. A comparable set of criteria to guide the design of trade policies that are consistent with legitimate national and international environmental policies, "greening the GATT", is beyond the scope of this chapter. The principles include:

(1) *Specify clear environmental objectives for the programs.* Objectives established by legislative or duly appointed administrative bodies satisfy the Green Box provision (a). Ideally, the objectives should be in terms of ambient environmental condition, such as permissible concentrations of pollutants, rather than recommended practices. If the farm practices are linked tightly with the environmental objective, then the linkage should be spelled out transparently as described below. Developed countries generally have not set environmental objectives for agriculture, a tradition that gives compensatory payments the appearance of general farm support, and hinders cost-effective program performance (Ervin *et al.* 1998).

---

[1] This section draws from Ervin (1999).

(2) *Clarify property rights in environmental resources to establish the applicability of payments, charges, and subsidies.* A frequent point of contention is whether a government payment for agri-environmental purposes distorts trade by giving the domestic farmer a competitive advantage. The definition of property rights in the environmental resource in question clarifies the farmer's rights and responsibilities for management, and thus makes clear whether a payment or charge is a legitimate or an implicit subsidy. The nature of nonpoint pollution from farming historically has stymied a clearer definition of responsibilities. However, improved science and shifts in public values to hold farming increasingly responsible for some forms of pollution have led to some clarifications.

(3) *Prefer the least trade-distorting agri-environmental management instrument.* Countries can choose from a wide array of measures to achieve their agri-environmental objectives, from compensatory payments, to trading pollution rights, to charges on pollution, such as for manure spills. Choosing the least trade-distorting scheme consistent with the specification of property rights and responsibilities ensures minimal adverse trade impacts. An example of this point is moving away from total land retirement schemes, such as the CRP, when some from of production can attain the environmental performance standard. Choosing an equally effective and less distorting measure is consistent with the GATT "equivalence" principle applied to SPS and other restrictions. The choice of instrument should be cast in a dynamic context. Choosing measures that foster reductions in long-run compliance cost to meet environmental objectives will affect trade less over time, although not necessarily in the short-run. The measures stimulate management and R&D innovations that improve cost-effectiveness.

(4) *Establish scientific linkage of the environmental objective with the program instrument.* This step ensures that clear theory and an empirical base exist to expect that the application of the instrument will help achieve the desired environmental objective. For example, if the objective is to improve the biological diversity of certain agricultural lands, then the establishment of *in situ* species through particular management agreements may be necessary. Perhaps due in part to the lack of clear agri-environmental objectives, the science to understand cause–effect relationships is weaker for agriculture than for other industries.

(5) *Implement monitoring and evaluation programs to document policy/ program efficacy.* To verify that the agri-environmental program is achieving its stated objective(s), evidence from monitoring and evaluation must be assembled that affirms its efficacy. Such evidence is not only good for program management, but essential to satisfy foreign nations that the program is a legitimate exercise. Environmental policies are institutions designed to achieve public objectives, and therefore, involve social values and social science. No amount of refinement of the physical and biological relationships between agricultural processes and environmental conditions will suffice to establish

"sound science" for some cases. For example, the public's desire for certain landscape attributes only associated with particular farming systems is a valid public objective, but cannot be supported by biological measurement. Economists can estimate the relative value of such outcomes through credible non-market valuation methodologies.

(6) *Apply equal treatment (for domestic products and imports) if applicable.* A standard requirement of GATT agreements is that domestic and foreign products must be treated equally to avoid discrimination. If imported foods are subject to control (e.g. to avoid excessive pesticide residues), then the same standards should apply to domestic agricultural products and vice versa. The case of restricting imports of GMOs promises to be contentious in this regard. Without a workable and enforceable labeling scheme, countries cannot implement such measures.

(7) *Ensure the transparency of agri-environmental measures.* Perhaps more than any other principle, the transparency of designing and implementing agri-environmental measures is necessary to bridge the cultures of trade and environment interests and build trust. The transparency condition is implicit in many of the above conditions, but deserves separate stand-alone emphasis. Ideally, this transparency will be matched by open trade-environment panel discussions and decisions.

The satisfaction of all seven conditions does not guarantee that agri-environmental programs will be immune from challenge in the WTO. However, they will reduce the likelihood of dispute. Two side benefits may accrue from pursuing the "good process" conditions. First, progress on harmonizing the national objectives of agricultural trade competitiveness and environmental protection may well follow. Second, agri-environmental programs that meet the criteria likely will be more cost-effective, sparing scarce public funds or extending the programs' reach.

5.5. CONCLUSIONS

Agricultural trade and environment issues have risen to the highest levels of international discourse over the last decade, and promise to play a major role in the next WTO round. Strong rhetoric about the manifold environmental benefits of liberalized trade competes with equally vehement assertions of widespread environmental risks from opening country borders to new trade flows. Experience and available science do not support either extreme view.

Broad conflict between agricultural trade and environmental protection in developed countries has not surfaced. Nonetheless, we cannot be sanguine about future conflicts. Major issues, such as trade in foods produced with GMOs, loom on the horizon. Little science is available to estimate the effects and construct welfare-enhancing measures. Such a meager scientific base should not be surprising. The sector has only recently undergone the first stages of trade liberalization. Also, environmental management programs for agriculture have been largely voluntary-payment approaches, with little unwanted cost imposed on producers.

The absence of serious conflicts has not dampened the rhetoric about potential future losses. Why? Quite likely, because strong business interest to expand trade and robust public demand for environmental protection could enlarge what have been until now narrow conflicts in specific regions or crops. Fears of rising future losses to trade and of spreading environmental degradation raise the stakes of developing policies that will avert major conflicts. This chapter presented evidence to support two general propositions about the linkages between agricultural trade and environmental protection, and offers design principles for building more GATT-legal environmental programs for agriculture.

First, little evidence exists to document that agricultural trade liberalization will significantly degrade or improve the environment in the short term. Neither large production nor large income adjustments are forecast for most developed countries. Some negative effects from increased transport and harmful nonindigenous species are noteworthy, and there are reasons to expect more conflicts in the long run, primarily due to rising global food demands and incomplete environmental programs. Agricultural policy reform and trade liberalization have been estimated to cause a net environmental improvement, but most studies estimate uneven and modest effects.

Perhaps the most worrisome environmental risks from agricultural trade liberalization lie in developing countries. If and when agricultural subsidies in developed countries are phased out or decoupled from production, the world supply of many commodities will fall and global prices will rise. Developing countries will step in to fill that supply and garner export earnings. But, serious environmental risks and damages from the increased production likely will follow due to immature agri-environmental management institutions in many developing countries. Some of those environmental effects, such as the loss of biological diversity, have relevance to developed countries. Institutions to resolve transboundary and global environmental issues related to agricultural trade are immature or nonexistent.

Second, economic theory and evidence do not suggest that environmental programs for agriculture have broadly affected food trade. This finding parallels the conclusions from a substantial body of research on the effects of environmental regulations on trade flows outside agriculture. Developed countries traditionally have used voluntary-payment approaches, such as cost-sharing for practice adoption, to control pollution and deliver environmental benefits from agricultural production. Pesticide regulations are an important exception to the voluntary approach, but they have not constrained production significantly as yet. Efforts to guard against environmental risks from trade flows, such as using phytosanitary rules to shield against importing harmful nonindigenous species, also have not exerted widespread effects on trade.

A continuation of these relatively small effects on trade is by no means assured. Some developed countries, such as New Zealand, Australia, and the US, are moving to more direct controls of environmental damages and risks caused by agriculture. The planned reductions in methyl bromide, an ozone-depleting gas, under the Montreal Protocol illustrate the growing tension. Thus, agri-environmental management reforms that foster improved protection while retaining competitiveness are of particular concern to the sector. More technical barriers to trade related to environmental purposes are also anticipated. Developing rules to clarify the appropriate implementation of such barriers for environmental purposes is a high priority.

Third, substantial progress can be made in developing agricultural trade and environmental institutions that work in harmony rather than conflict. Popularly referred to as "greening the GATT" and "GATTing the greens", both actions are necessary to improve global welfare. A skeleton framework for greening the GATT exists in the URA Agreement on Agriculture, but it needs much elaboration to guide public and private decision makers. The task of making agri-environmental programs less trade conflicting is more challenging. A set of principles is offered to guide the design and implementation of environmental programs for agriculture.

6

International Trade and Sustainable Forestry

EDWARD B. BARBIER

6.1. INTRODUCTION

This chapter is concerned with examining the linkages between the global trade in forest products and efforts to promote more *sustainable forestry*. Although this term has been defined in many ways in the literature, perhaps the most relevant interpretation is the general definition and criteria of *sustainable forest management* adopted by the International Tropical Timber Organization (ITTO): "Sustainable forest management is the process of managing permanent forest land to achieve one or more clearly specified objectives of management with regard to production of a continuous flow of desired forest products and services without undue reduction in its inherent values and future productivity and without undue undesirable effects on the physical and social environment."[1]

Both the potential obstacles to sustainable forest management posed by current trade policies and barriers and the possible positive reinforcement for sustainability through creating trade-related incentives will be explored. The much-touted contribution of certification and ecolabeling in ensuring that the trade in forest products is from sustainably managed sources will be critically evaluated. The chapter will also discuss the likely additional costs of achieving sustainable forest management, especially in developing countries, and the need to consider innovative mechanisms to finance these costs. A critical issue, which this chapter seeks to address explicitly, is the extent to which financing sustainable forest management should be borne by the trade in forest products or through additional finances provided from sources outside of the trade.

The following chapter will discuss mainly the linkages between the trade in forest products and the sustainable management of production forests for

I am grateful to comments on this chapter provided by G. G. Schulze, H. W. Ursprung and an anonymous reviewer.

[1] From the International Tropical Timber Council Decision 6(XI), Quito, 8th Session, May 1991. See also ITTO (1990) and Barbier *et al.* (1994).

timber, but is also concerned with the wider influences of the trade on timber-related deforestation and forest degradation. The main rationale for focusing primarily on the relationship between the international trade in timber products and sustainable management of production forests is that this relationship lies at the heart of the key trade and environment issues that are the focus of global concern currently.

However, it is also important to recognize that the analysis of trade and sustainable forestry linkages is subject to a few data limitations. Data availability usually allows examining these linkages only in terms of the trade in a few categories of wood products – roundwood, sawnwood, plywood, furniture, and pulp and paper products.[2] Lack of data has also precluded analysis of how trends in production of trade in higher processed products, such as furniture, doors, construction and joinery items etc., might impact on sustainable forest management. Finally, different sources refer collectively to the trade in the various products made from forest timber as the *forest products trade*, the *trade in wood products* and the *timber trade*. In this chapter, all three terms will be used interchangeably.

This chapter is organized as follows. The next section provides an overview of world production and trade in forest products, and of the forest resource base supporting it. Section 6.3 examines issues of market access and trade barriers in the forest products trade, concentrating particularly on those issues of most relevance to sustainable forestry. The following section elaborates further on the key debate concerning the role of certification and labeling of wood products as ensuring that the timber trade is from "sustainable" sources. Section 6.5 focuses on forest management issues, and in particular on the role of public policies in influencing timber management and the reforms required to implement sustainable forestry. Finally, section 6.6 discusses the issue of financing global sustainable forestry, and whether the additional costs required especially by poorer producer countries should be met solely from sources generated within the timber trade.

6.2. CURRENT TRENDS IN THE GLOBAL FOREST PRODUCTS TRADE AND DEFORESTATION

This section overviews briefly the current trends in the global production and trade in forest products. The current status of the global forest resource base supporting the trade in timber products is also reviewed.

---

[2] Most available information on the international trade in wood products has been assembled by the Food and Agricultural Organization of the United Nations (FAO), and it is this information that will be used primarily in this document. Following FAO convention, in the following document, wood or timber products will also be referred to as forest products generally.

6.2.1. Global Forest Products Trade Trends

Trade is still a relatively small proportion of global timber production. Approximately one quarter of wood based panels and paper products and one fifth of sawnwood and wood pulp are traded internationally. Only 6–7 percent of global industrial roundwood output is currently traded (Barbier 1996; FAO 1997).

At a global level, forestry contributes about 2 percent to world GDP and 3 percent of international merchandise trade. The regional pattern and direction of the global forest products trade has been fairly stable, with the global forest product market still largely dominated by developed countries, in terms of both exports and imports (FAO 1997). Nevertheless, two distinct trends have become discernible in recent years (Barbier 1996):

(1) First, the trade in forest products is highly regionalized within three important trading blocs, the Pacific Rim, North America and Europe (mainly Western Europe). Within each trading bloc the major importers are mainly developed countries, such as Japan, the United States, Canada and the European Union. However, in recent years developing countries particularly in Asia have been increasing their share of global imports (see table 6.1). Much of this demand reflects the increased growth in consumption of industrial wood products in developing countries. Newly industrialized countries with limited forest resources have also been increasing their imports of logs

Table 6.1. *The global forest products trade: major importers and exporters, 1995*

Importers	US$ mn	Exporters	US$ mn
United States	22,448	Canada	27,787
Japan	19,486	United States	18,148
Germany	10,948	Finland	11,953
Italy	8,637	Sweden	10,850
France	8,198	Germany	7,779
United Kingdom	8,084	France	5,851
Netherlands	5,163	Indonesia	4,728
Korea (Republic of)	4,972	Malaysia	4,226
Belgium-Luxembourg	4,066	Brazil	3,547
Taiwan	3,840	Austria	3,361
Spain	3,826	Russia	3,231
China	3,840	Netherlands	3,017
Canada	2,953	Italy	2,874
Switzerland	2,857	Belgium-Luxembourg	2,791
Hong Kong	2,796	Norway	2,179

Source: FAO (1997).

and semi-finished wood products as raw materials for the export-oriented processing industries.

(2) Second, the major global exporters of forest products still tend to be developed countries with temperate forest resources and processing industries. However, developing countries such as Indonesia and Malaysia have emerged as dominant world exporters of certain forest product exports, such as non-coniferous wood-based panels, logs and sawnwood (see table 6.1). Other developing countries, notably Brazil, Chile and the Asian newly industrializing countries, are beginning to have an impact on the international trade in wood pulp and paper products. In general, the trade in forest products has shifted towards value-added processed products.

However, the global forest products trade is still dominated by a few countries. In 1995, five countries accounted for 55 percent of world exports, and ten accounted for 70 percent. The top two exporters, Canada and the United States, were responsible for almost one third of the global export market. Similarly, five countries accounted for 48 percent of global imports, and ten countries for 66 percent. The top two importers, the United States and Japan, comprised nearly one third of the global import market (Bourke 1998 and table 6.1).

The trade in forest products has generally benefited from successive postwar GATT Agreements. Tariff barriers to forest products trade have continued to decline in recent years, particularly in the post-Tokyo Round era (Bourke 1988). The extent of the decline in tariffs differs with the market and product. With few exceptions, developed country markets tariff rates had fallen generally to very low levels even before the Uruguay Round schedules were agreed. For example, pre-Uruguay Round tariff rates for forest products in developed countries averaged 3.5 percent compared to 6.3 percent for all industrial products (WTO 1994).

The effect of the recently concluded Uruguay Round negotiations will be to reduce tariff rates on forest products further, including the phasing out completely of tariffs on pulp and paper products in major developed country markets. The extent of tariff escalation for forest products will be reduced in most importing markets, and many tariff rates will be bound. It is estimated that the likely gains in trade for major forest products could be in the region of US$460–593 million, but proportionately this amounts to a gain of only 0.4–0.5 percent of total forest product imports in the markets analysed (Barbier 1996, 1999). The implications of the Uruguay Round for the nontariff barriers increasingly faced by forest products is less clear. However, two special agreements, the Agreement on the Application of Sanitary and Phytosanitary (SPS) Measures and the Agreement on Technical Barriers to Trade (TBT), do provide the basis for tackling certain nontariff measures that have been used as trade barriers against forest products.

One of the current uncertainties for the global timber trade currently is how this trade is being affected by the recent economic and financial crises in East and Southeast Asia. Beginning with the severe Thai currency devaluation in June 1997, the crisis has spread rapidly through the region, affecting virtually every country. Not only have the major Asian timber producing countries – Indonesia, Malaysia, Papua New Guinea, Thailand, the Philippines, and the Solomon Islands – experienced major economic disruptions, but the crisis has affected several key Asian consumer countries as well – Japan, China, Singapore, South Korea, and Hong Kong. As indicated in table 6.1, some of the latter countries are major world importers of forest products, from both producers in Asia and elsewhere in the world. Thus the Asian economic crisis has certainly affected the regional forest products trade, and is starting to have an impact on the global trade as well.

For example, the combined effects of falling currencies and weakening market demand for forest products in Asian consumer countries has had a major impact on all producers from the region (Adams and Johnson 1998). With sharply falling exports and prices, the worst affected appear to be the two largest Asian exporters, Malaysia and Indonesia. However, there is also evidence that other major tropical producers outside the region, such as Brazil, are also being severely affected by the decline in demand from Asian consumers. Part of the problem is that, despite falling prices for most raw and semi-processed timber products, demand in other sizeable consumer markets, such as the European Union, appears to be too sluggish to absorb the excess supply of these products. Already, the impact on Asian forestry industries is becoming severe, with reports throughout the region of concessionaires ceasing operations, mills closing, and workers being made redundant. The long term implications for progress towards implementing sustainable forestry management in Asia are not easy to discern.

6.2.2. Global Forest Resource Trends

In terms of the status and management of global temperate and tropical forest resources, the most environmentally important resource is usually considered to be the closed forests (WRI 1992). In tropical countries closed forest resources have been subject to a higher rate of deforestation than in temperate countries. Reforestation is generally higher in temperate countries as well. Recent estimates of total forest loss (net of reforestation) by the FAO (1997) suggest that 12.26 million ha was lost annually over the 1980–90 period, and a further 13.03 million ha was deforested annually over 1990–95 at an estimated 0.65 percent annual deforestation rate. In comparison, all developed countries (excluding Russia) increased their total forest area by 8.78 million ha, or at an annual rate of 0.12 percent.

This changing pattern of global forest resources is thought to have two important implications for the trade in forest products (Barbier *et al.* 1994;

Sedjo and Lyon 1990):

(1) Declining tropical resources and expanding temperate resources will offset each other leading to stable prices for wood products generally, except for highly valued tropical woods.

(2) The shift to plantations and second-growth forests versus old-growth stands as the source of timber will continue. The long-term pattern will shift from the Pacific Northwest and the tropics, to plantations forests in North America and newly planted Southern Hemisphere forests. European forest resources are also projected to expand, at a net rate of around 1 percent annually (Pajuoja 1995).

Continued loss of old growth forests, and in particular tropical deforestation, will of course also have important environmental implications.

6.3. MARKET ACCESS AND NEW BARRIERS TO THE FOREST PRODUCTS TRADE

Despite recent progress in reducing tariff barriers to the forest products trade as a result of the Uruguay Round, there has been a proliferation in "new" tariff barriers in the forest products trade which could influence the adoption of sustainable forestry practices. The following section discusses these possible sources of restrictions to the trade. In addition, the relative competitiveness of forest products in influencing the long run returns to timber production are assessed, as is the development and exploitation of lesser known species. Finally, new issues have emerged, such as illegal logging activities and the use of Convention on International Trade in Endangered Species of Wild Fauna and Flora (CITES) to ban individual timber species, which may influence the relationship between trade and sustainable forestry.

6.3.1. New Trade Barriers

Section 6.2 noted the important progress that was made in further reducing tariffs on the global forest products trade as a result of the Uruguay Round. However, despite the substantial reduction in tariffs on the forest products trade, there has been widespread concern that new types of barriers to the trade, particularly nontariff barriers, could still impede market access for many products as well as lead to new forms of "protectionism". This concern is not misplaced, as in recent years there has been a proliferation of additional policies and regulations that have the potential of becoming "new" barriers to the forest products trade. These include:

- export restrictions by developing countries to encourage domestic processing of tropical timber for export;
- environmental and trade restrictions on production and exports in developed countries that affect international trade patterns;

- quantitative restrictions on imports of "unsustainably produced" timber products; and
- the use of ecolabeling and "green" certification as import barriers.

Although only the last two measures could be strictly defined as "new", all of these trade measures have been increasingly employed in recent years and have the potential to affect forest product trade flows significantly. The issue of certification and labeling will be addressed in more detail in section 6.4. The first three barriers will be briefly discussed below.

Developing countries are continuing to use export restrictions on wood in rough and semi-processed products to support domestic processing industries and improve export prospects for higher valued forest products. Several authors have recently reviewed the role of export taxes and bans in encouraging forest-based industrialization and sustainable timber management in tropical forest countries.[3] The general conclusion is that tropical timber export taxes and bans have proved only moderately successful in achieving the desired results in Southeast Asia. For example, although expanded processing capacity was established in Malaysia, the Philippines, and Indonesia, it was achieved at high economic costs, both in terms of the direct costs of subsidization as well as the additional costs of wasteful and inefficient processing operations.

Despite the losses in terms of economic inefficiencies and the implications for the management of their forest resource base, developing countries are unlikely to end such policies but may instead employ them more extensively. Many timber-producing countries see the use of export taxes and bans on raw or partially processed logs as the means to compensate domestic processing industries for import barriers faced in developed economy markets. However, with the post-Uruguay Round decline in tariff escalation and barriers generally in import markets for forest products, this argument is less valid.

Developed countries are increasingly employing a variety of environmental regulations in their forest industries – both alone and in conjunction with export restrictions – that may have significant trade implications. Whether or not such regulations are being used intentionally for this purpose, they may lead to trade distortions and discrimination. For example, the combination of trade and environmental restrictions on logging in the Pacific Northwest of the United States – such as the spotted owl reservations coupled with the state-level log bans – produced significant domestic and global trade impacts, including increases in global sawlog prices and regional shifts in production with related effects in major sawnwood and plywood markets (Flora and McGinnis 1991; Perez-Garcia 1991).

In many developed countries domestic policies to promote waste paper recovery and recycling have had important trade implications, particularly where they involve mandatory restrictions on the levels of virgin fibre and pulp

---

[3] See for example Barbier *et al.* (1994) and (1995); Vincent (1992).

use. For example, Bourke (1995) and Elliot (1994) discuss the trade implications for Canada – the world's largest producer and exporter of newsprint – of US state and federal recycled content laws for newsprint. In particular, the US recycled content laws may provide an unfair cost advantage to domestic producers because of the greater availability of used newsprint in the United States than in Canada. Similar problems apply to packing and reuse requirements, such as the recent European Union Packaging Directive and the Japan's regulations for recycling of paper, logging residues and dismantled houses. Such regulations all have the potential of being used as nontariff barriers to competing paper product imports, particularly if there are requirements on suppliers to recover packaging or to impose deposit and refund schemes (Bourke 1995; Weaver *et al.* 1995). Potential problems exist with other environmentally oriented regulations, such as the increasing restrictions on trade in wood panels using formaldehyde glue, regulations banning or controlling certain timber preservation processes and materials, and controls on processing materials, for example, the use of chlorine in bleaching pulp.

Although there are legitimate uses of all the trade policy measures discussed above, the rate at which they are being implemented and the frequency with which they have led to trade distortion and discrimination suggests that their use must be examined carefully. International agreements and rules governing their use should also be negotiated, and the interface and possible conflicts between multilateral environmental agreements and trade rules need to be explored through the auspices of the WTO. What clearly needs to be avoided is indiscriminate and widespread application of "new" barriers to the forest products trade that could easily override the gains in market access resulting from the recently concluded Uruguay Round.

6.3.2. The Relative Competitiveness of Forest Products

One reason why protectionism in the forest trade remains a persistent problem is that the trade remains both regionalized yet highly competitive (see section 6.2). In addition, the degree to which forest products from different regions compete among themselves and with nonwood substitutes for import markets is an important determinant to the long-run returns to forest products, and thus a potential influence on efforts to establish sustainable forest management. Changes in long-run returns may in turn influence the incentives for sustainable forest management.

The degree of substitution between tropical and temperate products in consumer markets illustrates the extent to which the markets for these two types of products are interrelated, or whether there are essentially two different markets for two distinct commodities. The available evidence indicates that the elasticities of substitution between temperate and tropical wood products are very low (Barbier *et al.* 1994). This suggests that there are two distinct markets, and tropical producers of these products would have difficulty in penetrating the larger temperate market. That is, temperate softwoods from different regions are still closer substitutes than softwoods and tropical hardwoods; equally, for tropical timber products there are strong substitution effects between products from different tropical regions or countries. In general, substitution by origin for tropical sawnwood and plywood in certain importing countries appears to be very high, especially for plywood. There is also evidence that in some major processing markets imports of tropical logs are subject to substitution by domestic softwood logs and by technical change.

Forest products may also be substituted by nonwood products in end uses and final markets. Although there is increasing anecdotal evidence of this occurring in many consumer markets, particularly in construction and furniture industries, estimating the magnitude or scale of this effect has proven more difficult. However, for specific products this substitution effect may be significant. For example, plywood is believed to face severe competition from solid synthetic panels, with price strongly influencing the choice of product in the construction industry. In addition, substitution may be more of a problem for wood-based composites, such as particle board, fiberboard and reconstituted panels, and wood pulp.

To summarize, empirical studies suggest that substitution between tropical and temperate timber products in importing markets has not been very significant. However, in response to export log bans by tropical producers, some importers are increasingly diversifying the source of their supply. Substitution of nonwood products for timber may be occurring, but the evidence is largely anecdotal. Substitution between tropical timber products originating from different countries or regions does appear to be very high, particularly for plywood. This would suggest that importers can substitute between sources of origin with relative ease but also that exporters can easily capture market share through price competition.

Thus the available empirical evidence indicates that producers *as a group* may enjoy significant market power. If all producer countries instigate sustainable forest management and this leads to higher prices for timber products across the board, then there may not necessarily be any significant loss of market share. However, if only a few producers instigate sustainable management and this leads to higher prices for their forest products, then substitution away from these products is more likely to occur.

6.3.3. Lesser Used Species

It has often been argued that, provided markets exist and exploitation can be conducted sustainably, there is substantial potential for expanding utilization of forest resources, particularly in tropical countries, by exploiting commercially lesser used species (LUS). For example, recent FAO statistics suggest that only about 26 percent of the potential standing volume in tropical harvest areas is being felled. In general, the forest products industry is based on the utilization of large sawlogs and logs for plywood and veneer. Over the 1988–92 period saw and veneer logs comprised 92 percent of the total industrial roundwood harvest in Indonesia, 97 percent for Malaysia and 93 percent for Papua New Guinea (Drake *et al.* 1995).

If LUS are to play an expanded role in both the temperate and tropical forest products industry, it is unlikely that this additional supply would be suitable for traditional solid wood products such as sawnwood and plywood. Instead, the likely use of LUS would be for wood pulp and reconstituted wood products such as fiberboard, particle board and reconstituted panels. At least one set of projections, for the Asia-Pacific region, suggest that the potential gains to producers of developing capacity to manufacture composite panel products and other engineered wood products might be substantial, and could offset the negative impacts on the forest products industry of declining supplies of forest resources for sawnwood, plywood and veneer (Drake *et al.* 1995). However, there are important considerations for any global strategy to promote LUS.

First, as discussed above, reconstituted wood products and wood pulp are currently some of the most competitive and volatile markets for timber products. Reconstituted wood panels are highly susceptible to substitution in import markets by semi-wood composites, such as cement fiberboard, composites made from agricultural and other recycled wastes, and a variety of nonwood substitutes. Any new form of virgin wood pulp would be competing in terms of quality and price with more traditional sources but also from recycled pulp. As noted above, the problem of market access for any new products generated from LUS is further exacerbated by the increasing proliferation of environmental, health, and other regulations specifying the composition and quality of both reconstituted wood and paper products in consumer markets.

Second, in many countries particularly in tropical regions, the potential for identifying let alone exploiting LUS is limited by lack of basic information on the availability and commercial viability of these species. This in turn reflects the limited human and technical resources devoted to inventorying and assessing the forest resource base of the timber industry. Forest Resources Assessment (FRA) has been promoted as a means for tropical producer countries to assess regularly their forests, including the identification of LUS for potential commercial exploitation through sustainable forest management (Kemp and Phantumvanit 1995). Basic assessment of this kind is essential if LUS are to be utilized more in the forest products industry.

Third, although it is generally assumed that exploitation of LUS is advantageous as it would increase stand harvesting yields and thus reduce per unit costs, and possibly lower regeneration costs by removing more of the residual stems, greater utilization of the standing volume of timber is not always compatible with sustainable forest management. A critical issue is whether there is a considerable loss of forest services, such as biodiversity and watershed protection, if more species and hence stems are removed. Clearly, more needs to be learned about the potential ecological effects on forests of increasing the rate of utilization of LUS and the resulting impacts of greater stem removal from stands. Again, an FRA might be a first step in determining the capacity for exploiting LUS through sustainable management practices. However, further analysis of market demand and the costs of exploitation and utilization is required to determine the commercial viability of these species.

Finally, exploitation of LUS must be considered in relation to the problem of increasing scarcity and depletion of key timber species, and the role of forestry policy and management in controlling this problem. This issue will be discussed further in section 6.5.

6.3.4. Illegal Logging and Trade

A growing concern in recent years has been the growth in the illegal trade in timber products, especially logs. This trade is thought to be directly related to restrictions on log exports imposed by producer countries, as well as tighter controls imposed on harvesting practices. Recently, international bans on trade of endangered species may have also incited an illegal trade in these species. Illegal activities in the timber trade cover a number of widespread practices involving illegal logging, illegal trading (i.e., "timber smuggling") and illegal pricing and classification of timber (see table 6.2). For obvious reasons, assessing the extent of these activities is extremely difficult, and estimates can vary widely. Yet the continuing and widespread growth in this trade poses a grave threat to implementing more sustainable forestry practices through legal trade-related incentives.

A recent review of studies by the Environmental Investigation Agency (EIA 1996) noted that the World Bank has estimated that on a global scale 5,000 km^2 of tropical forests were being logged each year during the early 1990s, an area nearly the size of the Indonesian island of Bali. However, a report by the Worldwide Fund for Nature (WWF) claims that virtually all logging for export currently taking place in India, Laos, Cambodia, Thailand, and the Philippines is illegal; and much logging in Malaysia and Indonesia could be classified as illegal. Although most of the recent concern has been with illegal activities in tropical countries, and the consequent loss and degradation of tropical forests, there is also evidence of significant illegal harvesting and other practices in temperate countries as well.

Although difficult to verify conclusively, there is some evidence that the scale of illegal trade in logs is continuing unabated and possibly even increasing. For example, ITTO's Annual Review of member countries' trade has indicated a perpetual problem of under-reporting of log exports, that is, some countries reported exports of logs are consistently and significantly below the corresponding level of imports reported by trading countries. Trade statistics for some importing countries have indicated sizeable log imports from countries where a complete log export ban is supposed to be in force. Publicizing such

Table 6.2. *Illegal practices in the forest products trade*

1. Illegal logging
   - Logging timber species protected by national or international law, such as the Convention on International Trade in Endangered Species of Fauna and Flora (CITES)
   - Logging outside concession boundaries
   - Logging in protected areas such as forest reserves
   - Logging in prohibited areas such as on steep slopes, river banks and catchment areas
   - Removing under/over-sized trees
   - Extracting more timber than authorized
   - Logging without authorization
   - Logging when in breach of contractual obligations
   - Obtaining concessions illegally

2. Illegal trading
   - Export/import of tree species banned under national or international law
   - Illegal log export/import in contravention of national bans

3. Illegal timber pricing and classification
   - Transfer pricing (i.e. internal over or under pricing of timber products) by a company to transfer profits abroad, usually to avoid tax in the country where a timber operation is occurring
   - Under-grading, under-valuing and misclassification of species by a company to avoid royalties and duties by declaring a lower value for timber extracted from concessions

Source: adapted from EIA (1996).

discrepancies can have an impact. For example, as a result of the evidence of under-reporting in ITTO's Annual Review, Papua New Guinea has undertaken to tighten controls over its log exports.

To some extent, the growing illegal trade in logs is a consequence of many developing countries implementing export bans and restrictions, as well as prohibitively high tariffs, as a means of encouraging a shift in the composition of their exports away from logs to more highly valued processed timber products. The loss of export earnings and government revenue from any resulting increase in illegal log exports is yet another significant cost of such policies.

However, illegal logging and trade is also a serious setback for promoting sustainable forest management practices. By their very nature, such activities involve destructive and short-term practices that are damaging to forests. In addition, the loss in export duties, timber royalties, and income taxes to developing country governments means less revenues available to promote sustainable forest management and improve forest departments and institutions.

6.3.5. Endangered Species and CITES

According to the 1994 IUCN threat categories and criteria, to date over 7000 tree species have been documented as threatened. Nearly half of these species are considered vulnerable, and a particularly wide range of tropical tree species fall in this category as a result of increased tropical deforestation over the past 150 years.[4]

Because of the concern over these trends, a number of developed countries, strongly supported by conservation groups, have attempted to have various commercially important tropical tree species placed in one of three appendix listings of the Convention on International Trade in Endangered Species of Wild Fauna and Flora (CITES). Appendix I listing essentially prohibits trade in a species; Appendix II listing requires an export permit to be issued by the CITES Authority in the country exporting the species; and Appendix III listing permits commercial trade provided that export permits are issued for the country or countries listing the species, or a certificate of origin is issued from other countries that reexport the species. Table 6.3 shows the timber species currently listed in the CITES appendices. Only two are traded internationally in significant volumes: *Pericopsis elata* (Appendix II) and *Swietenia macrophylla* (Appendix III).

The recent attempts to have more commercially important tropical species listed have been controversial. Many trade and forestry interests, including exporting countries affected by the listings, have questioned whether some of the species proposed for listing are really endangered or threatened, and

Table 6.3. *Timber species listed in CITES appendices*

Appendix I
Araucania araucania (Argentina), *Fitzroya cupressoides, Pilgerodendron uviferum, Dalbergia nigra, Abies guatemalensis, Podocarpus parlatorei, Balmea stormiae*

Appendix II
Araucania araucania (Chile), *Caryocar costaricense, Oreomunnea pterocarpa* (also referenced as *Engelhardia pterocarpa*), *Pericopsis elata, Platymiscium pleiostachyum, Pterocarpus santalinus, Swietenia humilis, Swietenia mahogani, Prunus africana, Taxus wallichiana, Aquilaria malaccensis, Guaiacum officinale, Guaiacum sanctum*

Appendix III
Talauma hadgsonii, Swietenia macrophylla, Podocarpus nerifolius, Tetracentron sinense

Source: FAO (1997).

---

[4] Personal communication, Sara Oldfield, World Conservation Monitoring Centre, Cambridge, UK.

whether the procedures used to determine this are sound, given the generally poor information on the inventory of species in tropical forests and on the actual species composition of traded timber products. Given that the listing of a species – even in Appendix III – is considered to have a substantial negative impact on trade in that species, there is general concern that CITES is being used as a means to control, if not limit, the trade in tropical timber products. Because of these concerns, CITES has recently established a Timber Working Group to monitor and make recommendations concerning both the listing of timber species as well as the implementation of the appropriate export controls on a listed species (FAO 1997).

However, a further problem with a CITES listing, as noted in the previous subsection, is that it can create inducements to increase illegal trade in a species. A CITES listing places a large burden on the CITES Authorities in countries involved in the trade of the listed species to implement the required controls. Without significant support by the relevant governments and trade interests, CITES controls on trade may be less effective.

For example, a recent review of the implementation of CITES Appendix III for *Swietenia macrophylla* (big-leafed mahogany, native to Central and South America) found that the controls have been somewhat effective with regard to trade between the key producer and consumer countries (Buitrón and Mulliken 1997). However, the failure of several range states to implement CITES import controls have undermined their effectiveness. In addition, the Appendix III listing does not specifically require certificates of origin or reexport to be presented at the time of export, thus limiting its usefulness with regard to export controls. Although the Timber Working Group recommended changes in this scheme, it did not ask that the name of the exporter be included on the certificate of origin. Thus, the system of controls is vulnerable to abuse, such as the transfer of CITES documents from one exporter to another without the CITES Authority granting permission.

Any attempt to use CITES to control or limit the forest products trade is simply not legitimate. There is a genuine need to control trade in endangered species. However, it is important to ensure that any trade restrictions imposed are, first of all, necessary to guarantee protection of an endangered species, and second, capable of improving the survival of a species that would otherwise be endangered by commercial trade. It is not obvious that many timber species fall into both of these categories.

6.4. CERTIFICATION AND LABELING: CAN THEY PROMOTE SUSTAINABLE FORESTRY?

The previous section noted that there is concern that ecolabeling and green certification could be used potentially as a means to protect domestic markets in consumer countries through restricting imports. However, proponents of timber certification and labeling argue that these instruments are both essential and necessary to reassure consumers that their wood products are obtained from genuinely sustainable sources, and that in the long run the timber trade will gain from guaranteed, higher valued markets. This section assesses progress to date on forest product certification and labeling schemes, and discusses their potential contribution to improved sustainable forestry globally. The evidence suggests that progress has been slow, and the contribution is still small. Efforts to make global forestry more sustainable will require a more concerted national policy level effort backed by international negotiation, or "country" certification.

6.4.1. Timber Certification

The number of ecolabeling and certification initiatives applied to the forest products trade has increased dramatically in recent years. Generally, the aim of these initiatives is to distinguish "sustainably" produced forest products or to ensure that forest product imports conform to domestic environmental standards and regulations. Understandably, there is considerable concern among producer countries and forest-based industries that certification and labeling will be used as nontariff barriers limiting access to key import markets. Provided that such regulations and schemes are nondiscriminatory, transparent and justified, are agreed mutually between trading partners or through multilateral negotiations, comply with GATT/WTO rules and conform with internationally recognized guidelines, then their potential use as trade barriers will be drastically reduced. However, the possible role of voluntary and nondiscriminatory timber certification in promoting sustainable forest management on a significant scale globally is, at best, still not clear.

The term "certification" has been used indiscriminately to cover a wide range of processes. In this document the term *timber certification* will be used to mean a process which results in a written statement, that is, a certificate, attesting to the origin of wood raw material and its status and/or qualifications, often following validation by an independent third party (Baharuddin 1995). To be effective in reassuring consumers that wood products originate from "sustainably managed" sources, timber certification requires both certification of the product process and certification of the sustainability of forest management practices. The latter requires verification of the forest management system in the country of origin, including the environmental and social impacts of forestry practices, against specified sustainable management criteria and standards. The former involves inspection of the entire product processing chain of supply from the forest to final product, through domestic and export markets if necessary.

Proponents of certification argue that it can assist potentially in promoting sustainable forest management while simultaneously reassuring consumers. A properly designed, voluntary and independently accredited certification scheme at the global level can be a means by which the various stakeholders can hold producers accountable; it can provide a market-based incentive to the individual producer to improve management; it can meet consumer demands for wood from well-managed forests without creating trade discriminations; and it can be a mechanism for monitoring multiple factors involved in forest use (Dubois, Robins and Bass 1995).

However, others suggest that the evidence for considerable additional demand for certified wood products is unproven and only in certain small "niche" markets may customers be willing to pay more for certified timber (Varangis, Crossley and Primo Braga 1995). In fact, there is concern that the impacts of certification on production and distribution costs might reduce the competitiveness of wood products in consumer markets. It is also argued that, although certification requires sustainable forest management as a necessary prerequisite, implementation of sustainable forest management does not require certification to take place (Kiekens 1995, 1997). The promotion of certification globally should not either displace or divert resources from ongoing efforts in the major timber supplying countries to implement national forest policies, regulations and standards in accordance to international and national commitments to sustainable forest management. Finally, it is argued that the necessary but stringent conditions required for an accredited global certification scheme are bound to have only a limited impact on a small proportion of global timber production, and equally, on the sustainable management of a limited area of forests (Baharuddin and Simula 1994; Kiekens 1995).

Recent estimates of the total global forest area certified are depicted in table 6.4. Total production amounts to around 3.5 million m^3 from about 5.1 million ha of certified forests. In fact, certified production accounts for only 0.23 percent of the world's industrial roundwood production. It is unlikely that the supply of certified timber and timber products will expand very fast. Even under optimistic projections, only 15 percent of traded wood products are expected to be affected by certification by the year 1999 (Baharuddin and Simula 1996). In fact, recent progress in certification appears to have slowed. A survey in March 1997 revealed that, although the number of certified forests had nearly doubled from 25 to 58, the total area of certified forests had fallen from 5.1 to 3.1 million ha. The main reason for this decline is that a number of large forest concerns did not have their certificates renewed. On the other hand, total production of roundwood estimated from certified forests is now estimated to be 9.5 million m^3 annually (Baharuddin and Simula 1997).

One of the additional benefits of certification might be that it will allow timber exporters to avoid losses of market share and revenues. An attempt to estimate these benefits for tropical timber certification has been conducted by Varangis, Crossley and Primo Braga (1995). The results are depicted in table 6.5. The total gain from certification of tropical timber products is US$428 million, or 4 percent of current developing country timber product exports. An interesting aspect of this estimate is that the vast majority of the

Edward B. Barbier

Table 6.4. *Certified global forests*[a]

Region	Production (m^3)	Area (ha)[b]	Percentage of total production[c]	Percentage of total area[d]
Africa[e]	NA	6,000	NA	0.45
Plantation	NA	6,000		NA
North America	502,408	617,000	0.09	0.14
Natural	502,408	617,000		0.14
Latin America[f]	101,350	672,940	0.12	0.10
Natural	101,350	597,000		0.09
Plantation	NA	75,940		1.05
Asia[g]	738,810	2,847,394	0.85	1.66
Natural	1,310	12,610		0.01
Plantation	737,500	2,834,784		31.8
Europe[h]	2,135,000	916,720	9.70	8.43
Semi-natural	2,135,000	916,563		8.43
Plantation	NA	157		NA
World Total	3,477,568	5,060,054	0.23[i]	0.14[j]
Natural/semi-natural	2,740,068	2,143,173		0.06
Plantation	737,500	2,916,881		10.12

[a] Based on Baharuddin and Simula (1996), unless specified otherwise.

[b] In some cases includes nonforest land as well.

[c] Percentage of total industrial roundwood production in 1993 for the countries specified, from FAO (1995).

[d] Percentage of total forest area, natural forest area and plantation area for the countries specified, from WRI (1994).

[e] Certified forest is located in South Africa.

[f] Certified forests are located in Costa Rica, Brazil, Guyana, Honduras and Mexico.

[g] Certified forests are located in Indonesia, Malaysia, Papua New Guinea and Solomon Islands.

[h] Certified forests are located in Poland and the United Kingdom.

[i] Total production from certified forests expressed as a percentage of total world industrial roundwood production in 1993, from FAO (1995).

[j] Total certified forest, natural/semi-natural forest and plantation areas expressed as a percentage of total world forest, natural forest and plantation areas in 1980, from WRI (1992). Total world plantation area excludes plantation area from Europe, North America and the former Soviet Union.

NA = not applicable or not available.

gains from certification arises from avoiding losses in markets and revenues in the absence of certification and not from the additional gains of any "green premium" – despite the generous assumption that the latter might be as high as 10 percent and with no substitution effects.

It has been suggested that if both tropical and timber products were certified, more markets would certainly be affected, and as a result around 15–25 of the total share of global forest trade could be influenced by certification (Baharuddin and Simula 1996). However, critics argue that the growing importance of domestic markets in tropical countries and the dominance of

Table 6.5. *Estimated revenues from tropical timber certification in European and North American markets*

Source of revenue	Export revenues gained (US$ mn)	Share of developing country wood Exports[a] (%)
1. Incremental revenue from European and American 'niche' markets[b]	62	0.58
2. Avoidance of net revenue losses without certification (i − ii)	366	3.43
(i) Loss of European and North American markets[c]	622	5.83
(ii) Gains from diversion to other markets[d]	256	2.40
3. Total net revenue gain (1 + 2)	428	4.02

Notes: Based on 1991 trade flows and values.

[a] Developing countries excluding China, Argentina, Chile and Near East countries. Total value of forest product export revenues in 1991 estimated at around US$10.66 billion, of which US$9.02 billion from nonconiferous logs, nonconiferous sawnwood and wood-based panels and US$1.64 billion from furniture and processed wood products.

[b] Assumes a 10% increment in revenues due to a "green premium" from certification affecting 10% of the North American and 20% of the European markets.

[c] Assumes that the markets in [a] will be lost in the absence of certification.

[d] Assumes that developing country exports would increase by 8.8% to non-European/North American markets in the absence of certification.

Source: Varangis, Crossley, and Braga (1995).

markets where consumers are less interested in certified timber, limit the potential market effects of this instrument for tropical timber products (Bourke 1998; Kiekens 1997). In the case of European temperate forests, it is argued that the only forests suitable for certification are public forests and a few large-sized private forests, mainly located in Sweden, but to date any certification of these forests has not led to any improvement in forest management or improved market access for certified products (Kiekens 1997).

There is also the issue as to whether timber certification inevitably leads to higher prices for timber products in final consumer markets. This issue has proven difficult to resolve conclusively, and it relates to the evidence concerning the overall costs of certification. It is useful to distinguish two costs: the *direct costs* of certification in terms of implementing such schemes and the *indirect costs* of certification through any trade losses and diversion in final consumer markets as a result of substitution between certified and noncertified products.

For example, for tropical products the costs of assessing or auditing have been estimated at about US$0.3 and US$1 per hectare per year in developing countries, and the costs of certifying the chain of supply for processing could be up to 1 percent of border prices (Baharuddin and Simula 1994). For temperate products in developed countries, cost estimates for certifying forests are roughly similar – US$0.3–0.6 per hectare (Varangis, Crossley and Primo Braga

1995). Dubois, Robins and Bass (1995) suggest as a rough approximation the minimum costs of certifying forest management would be a fixed assessment cost of US$500 plus US$0.40 per ha for the initial assessment and US$0.15 per ha for each subsequent visit. In addition, to meet established certification criteria forest management costs in North American temperate and boreal forests will rise by 20–30 percent, and could be as high as 100 percent. Finally, there are likely to be some indirect costs in terms of trade losses and diversion for certified timber products in importing markets – although the precise magnitude of these indirect costs of certification are difficult to determine presently.

There is now an emerging international consensus that an adequate international framework is needed both to ensure harmonization and mutual recognition of certification systems and to ensure an effective international accreditation process of certification bodies. The important criteria for any internationally accredited certification body is that it is independent, impartial, and able to demonstrate that its organization and personnel are free from any commercial, financial or other pressure (Dubois, Robins and Bass 1995). Equally, to achieve harmonization and mutual recognition, a voluntary international certification system must: (i) be comprehensive and cover all types of forests and wood products; (ii) be based on objective and measurable criteria; (iii) produce reliable assessment results and thus be fully independent from any vested interests; (iv) be transparent and involve a balanced participation of the interested parties and stakeholders thereby ensuring their commitment; (v) represent all involved parties; and (vi) be goal oriented and cost-effective (Baharuddin and Simula 1996). A major problem affecting progress on certification appears to be the duplication and poor coordination of efforts on developing internationally recognized criteria and indicators of sustainable forest management (Baharuddin and Simula 1997). There are at least eight intergovernmental processes, some closely linked and others overlapping, that are attempting to derive such criteria. In addition, there are numerous national-led initiatives that are being developed more or less independently and without necessarily conforming to internationally agreed common criteria. Only one recent initiative, implemented by the Centre for International Forestry Research (CIFOR), appears to be evaluating and developing criteria and indicators for sustainable forest management at the forest or forest management unit level, and which could be used as the basis for assisting various national efforts in determining how close their forest management efforts are conforming to sustainability standards.

Establishment of an international framework covering all existing and proposed timber certification schemes, as well as a common core set of criteria and indicators for sustainable forest management on which such schemes can rely, is clearly a long term process. As a multinational forum, the IFF could endorse this process and encourage the parallel and cooperative development of existing and proposed international schemes, as well as the related national and regional schemes, with the overall objective of achieving international harmonization and mutual recognition of standards. In addition, the international community should support WTO's efforts to ensure that existing and new certification and ecolabeling schemes for wood products are not used in a discriminatory way as a form of "disguised protectionism". The purpose of timber certification should be to reinforce the positive incentives for sustainable forest management and not to penalize or restrict production and trade in timber not meeting standards.

However, there is also the issue as to whether the international community should consider a different sort of certification scheme altogether, such as a country-level certification scheme that has been proposed recently (Barbier *et al.* 1994). The latter "certification" approach will now be discussed in more detail.

6.4.2. Country Certification

As timber certification is currently influencing only a very small proportion of the global trade in forest products and an equally limited area of the world's production forests, it cannot be considered the main instrument for promoting sustainable forest management globally. Given the pressing need to promote sustainable forestry, it is imperative to develop urgently other instruments complementary to timber certification that are more directly aimed at wholesale and timely improvements in forestry management policies and regulations in producer countries.

One such approach is the concept of *country certification*. Originally proposed in a report to ITTO (Barbier *et al.* 1994), country certification involves certifying through explicit bilateral or multilateral recognition all timber products from a country that can prove it is complying with an internationally agreed objective, such as a sustainable forest management target. Such a scheme could be enacted for all timber producer and consumer countries through an international agreement on global forests. It may also require additional assistance for poorer countries with inadequate financial resources to achieve, implement, and monitor the key policy objectives. To be effective, country certification would require two broad sets of policy commitments from timber producing and consuming countries respectively.

The first set of policies would require producer countries to undertake substantial reviews of their forest sector policies and to correct those policy distortions that work against sustainable timber production objectives, as such distortions are believed to be at the heart of inefficient and unsustainable forest sector development and timber-related deforestation (see section 3). The second set of policies would require a commitment by consumer countries to remove any remaining tariff and nontariff barriers to timber imports into domestic markets, for those producer countries that demonstrate a commitment to forest sector policy reform, to promote actively the use of tropical timber imports from exporting countries that are implementing "sustainable management" policies, and to remove any of the "new barriers" identified in section 3 to the imports from participating producer countries.

As with timber certification, any country certification scheme needs to be voluntary and internationally agreed. If poorly implemented without sufficient international transparency, recognition or commitment, a country certification scheme would have little impact on improving sustainable forestry management globally. It would neither take advantage of the trade-related incentives needed for encouraging sustainable management of forests nor provide the stimulus for fostering further cooperation in related areas, such as timber certification. In addition, an important precondition for implementing country certification would be the development of internationally recognized and agreed criteria and indicators of sustainable forest management, so that all producer countries would have a common core set of criteria and indicators for assessing how well their forest management efforts and policies are measuring up to agreed sustainability standards and targets.

The need for policy-based instruments such as country certification is essential if any multilateral agreement or commitment to ensure that all forest product exports are from sustainably managed sources is to be realistically implemented. At the same time, however, the implementation of any such scheme clearly requires greater progress than achieved by the international community to date in implementing international obligations and agreements to ensure long-term commitment of major producers and consumers in the timber trade to sustainable forestry.

6.4.3. International Obligations and Agreements

In recent years, the possible development of an international and non-discriminatory agreement for forest products from all types of forests has been debated in a variety of global fora. In particular, one proposal suggests extending the Year 2000 objective of the International Tropical Timber Agreement (ITTA) for all types of forests.

The ITTA was adopted in November 1983 and came into force on April 1, 1985, and its Secretariat, the International Tropical Timber Organization (ITTO), was established in November 1986. The ITTO has a current membership of 53 countries, of which 26 are tropical timber producer countries and 27 are consumer countries representing the world's major importers of not just tropical timber but all forest products. Since 1990, the main focus of ITTO's work programme has been the "Year 2000 objective", in which all producer members committed themselves to ensuring that all their tropical timber exports come from sustainably managed forests by the year 2000.[5]

---

[5] The actual target year 2000 has been unrealistic for many poorer producer countries to attain; thus the "Year 2000 objective" has been more appropriately interpreted as requiring all producer members to achieve exports of forest products from sustainably managed sources as rapidly as economic and financial conditions will allow them to fulfill this objective.

The new ITTA was agreed in 1994, and became operable on January 1, 1997 for an initial four years, with a possible extension to ten years. During the process of negotiating the new agreement, the issue as to whether the ITTA should be broadened to cover all the world's forests was debated. Although the new ITTA has not been strictly altered beyond its original mandate to cover just tropical forests, the terms of the new agreement does allow some scope for such broadening of the ITTA mandate beyond tropical forests. For example, in several key paragraphs, the new ITTA replaces the term "tropical forests" with "timber producing forests", and stresses that some "obligations" under the treaty are required by "all members" and not just "producer members". Thus, the new statement on the "Year 2000 objective" commits the ITTA " ... to enhance the capacity of members to implement a strategy for achieving exports of tropical timber and timber products from sustainably managed sources by the Year 2000", and "to encourage members to develop national policies aimed at sustainable utilization and conservation of timber-producing forests." Already, many consumer member countries – who are also major exporters of temperate forest products – have indicated that they are willing to follow similar commitments as the "Year 2000 objective" to ensure sustainable management of their forests for the timber trade.

The issue as to whether the ITTA, or any other international forum on forests, should eventually evolve a multilateral agreement to cover all forests, and in particular, to commit all producer and consumer countries to ensure that the global forest products trade is derived from sustainably managed sources, is still an important proposal that the international community must address. In particular, the issue as to what international and national policies are required to facilitate the long-term management, conservation, and sustainable development of all types of forests that are sources for the global forest products trade, and whether these policies need to be endorsed through multilateral agreement and commitments, is an important area for the international community to consider.

6.5. IMPLEMENTING SUSTAINABLE FORESTRY

It is generally assumed that most of the world's forests are not being harvested on a sustainable basis, and that the cost of timber produced under sustainable forest management will be generally higher than for timber produced under current, less sustainable practices. Although this appears to be a reasonable assumption, given the additional harvesting regulations and management practices that need to be implemented to improve the sustainability of forest operations, assessments of the additional costs of implementing sustainable management are notoriously unreliable. There also appears to be very little discussion as to why "unsustainable" practices are occurring in the first place, or why forestry and other policies in many producer countries fail to encourage more sustainable management of timber resources.

To shed light on these issues, the following section assesses the role of public policies in influencing timber management, and the reforms required to internalize the costs of unsustainable management.

6.5.1. Public Policies and the Incentives for Sustainable Forest Management

Many of the direct and indirect impacts of unsustainable timber extraction involve losses of environmental goods and services that are essentially non-marketed and would not be normally taken into account by the private individuals making the harvesting decisions. Although some of the resulting environmental values lost – such as declining biodiversity, loss of carbon storage and even damage to watershed protection function – may benefit individuals in other countries, many of the foregone values can potentially affect the welfare of domestic populations within timber producing countries. It is the role of public policy in producer countries to ensure that the latter welfare impacts are fully incorporated, or 'internalized', in the timber harvesting decisions undertaken by private individuals. Unfortunately, in many producer countries, wider economic and forestry policies not only fail adequately to "correct" economic incentives to account for the costs of unsustainable forest management but actually encourage such practices through distortionary influences on markets.

As depicted in figure 6.1, there is a wide range of economic and forest sector policies that can influence timber management. Although forest exploitation is directly affected by economic policies aimed specifically at the forestry sector, more general "economy-wide" policies (e.g. fiscal and monetary policy) can affect general economic conditions, with indirect knock-on effects in the forestry industry.

Many domestic forestry policies do not even begin to approximate the appropriate incentives required to achieve a socially efficient level of timber harvesting that accounts for all environmental externalities. More often than not, pricing, investment and institutional policies for forestry actually work to *create* the conditions for short-term harvesting by private concessionaires, and in some instances, even *subsidize* private harvesting at inefficient and unsustainable levels.

Short-term concessions and poor regulatory frameworks coupled with inappropriate pricing policies often contribute to excessive rent-seeking behavior in tropical timber production (Barbier *et al.* 1994; Gillis 1990; Repetto and Gillis 1988). That is, concessionaires have an incentive to open up additional stands for harvesting in order to "mine" timber for high short-term profits. By not charging sufficient stumpage fees and taxes or by selling harvesting rights too cheaply, by and large most governments have allowed the resource rents to flow as excess profits to timber concessionaires and speculators, often through short-term harvesting operations. For example, in the

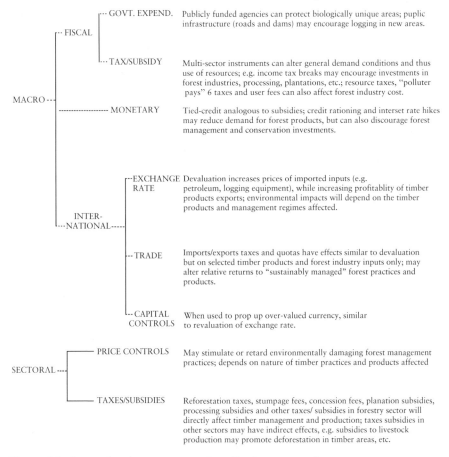

Figure 6.1. *Sectoral and macroeconomic policy impacts on forest management.*
Source: Barbier *et al.* (1994).

Philippines, the social gain from logging old-growth forest was found to be negative, once the social costs of timber stand replanting, the costs of depletion and the costs of off-site damages were included (Paris and Ruzicka 1991).

Hyde and Sedjo (1992) point out the difficulty experienced by forestry departments, particularly in developing countries, in administrating and collecting various timber fees and taxes. Much of the problem may have to do with the complexity of fees and concession arrangements, which makes enforcement and supervision of revenue collection difficult. In a review of forest pricing and concession policies in West and Central Africa, Grut, Gray and Egli (1991) suggest replacing the multiplicity of forest fees with an annual concession rent, set by competitive bidding, and replacing logging concessions with forest management concessions that should be regularly inspected.

Public policies also have an important influence on the pattern of forest-based industrialization and its implications for long-term economic development and deforestation. Vincent and Binkley (1991) note that stumpage prices (e.g. the prices of harvested logs at the stand) have a crucial role to play in the interrelated dynamics of timber reserve depletion and processing expansion, particularly in facilitating the transition of the forest sector from dependence on old-growth to secondary-growth forests and in coordinating processing capacity with timber stocks. Unfortunately, in most developing countries, stumpage prices tend to be administratively determined rather than set by the forces of supply and demand, thus understating stumpage values and failing to reflect increasing scarcity as old growth forests are depleted.

Improper policies also have an impact on timber forest management and its environmental effects in industrialized (OECD) countries. Logging fees or royalties for timber harvested from public lands are also based on administrative pricing. The standard calculation is to take the short-run derived demand lumber price at the mill minus harvest, extraction and (log to lumber) conversion costs in order to determine the royalty (Hyde and Sedjo 1992). Such pricing methods are not related to long-run "user" costs or environmental values, and in many instances do not even approximate market and economic scarcity values for timber. For example, in Australia old-growth forest hardwood sawlogs and softwood sawlogs were generally priced below market price; pulp log royalties were found to be both above and below market price (Resource Assessment Commission 1991).

As outlined by Wibe (1991), other problems also exist with regard to ensuring that private investors and concessionaires in OECD countries produce timber at a long-run privately efficient level. First, markets for forest land in these countries are far from perfect and free, preventing any investment in forestry from being fully capitalized through selling the standing timber or planted stand. For example, in the Nordic countries, Germany, and France restrictive regulations exist on the market for forest lands. In addition, regulations on the buying and selling of forest land usually imply large transaction costs, especially when holdings are small, which is normally the case. The result is that private forest owners tend to invest too little in regeneration and/or reforestation. On publicly owned forest lands there are often problems in securing efficient contracts with private forestry activities.

Subsidies in OECD countries, particularly for plantation establishment, are now recognized to have direct and indirect environmental impacts, as several case studies have revealed (Jones and Wibe 1992; Wibe 1991). For example, in Sweden subsidization of forest land drainage to increase timber production has led to the loss of over 30,000 ha of wetlands annually. In the United Kingdom in the 1980s, tax concessions on afforestation led to the location of coniferous plantations on land of poor or negligible agricultural value, such as wetlands, heath, moorland, but which have high environmental value as natural wildlife habitat and for other amenities.

6.5.2. Internalizing the Costs

The above "snapshot" review of forestry policies in producer countries suggests that, to a large extent, the unsustainable management of forests has become "institutionalized" in many countries and regions as a result of long-standing and widespread policy distortions leading to inappropriate incentive mechanisms. Several important points for the debate concerning the relationship between trade and sustainable forestry emerge from this review.

First, inadequate and often distortionary public policies are a major barrier to sustainable forest management in producer countries. The result is inappropriate economic incentives at the stand level that lead to inefficiencies in timber harvesting and create the conditions for short-term extraction for immediate gain, while at the same time failing to "internalize" the direct and indirect environmental impacts of forestry operations. Improper policies also have a more long-term and wide-scale effect on the pattern of forest-based industrialization and its implications for the management of the forest resource base as a whole, including the conversion of forest land to agriculture and other uses.

Thus policy reform to improve sustainable forest management may not only reduce the direct and indirect environmental impacts of forestry operations but may also be justified on economic efficiency grounds for long-term development of the forestry industry and the use of forest resources. The result is that producer countries may incur significant short-term costs from encouraging policy reforms and regulations to encourage sustainable forest management, but they are also likely to gain substantially in the long run from a more efficient forestry sector. Even in the short run, the reduction in subsidies, preferential tax breaks and other inducements may be an additional financial benefit of policy reform.

Equally, the transition to sustainable forest management may impose additional costs at the stand level for residual stand management and increased environmental protection. For example, Baharuddin and Simula (1996) suggest that the increased costs may derive from five different sources:

1. setting aside of areas;
2. lower harvesting yields;
3. additional silvicultural and harvesting costs;
4. additional costs of planning and monitoring; and
5. different distribution of costs and benefits over time.

Low-intensity harvesting will generally mean less timber extracted per hectare in the short term. However, these costs can be at least partially offset by improved harvesting techniques and better planning that lower operating costs. In addition, the current income foregone with reduced yields initially may be more than compensated over the long run from improved stand productivity and yields as a result of reduced residual damage and better stand regeneration and recovery. Too often, assessment of the costs of sustainable forest management focuses on the short-term costs of implementing improved management and fails to take into account the potential long-term gains in stand productivity and income.

Estimating the additional costs to both timber operations at the stand level and forestry industries at the national level of implementing sustainable forest management practices is therefore extremely difficult. However, the available evidence does suggest that on the whole:

- the transition to sustainable management is likely to impose some increase in production costs in the short term, both at the industry-wide level and the stand level;
- the additional costs may be higher for tropical than temperate countries,
- it may no longer be economically worthwhile to harvest some forests, and large areas of some countries' forest resource base may have to be "set aside" from production which could result in some income losses; and
- increases in costs and stumpage prices at the stand level do not necessarily mean significantly higher prices for final forest products.

The costs of implementing sustainable forest management are likely to vary significantly across forests, countries and regions. A range of 5 percent to 50 percent additional production costs is possible (Baharuddin and Simula 1996). For temperate and boreal forests, the available estimates suggest generally an increase of around 20–30 percent in costs (Dubois, Robins and Bass 1995). For tropical countries, the variation in estimates is much wider, but on average, higher than for temperate regions. Most estimates suggest that the costs of sustainable forest management per cubic metre of log produced lies between 10–20 percent of the current average international tropical log price of about US$350 (Varangis, Crossley and Primo Braga 1995).

The higher costs of sustainable forest management on overall timber production are likely to make it infeasible to harvest some forest areas that would have otherwise been logged. This makes perfect sense in cases where the failure to "internalize" the environmental and long-run costs of timber operations has meant that these operations remain financially profitable even though they are socially inefficient. A comparison of the private and social returns to selective logging on steeply-sloped (30–50 percent) old-growth forest in the Philippines illustrates this point (see table 6.6 and Paris and Ruzicka 1991). The magnitude of the estimated damage to downstream activities indicates that the Philippines would be better off by not harvesting old-growth forests on such steep slopes, even though the private concessionaire would gain financially from unsustainable harvesting on the steep terrain.

Table 6.6. *Private and social efficiency: logging in the Philippines (US$1 = 27.7 Philippino Pesos (P))*

(Figures in Philippino Pesos per ha per year)	Sustainable management scenario[a]
Value of log harvest[b]	5720
Road building, harvesting and transportation costs	− 2369
Net financial return to short run timber harvesting	3351
Internalizing cost of sustainable timber management	
Cost of protection, timber stand improvement and enrichment planting	− 1000[c]
Net returns to "privately efficient" timber harvesting	2351
Shadow pricing adjustments	
Adjustment to market price of logs[d]	572
Adjustment to local harvesting costs[e]	474
Social costs of degradation of nontimber values	
Cost of marginal offsite damage to downstream activities	− 6245[f]
Net returns to "socially efficient" timber harvesting[g]	− 2848

Notes:

[a] Old growth forest selectively logged and subsequently protected.

[b] Legal operations using existing selective logging systems assumed. Private profits of illegal operators will be higher. Different combinations of yield and price are possible to capture the variations in the quality of standing forest. Assumption is that one hectare of old growth forest of 30–50 percent slope sustainably yields 100 cum every 35 years or 2.86 cum per year. Market price is 2000 Philippino Pesos per cum.

[c] P1000/ha for one year to ensure sustainability of production on the one hectare in question.

[d] Equal to market price adjusted upwards by 10 percent to account for low cost illegal supplies.

[e] Standard conversion factor 0.8 applied.

[f] P2600 for 3 years discounted at 12 percent. Off-site damage assumed to be limited in duration instead of being sustained in perpetuity. Damage resulting from the selective logging of a single plot. The high magnitude of the estimate is a result of logging on very steep slopes.

[g] The overall returns are only a rough indication and are based on estimates prepared in 1990 as part of the formulation of the Philippine Master Plan for Forestry Development.

Source: Adapted from Paris and Ruzicka (1991).

On the other hand, the widespread implementation of sustainable forest management across many regions in the country could result in the removal of many forest areas from potential production. Although there would no doubt be substantial environmental gains, the economic costs to producer countries could be significant, particularly for tropical timber exporting countries. This was demonstrated in a recent policy simulation that indicated the additional economic impacts to tropical forest countries of "setting aside" 10 percent of their forest resource base (Perez-Garcia and Lippke 1993). The simulation indicates that such reductions in supply would result in a loss of wealth for tropical timber producing countries. Over the long run, permanent set asides would mean that the remaining production forest inventory could not support as high a level of sustainable harvest as under base case projections.

Equally, although there is often a presumption that sustainable forestry involves "low-intensity" harvesting of natural forests, there is also a strong case to be made for the establishment of intensively harvested pure stand plantations for commercially valuable tree species. For example Browder *et al.* (1996) have shown that, based on financial criteria alone, establishing pure stand mahogany plantations outside the tropical forest zone in the Brazilian Amazon may be an important means for exploiting the lucrative international trade in mahogany while simultaneously reducing the pressure for selectively extracting mahogany from a wider area of natural forest – often with environmentally damaging consequences. Thus the move from extensive forest exploitation for a few commercially viable species to a more intensive management regime of regeneration and harvesting of these species in designated production and plantation areas has a potentially important role to play in a national sustainable forest management strategy.

Finally, it is sometimes argued that the higher additional costs of sustainable forest management will make many timber products uncompetitive in final markets. However, although harvesting costs are often a large proportion of the stumpage value of logs, for most processed forest products the costs of the wood raw material is a small proportion of the total costs of harvesting. This is particularly the case for products traded globally; for example, typical stumpage values in tropical countries of US$6–30 per cubic meter of log equivalent end product often represents less than one percent of the final value of the product being sold in foreign consumer markets (Barbier *et al.* 1994). As a consequence, even reasonably large increases in harvesting costs and the stumpage value of timber can have little or only a modest impact on the final product price in consumer markets. Thus the evidence for both traded tropical and temperate wood products suggest that a doubling of harvesting costs may lead to an increase of 10–15 percent of the costs at the importer or wholesaler level and less than a 10 percent increase in the retailer's cost (Barbier *et al.* 1994; Dubois, Robins and Bass 1995).

A model of Indonesia's forestry sector developed by Barbier *et al.* (1995) was used to simulate a policy initiative by Indonesia to implement more "sustainable" management of its remaining production forests. As shown in table 6.7, scenarios depicting 25 percent and 50 percent increases in harvesting costs across Indonesia's forestry sector were examined. Although domestic log prices are affected significantly by the increased harvest costs, any resulting impacts on the rest of Indonesia's forestry sector seem to be somewhat dissipated.

Table 6.7. *Indonesia: timber trade and tropical deforestation simulation model. Policy scenario: sustainable timber management (% change over base case)*

Key variables	25% rise in harvest costs (%)	50% rise in harvest costs (%)
1. Prices (Rp/m^3)		
Log border-equivalent price (unit value)	41.59	83.06
Sawnwood export price (unit value)	4.04	8.09
Plywood export price (unit value)	2.86	5.72
2. Quantities ('000 m^3)		
Log production	− 0.94	− 1.87
Log domestic consumption	− 1.37	− 2.73
Sawnwood production	− 1.89	− 3.77
Sawnwood exports	− 1.03	− 2.05
Sawnwood domestic consumption	− 2.28	− 4.55
Plywood production	− 0.87	− 1.73
Plywood exports	− 0.38	− 0.75
Plywood domestic consumption	− 3.12	− 6.24
3. Deforestation (km^2)		
Total forest area	0.02	0.04
Annual rate of deforestation	− 2.28	− 4.23

Source: Barbier *et al.* (1995).

Indonesia's sawnwood and plywood exports seem to be the least affected by the increased harvest costs, which would suggest that external demand factors exert an important counteracting influence.

6.6. FINANCING SUSTAINABLE FORESTRY

Although in the long run producer countries ought to obtain sufficient returns from timber production, value-added processing, and exports to cover the additional harvesting costs and other economic impacts of sustainable timber management, in the short and medium term the transition to sustainable forest management may be costly for many producer countries, particularly developing economies. In addition, some of the additional costs that producer countries may be asked to bear would be associated with maintaining certain forest services – such as carbon storage, biodiversity preservation and watershed protection – that yield mainly global and regional benefits. Finally, to undertake some of the required basic improvements in forest resource assessment, monitoring and planning of forest management, chain-of-supply inspections of processing activities, independent certification and even basic forest sector policy analysis may require a transfer of resources, skills and even technology to build up local capacity in many developing countries.

It is now generally accepted, as well as enshrined in the Forest Principles accord of the 1992 UNCED Conference,[6] that compensating developing countries for their role in maintaining forest resources that have value on a *global level* is a fundamental basis of multilateral policy action. It should also be recognized that compensation is needed by developing countries for the income they may forego in protecting their forests and for the additional costs incurred in implementing sustainable management practices for their production forests.

However, actual assessment of the additional financial and technological assistance required to compensate developing countries is extremely difficult to undertake empirically. Based on broad estimates made for ITTO and UNCED, additional funds required by all producer countries to implement sustainable management of their tropical forest resource could be anywhere in the range of *US$0.3 to 1.5 bn annually* (Barbier *et al.* 1994).

Although these figures would suggest the need for additional financial assistance for producer countries, the real issue is whether the financing ought to be raised from the tropical timber trade or from other sources. There are essentially three policy options available:

1. redirection of existing revenue from the trade;
2. appropriation of additional revenue from the trade; and
3. additional funding from sources external to the trade.

Preliminary estimates suggest that just under US$1.5 bn in additional funds could be raised through a *tax transfer* of revenue from the trade between consumer and producer countries – an amount closer to the "upper bound" of the estimate financing for sustainable management required by these countries (Barbier *et al.* 1994; OFI 1991). However, consumer countries are likely to be concerned about the fiscal and political implications of a tax transfer scheme. Based on the above calculations, their governments would have to forego nearly US$3.7 bn in tax revenues from the trade – more than 2.5 times what producer country governments gain in increased revenues (Barbier *et al.* 1994). Rather than lower their VAT or other taxes on the trade in timber products from developing countries, consumer country governments could instead transfer directly some proportion of the revenue raised through these taxes to developing countries. Although the consumer country governments would still forego substantial revenues, they would most likely have to forego *less* revenue if it were directly transferred to developing countries to help them meet their US$0.3–1.5 bn target than under the tax transfer scheme (Barbier *et al.* 1994).

In recent years there has also been renewed interest in the use of trade instruments – such as a small 1–5 percent surcharge on tropical timber imports into developed country markets – to appropriate additional revenue from the trade for sustainable forest management. If endorsed by a global forestry

---

[6] United Nations Conference on the Environment and Development.

agreement, an import surcharge would be within GATT/WTO rules through subarticle XX(h). A differentiated surcharge could also be imposed so that imports of processed tropical hardwood products face less discrimination than logs, thus reducing any remaining distortions from escalating tariffs. The funds raised could most likely be transferred to a mutually recognized international body, such as the ITTO, for distribution to tropical producer countries – possibly through specific projects and programs.

One of the major concerns of developing countries is that any revenue-raising import surcharge, even at very low levels, would be distortionary. In particular, if the tax was levied by all importers on a wide range of tropical timber products, then there could be a more significant impact on total world trade in these products through substitution effects (Barbier *et al.* 1994). Moreover, such a tax would discriminate in favor of temperate forest-based industries of the developed market economies. Thus as an alternative, developing countries may prefer a tax on exports rather than an import surcharge. This would give producers more direct control over the proceeds of the tax. In addition, an export tax would affect all import markets rather than just one, thus spreading the costs of sustainable management to all producers and consumers (Barbier *et al.* 1994; Buongiorno and Manurung 1992).

Finally, there is the issue of whether the amount of funds raised through any trade surcharge would be adequate for the task. The studies undertaken so far suggest that the amount of net funds raised from a trade surcharge of 1–5 percent may fall short of the approximate target of US$0.3–1.5 bn required annually by developing countries as additional resources for sustainable forest management (Barbier *et al.* 1994; Buongiorno and Manurung 1992; NEI 1989).

All the above proposals for raising the additional funds for sustainable forest management from the global forest products trade are clearly controversial and pose great difficulties for comprehensive international agreement on mutually recognized and transparent schemes. It is also unclear why the international trade in forest products should be used to raise revenues to cover the costs of sustainable forest management, when the vast majority of timber production worldwide does not even enter into trade (Barbier *et al.* 1994). Moreover, given that commercial logging is not the primary cause of deforestation globally, schemes to raise revenues from the world forest products trade to provide financial assistance for sustainable forest management would involve unnecessary, and possibly inappropriate, discrimination against the timber trade.

Thus there is a strong rationale for additional funds to be made available to developing countries for sustainable forest management from sources outside the forest products trade rather than raising revenues through redirecting existing or appropriating additional revenues from the trade. Comprehensive international agreements, targeted financial aid flows and compensation mechanisms to deal with the overall problem of sustaining global forest resources may ultimately eliminate the need to consider interventions in the forest products trade to secure funds for sustainable forest management.

It is unlikely in the current global economic climate that there will be a concerted international effort to substantially increase bilateral or multilateral aid flows for sustainable forest management globally. Nevertheless, there still remains the possibility of designing new sources of financial assistance that are separate from existing developing country aid budgets. The Forestry Principles signed at UNCED are effectively a step in that direction, and international commitments through the National Forestry Action Plan (NFAP) and Global Environmental Facility (GEF) continue to reinforce the global interest in forestry and biodiversity protection. The case could be made that a comprehensive international agreement on global forest management should include provisions for additional funds and technical assistance over the short and medium term to support the transition to sustainable forest management in developing countries. Such assistance could take the form of compensation payments made from an international "rain forest fund" (Amelung 1992). Alternatively, such payments could arise through the establishment of a global system of marketable forest protection and management obligations (FPMOs) that could be initially limited to mainly bilateral agreements with little trading before being implemented globally (Sedjo, Bowes and Wiseman 1991). Other schemes that essentially provide the mechanism for trade in forest services include debt-for-nature swaps, carbon offsets, internationally tradeable carbon dioxide permits and tradeable development rights.

6.7. CONCLUSION

It is now generally accepted that the international trade in timber products is not the major source of global deforestation. However, there is evidence that unsustainable harvesting practices, illegal logging, and poorly developed forestry policies are rife in many producer countries, and that timber-related deforestation is much greater than it needs to be. In addition, there is evidence that consumers in the major importing markets for forest products are sufficiently concerned that there is a growing movement to ensure that products in these markets can be "certified" as coming from "sustainably managed" sources of timber supply.

This chapter has sought to review some of the critical issues linking trade and sustainable forestry. The crux of the problem of unsustainable management of global timber resources is the failure of many producer countries to implement forest management, trade and other policies that encourage the long-term and sustainable exploitation of their own timber resource base. The result is inappropriate economic incentives at the stand level that lead to inefficiencies in timber harvesting and create the conditions for short-term extraction for immediate gain, while at the same time failing to "internalize" the direct and indirect environmental impacts of forestry operations. Improper policies also have a more long-term and wide-scale effect on the pattern of forest-based industrialization and its implications for the management of the forest resource base as a whole, including the conversion of forest land to agriculture and other uses.

Thus policy reform to improve sustainable forest management may not only reduce the direct and indirect environmental impacts of forestry operations but may also be justified on economic efficiency grounds for long-term development of the forestry industry and the use of forest resources. The result is that producer countries may incur significant short term costs from encouraging policy reforms and regulations to encourage sustainable forest management, but they are also likely to gain substantially in the long run from a more efficient forestry sector. Even in the short run, the reduction in subsidies, preferential tax breaks and other inducements may be an additional financial benefit of policy reform.

Producer countries that take a long-term perspective on the development of their forest industries and through sustainable management of their production forests are likely to gain substantially from the expanding international trade in timber products. These countries are also likely to benefit from the nontimber values of their remaining protected forest areas. The international community could assist many producer countries, especially low income economies, to make such adjustments through establishing a comprehensive international agreements on, in the first instance, the global trade in timber products, and possibly, on global forest management generally. As discussed in this chapter, however, an international timber agreement will only be effective if there is a commitment of both consumer and producer country signatories to commit to using policy-based instruments, such as country certification, as part of any multilateral agreement, so that genuine progress to ensuring that all forest product exports are from sustainably managed sources can be realistically implemented.

7

International Trade in Hazardous Waste

MICHAEL RAUSCHER

7.1. INTRODUCTION

In an internal memo, which was later published by the *Economist* magazine (*Economist* 1992), the chief economist of the World Bank, Lawrence Summers, posed the question whether it would be advisable to dump toxic waste in countries with low wages and low population densities. Low population density implies that the number of people affected by environmental risks is small. Low incomes result in a low willingness to pay for environmental quality or – to put it the other way around – high degree of tolerance to environmental hazard. Moreover, low wages are said to indicate a low economic value of human life and health and, therefore, relatively small costs of health-impairing pollution. This memo evocated a public outcry because of its immoral or cynical line of arguing. Nonetheless, the memo was an attempt to address an important issue: the measurement of environmental damages, which are often considered to be unmeasurable. The drastic example of low wages being an indicator of low impacts of environmental disruption was a way to make the implicit value judgments contained in the neoclassical view on comparative advantage more explicit. According to the law of comparative advantage, toxic waste should be stored or treated where the environmental costs are low, that is, *ceteris paribus*, in low-income under-populated countries. Free trade between countries with low and countries with high environmental costs is then said to be beneficial for all parties involved. First, it is based on voluntary exchange. Thus, if a country did not benefit from trade, it would not trade. Second, the international division of labor improves the efficiency of allocation since it makes factors of production move into their most productive utilizations. Applying these arguments to hazardous waste, one would come to the conclusion that international trade in this commodity is a good thing.

Environmentalists have a completely different view on international trade in hazardous waste. They argue that much of this trade involves an unequal

I am indebted to two referees for helpful comments and critique.

exchange between the North and the South. Industrialised countries export their domestic environmental problems to the developing world, which faces huge problems with the disposal and processing of hazardous substances. A ban of trade in hazardous waste would bring relief to the South and would coerce the North to either avoid the generation of this waste or develop environmentally sound methods for its disposal. See Moyers (1990) and Daly and Goodland (1994) for such arguments. In reality however, trade in hazardous waste is neither completely liberalized nor completely prohibited. Transboundary movements of hazardous substances are possible but they are highly regulated and restricted. The most important international agreement in this respect is the Basle Convention on the Control of Transboundary Movements of Hazardous Waste and Its Disposal (hereafter: Basle Convention). It prohibits purely private transactions in hazardous waste and exports to non-parties and it requires reimports of exported toxic substance in special cases.

This paper intends to give an overview of the economics of international trade in hazardous waste. It is organized as follows: The next section will briefly discuss the extent of trade in hazardous waste and it will present the institutional framework as given by the Basle Convention. Then we will deal with determinants of the patterns of trade. Who is an exporter and who is an importer of toxic waste? Afterwards, the paper will deal with welfare effects of trade in hazardous waste. It will be seen that in a first-best world, trade is indeed beneficial to all parties involved although the object of the trade consists of dangerous substances. However, if there are distortions such as insufficient environmental policies, then international trade may be harmful. After a short section on the effects of trade liberalization on environmental regulation, I will consider the impact of hazardous waste regulation on trade. Are there good reasons to modify environmental regulation in order to achieve trade-related objectives? Must we expect a harmful process of environmental deregulation? In a next step, imperfections such as regulatory and enforcement deficits as well as asymmetric information will be considered. Afterwards, international externalities will be considered in an interjurisdictional-competition framework. Section 7.7 deals with imperfect competition, that is, with the question whether a monopolistic supplier of hazardous waste can exploit importing countries. In section 7.8, trade barriers will be considered and the final section will summarize the results.

7.2. TRADE IN HAZARDOUS WASTE: SOME STYLIZED FACTS AND THE INSTITUTIONAL FRAMEWORK

Hazardous waste is waste that poses a threat to human health or the environment and therefore requires special care during transportation, storage, treatment, and disposal. (Asante-Duah and Nagy 1998: 1–2). The statistical records on generation and, particularly, trade of hazardous waste are often incomplete and sometimes inconsistent. Thus, it is difficult to provide reliable numbers. However, what is known is that hazardous waste has become a large-scale problem only in the recent past. The generation of hazardous waste in the US for instance has increased by a factor of ten from 1980 to 1990. The most important producer of hazardous waste is the US, which alone accounts for some 75 percent of OECD waste output. OECD countries on average import and export some 1 percent of their hazardous waste output. Countries with export shares above 10 percent are the Netherlands, Switzerland, Austria, and Australia. Countries importing more than 10 percent of their domestic outputs are again Netherlands, Austria, and Denmark (Asante-Duah and Nagy 1998: 70–3). There appears to be some intra-industry trade in hazardous waste in Europe. Records on North-South trade are even less reliable that those for industrialized countries. In many cases, transports have been illegal or they have never been recorded by official authorities. Asante-Duah and Nagy (1998: 75–80) present some anecdotal evidence concerning imports of hazardous waste by developing and transition countries. It is seen that these importing countries in many cases lack the capability of environmentally sound treatment or disposal. Nevertheless, they are willing to accept these substances for rather low compensation payments, that save the exporters a substantial amount of money.

International trade in hazardous waste is regulated by the Basle Convention. For surveys see Douma (1991), Rauscher (1997: 300–1), and Asante-Duah and Nagy (1998: 98–103). This Convention is an international environmental agreement, which came in to force in 1992 and to which most industrialized and many developing countries are signatory parties. It requires them to minimize the generation of hazardous waste, to ensure that adequate disposal facilities are available, to control and reduce international movements of hazardous substances, and to prevent illegal traffic. The particular rights and obligations of the signatory parties are specified in Article 6 of the Convention. International trade in hazardous waste is to be supervised by the authorities of the states involved with a transboundary shipment. The exporter of hazardous waste is required to notify the authorities of the exporting, importing, and transit states and written letters of consent by the latter two are necessary for the transactions to take place. This means that purely private shipments of toxic waste without an involvement of government authorities are prohibited. Moreover, every country has the right to refuse imports of hazardous waste. Another restriction of trade is contained in Article 4(5) of the Basle Convention: exports of toxic waste to nonparties are prohibited. However, this constraint is mitigated by Article 11, which allows signatory parties to export toxic waste to countries with which they have agreements that are at least as stringent as the Basle Convention. Moreover, the Convention demands that under certain circumstances hazardous waste be reimported by the exporting country. In particular if the movement of toxic waste is not completed in accordance with the terms of the contract and possibilities for an environmentally sound disposal in the transit or importing country do not exist, the exporting country is held responsible for reimporting the waste. Besides simply limiting the volume of trade in toxic waste, the Convention has an incentive effect. Due to the threat of reimports, the exporters care about sound environmental regulation in the importing countries, and countries wanting to exploit their comparative advantage in the disposal and processing of toxic substances tend to implement stricter standards in these industries.

From the point of view of a free-trade advocate, the regulations contained in the Basle Convention are severe trade restrictions that prevent an efficient and welfare-enhancing division of labor between potential exporters and importers of hazardous waste. The mandatory involvement of government authorities subjects the hazardous waste trade to bureaucratic discretion. The obligation to reimport shifts the risk of insufficiency of storage or processing facilities from the importer to the exporter. The right of a country to ban hazardous waste imports is a severe trade impediment. Finally, the prohibition of trade with nonparties in an obvious discrimination. These features of the Basle Convention do not only contradict the view of dogmatic free-traders, they are also inconsistent with the spirit of General Agreement on Tariffs and Trade (GATT). The GATT defines circumstances under which trade interventions are justified. In such cases, trade restrictions must be nondiscriminatory, domestic like-products must be treated in a similar way as foreign goods and no other instruments that are less trade-distorting should be available.[1] The measures taken by the Basle Convention do not satisfy these criteria.

Environmentalists in contrast complain that the Basle Convention is not strict enough. The principle of trade only among parties is diluted by Article 11. Moreover, the definition of what constitutes hazardous waste which is to be regulated by the Convention is considered to be insufficient or too vague. Besides the Basle Convention, there are numerous other multilateral and bilateral international agreements on hazardous waste trade, for example on the EC and the OECD levels but also by groups of developing countries, such as the Organization of African Unity, OAU. See Asante-Duah and Nagy (1998, chapter 5).

7.3. FACTOR ABUNDANCE AND THE PATTERNS OF TRADE

International trade in toxic waste is trade in a commodity with particular characteristics and it is in many cases performed between countries that are rather dissimilar, that is, developed and less developed countries. Although there appears to be some intra-industry trade, particularly among industrialized countries, a substantial part of the toxic-waste trade may be explained by Heckscher–Ohlin theory. International differences in endowments with

---

[1] On environmental aspects of the GATT, see Rege (1994), Esty (1994: 46–52), and Rauscher (1997: 301–3).

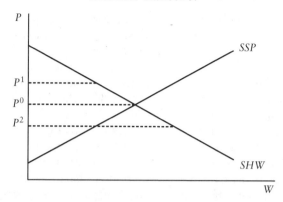

Figure 7.1. *Supply and demand of hazardous waste.*

factors of production are then viewed as being the reason of specialization and trade. We will present an intuitive version of this theory based mainly on a diagrammatic approach. A similar approach has been chosen by Berger (1998). For a more formal approach, see Rauscher (1997: chapter 4). First of all, it should be noted that hazardous waste is a commodity with a negative value; it is a bad rather than a good. The corresponding good with a positive value is the importer's service to help the exporter to get rid of the hazardous waste. This is depicted in figure 7.1. It represents a country's market for hazardous waste. W denotes the quantity of hazardous waste and P the price to be paid by the producers of the waste to the waste-management sector. SSP denotes the supply of storage and/or processing services and SHW is the supply of hazardous waste or the demand for storage and processing services. The SSP curve is positively sloped because the treatment of waste utilizes scarce resources. The SHW curve is negatively sloped since a high cost of waste disposal is an incentive to avoid waste in the production process. The equilibrium price is P^0. If the price on the world market is larger than the domestic autarky price, for example, P^1, then the country will be an importer of toxic waste. If it is lower, for example, P^2, it will be an exporter.

It follows that the patterns of trade are determined by the shapes of the SPP and SHW curves in different countries. An upward shift in the SPP curve is explained by tighter regulation of the waste treatment industry, in particular by tighter environmental standards. An upward shift in the SHW curve may be due to lax environmental standards in the waste-generating production sectors and by the size of these industries. From the effects of these shifts on the domestic price, it follows immediately that a country is an exporter of toxic waste if, *ceteris paribus*,

- the waste management sector is subject to tight standards,
- the waste-generating sector is subject to lax standards,
- the waste-generating sector is large.

In Rauscher (1997: 95) the size of the waste generating sector is positively related to the economy's stock of physical capital. Thus, capital-rich countries such as industrialized countries tend to export hazardous byproducts of their manufacturing output. What determines the strictness of environmental regulation? In the context of our model, all environmental externalities can be internalized by an appropriate regulation of the waste management sector.[2] The strictness of this regulation depends on variables such as availability of safe deposit sites and on the willingness of the voter to pay for environmental quality. This in turn is influenced by the degree of environmental concern and by the ability to pay, that is, on per capita income. This implies that, everything else being equal, low-income countries with low ability to pay tend to accept toxic waste at lower compensation payments and, therefore, to become waste importers. Finally, a Ricardian element determines the price of toxic waste. Technological progress shifts the *SSP* curve downwards and increases domestic waste processing.

7.4. GAINS AND LOSSES FROM TRADE

In mainstream economic theory, international trade is in general advantageous to all parties involved. The reason is that transactions would not be made if people acting in their self-interest did not benefit from them. The exception to this rule is second-best theory. If there are distortions in the economy, it is possible that these are partially corrected by barriers to trade. The removal of trade barriers then leads to welfare losses. In the context of environmental economics, distortions are omnipresent. Let us assume that there are non-internalized environmental externalities, for example, in the case of insufficient environmental regulation. In our model framework, we will assume that the waste-treatment sector of the economy is not appropriately regulated such that the marginal social cost of waste disposal exceeds the marginal private costs.

In figure 7.2, this is depicted for a small waste-exporting country. Again *SHW* is the hazardous waste supply curve, *SSP* is the supply curve of the waste disposal industry under perfect regulation, and *SSP'* is the same curve if this industry is insufficiently regulated and can shift some of its cost to the rest of society. In the case of an autarkic economy, the price is determined by the intersection of the demand and supply curves. This gives P^A. P^W is the resulting world market price after trade liberalization. In the case of optimal environmental regulation, the gains from trade consist of an increase in the production sector's surplus and a reduction in the waste processing industry's surplus.[3]

---

[2] Sources of environmental disruption other than hazardous waste are assumed to be exogenous and constant.

[3] The production sector's surplus is the share of the area under the *SHW* curve which is located above the relevant market price. This corresponds to consumer surplus in the usual microeconomic analyses. The processing industry's surplus is the area between the relevant market price and the *SSP'* curve. This is a conventional producer surplus.

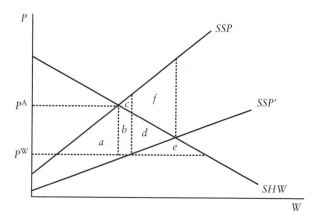

Figure 7.2. *Gains from trade in the exporting country.*

The net effect is a welfare gain of $a + b + d + e$. In the case of insufficient regulation, the autarky level of hazardous waste output is larger than in the case of perfect regulation. Thus, the private gain due to trade expansion is reduced to e. In addition to this private gain, however, there is an environmental benefit of magnitude $d + f$. This is due to the fact that noninternalized environmental damages are reduced since hazardous waste is exported. In the diagram this is measured by the area between the *SSP* and *SSP'* curves. It is seen that the total welfare gain, $d + e + f$, may turn out to be larger than in the case of perfect environmental regulation.

Next consider the importing country. The difference in figure 7.3 compared to the previous diagram is that the world market price is above the autarky price level. The variables are indicated by asterisks now. In the case of a perfectly internalizing environmental policy (*SSP* curve), the gains from trade are measured by area a. If the environmental regulation is insufficient (*SSP'* curve), the private gains from trade are larger: $a + b + c$. However, there is a noninternalized environmental effect $-c - d$. The net welfare effect is $a + b - d$, and this may well be negative. The likelihood of welfare losses becomes even larger if the deviation of environmental regulation from its optimal level is so large that a potential exporter of hazardous waste is turned into an importer.

In a next step, let us consider the interaction of a group of industrialized countries with adequate, that is strict, environmental laws and a group of developing countries that do not generate toxic waste themselves but act only as recipients of waste shippings. We will consider the effect of a regulatory deficit in the developing countries in figure 7.4. *EHW* is the export supply of hazardous waste, and *SSP** and *SSP*$^{*'}$ are the import demand functions of the developing countries in the cases of adequate and inadequate policies, respectively. P^0 is the world market price in the case of perfect regulation, P^1 is the price if developing countries have insufficient environmental policies. In a

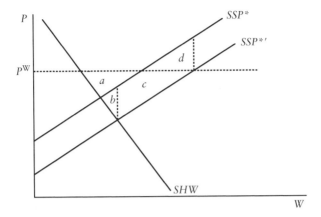

Figure 7.3. *Gains from trade in the importing country.*

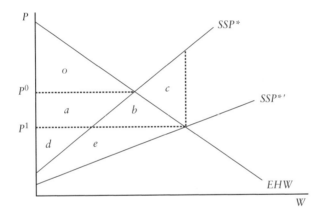

Figure 7.4. *Gains from trade in a North–South model.*

first step, consider the case of perfect environmental regulation. Compared to the autarky situation, the waste-exporting countries gain o and the importing countries gain $a + d$. The move from SSP^* to $SSP^{*'}$ reduces the world market price. This is an improvement of the terms of trade of the waste-exporting countries and results in an increase in the gains from trade by $a + b$. Trade is beneficial to the waste exporters and the benefits are increased by environmental laxity in the importing countries. Now consider the importing country. In the case of insufficient internalization the private sector's gain from trade is $d + e$ whereas society as a whole experiences a loss of environmental quality of $-e - b - c$. The total effect of international trade from the importers' point of view is negative, $d - b - c$. The global welfare effect of international trade is $o + a + d - c$, which is positive in figure 7.4, but may also be negative in the case of a very substantial deviation of the insufficient from the optimal regulation level.

The results can now be summarized as follows:

- A regulatory deficit in the waste disposal and processing industries of the importing countries is beneficial to the exporters of toxic waste.
- It is harmful to the importers and may cause welfare losses from trade liberalization for them.
- The gains from trade may be negative for the world as a whole.

It should be noted that these results have been obtained under two implicit assumptions. First, we have neglected transport externalities, that is, pollution spills that occur during the transportation from the exporting to the importing country. Second, we have neglected transfrontier pollution spill-overs. When environmental costs of the transportation of toxic waste are taken into account, then the gains from trade may become negative for all parties involved unless these costs are adequately internalized. With the internalization of these costs, however, transportation may become so expensive that trade in hazardous waste is not profitable any more. Transfrontier spill-overs of pollution imply that the exporting country has to bear some of the environmental cost of storage and processing of hazardous waste even if these activities are taking place in another country. A historical example for this is the trade in toxic waste between the Federal Republic of Germany (FRG) and the German Democratic Republic (GDR) in the 1970s and 1980s. Much of this waste was disposed of under rather dubious conditions on a site called Schönberg in the north-west of the German Democratic Republic right at the border between East and West Germany. See Nunez-Muller (1990). In a case of leakage, West Germany would probably have suffered severe environmental harm. Thus, it is not always true that exporting toxic waste really means exporting the problem.

7.5. CHANGES IN ENVIRONMENTAL POLICY FOLLOWING THE LIBERALIZATION OF TRADE IN HAZARDOUS WASTE

The possibility of trading toxic waste should affect the design of environmental policy. Usually, environmental economists make the assumption that marginal environmental damage is an increasing function of pollution that is, the environmental cost of pollution is rising more than proportionally if the level of pollution is increased. This is sensible because in most cases the assimilative capacity of ecosystems is a declining function of pollution.[4] Since trade in toxic waste exports a part of the pollution from the country where it has been generated to the country where the waste is stored or processed, the marginal

---

[4] An exception is the case of hot spots, i.e. areas that are polluted so much that additional pollution does not matter any more.

environmental costs in the two countries are changed. Since these costs equal optimal environmental taxes, this should have an impact on environmental policy. One can expect that (explicit or implicit) environmental taxes are reduced in the exporting and increased in the importing country. In reality, however, it is difficult to observe such changes in environmental policy since (i) real-world environmental regulation is often governed by political rather than economic objectives, and (ii) there are various variables influencing environmental regulation and in practice it is difficult to isolate the effects of trade liberalization.

Moreover, transportation externalities must be taken into account. An environmental policy aiming at internalizing the environmental externalities from transportation should introduce trade taxes or otherwise regulate the transboundary movements of hazardous substances. Note that these trade taxes do not only depend on the type of the waste but also on the way it is transported.

7.6. EFFECTS OF CHANGES IN ENVIRONMENTAL REGULATION AND THE DESIGN OF OPTIMAL ENVIRONMENTAL POLICIES

The literature on international trade and the environment has established that environmental policies in open economies may, under certain circumstances, differ from policies in closed economies. The underlying reason is that environmental policies can be used to achieve trade-related objectives. This is not the case if the country under consideration is small and, thus, cannot influence the world market. Then tighter environmental regulation in the waste sector leads to an increase in exports or a reduction in imports. Optimal regulation is governed by the rule that the tax to be charged for the polluting activity should equal the marginal environmental damage.

Matters are different if we consider a large country with market power in the market for hazardous waste.[5] Tighter environmental regulation of the waste sector leads to an increase in the price for waste disposal. This is good for the waste-importing country and bad for the waste-exporting country. Tighter environmental policy in the exporting country and laxer environmental policy in the importing country increase the volume of waste traded and, therefore, the environmental cost of transportation. Moreover, in the case of transfrontier pollution, a leakage effect has to be considered. With tighter environmental regulation in one country, the quantity of waste to be disposed of in the rest of the world is increased. If this country is an importer, its imports are reduced. If it is an exporter, its exports are increased. In both cases the risk of transfrontier

---

[5] The large-country effect on environmental policy has been considered in Markusen's (1975a) seminal paper on trade and the environment. For some extensions see Krutilla (1991) and Schulze and Ursprung (this volume: Chapter 2).

pollution rises. As an example consider a situation where toxic waste is stored very close to the common border of the trading countries, like in the case of the Schönberg site in Germany, which has been mentioned before. Tighter standards in the Federal Republic of Germany (FRG) led to increased exports to the GDR. Since standards in the GDR were rather lax, there was an increasing risk of transfrontier pollution. Thus, exporting toxic waste does not always mean that environmental risks are exported as well.

From these considerations, it follows that an optimal environmental tax to which the waste sector is subject consists of four parts if the country under consideration is large:

(1) The first part is the domestic marginal environmental damage.
(2) The second part is a terms-of-trade term. It is positive if the country is an importer and negative if it is an exporter of toxic waste.
(3) The third part is a term covering the transportation problem (provided that the transportation externality affects the country under consideration). As the terms-of-trade term, it is positive for the importing and negative for the exporting country.
(4) Finally, there is a leakage part. It is negative for both countries since laxer environmental policies at home reduce the transfrontier spill-over from abroad.

See Rauscher (1997: chapter 4) for an algebraic derivation of these results.

These trade-driven changes in environmental policy in one country have external effects on the rest of the world. Terms-of-trade improvement for one country means terms-of-trade deterioration for its trade partner. Reduction of trade for reasons of transport externalities is beneficial for other countries that suffer from these externalities as well. And a laxer domestic regulation increases the transfrontier pollution problem faced by other countries. These externalities lead to deviations of national policies from the cooperative first-best optimum. We can distinguish three cases:

1. Terms-of-trade effects dominate. In this case the importing country chooses a tighter-than-optimal level of regulation. The exporter's environmental regulation is too lax.
2. Transfrontier pollution dominates. Both countries choose too-lax regulation of their waste sectors.
3. Transport externalities dominate. The importing country's regulation is too strict, the exporting country's regulation is too lax.

7.7. IMPERFECTLY COMPETITIVE MARKETS

It is often argued that the laxity of environmental regulation in developing countries is to be explained by their dependence on large multinationally operating firms. These multinationals can exploit their market power *vis-à-vis*

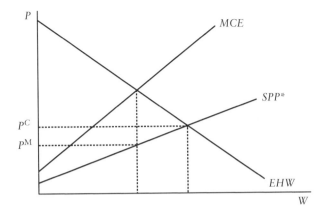

Figure 7.5. *Monopolistic supply of hazardous waste.*

small countries that compete against each other. A scenario showing such a situation is depicted in figure 7.5. As in figure 7.4, the *EHW* and *SSP* curves denote the export supply of hazardous waste and the supply of storage and processing, respectively. To a monopolist, however, the marginal cost of buying waste-management services is larger than the market price. An additional unit of waste leads to a price increase which makes the existing exports more expensive. This is denoted by the marginal-cost-of-exports (*MCE*) curve. The monopolist's optimum is to restrict exports until the marginal cost of export equals the willingness to pay for waste disposal. This reduction in waste exports leads to a price reduction. The price, P^M, is less than the competitive price, P^C. This implies that the regulation of the waste sector in the developing country has indeed been relaxed. Monopoly power on the hazardous-waste supply side of the market leads to deregulation by the importers.

A monopolist has a positive effect on the environment. The reduced supply of toxic waste reduces environmental risks. The deregulation effect is a consequence of the convexity of the environmental-cost function. With less pollutants, the marginal damage is reduced and so is the (implicit) tax rate. Thus, albeit domestic waste generation may be increased after the reduction of the price, the net effect on the quantity of hazardous waste to be disposed of is negative.

7.8. RESTRICTIONS ON TRADE IN HAZARDOUS WASTE

It has been seen that trade in hazardous waste may be beneficial in some situations and welfare-reducing in others. Moreover, it was shown that the regulation of the waste treatment and disposal sectors can be used to achieve trade-related objectives. The question now is whether environmentally motivated objectives justify trade restrictions. Three kinds of arguments can be made on theoretical grounds in favor of such measures. The first one is the internalization of transport externalities. If transport is environmentally disruptive, this activity should be restricted, either by means of taxation or by other environmental-policy instruments. Since this is rather obvious, we will not dwell on this any further. The other two problem areas require more attention. On the one hand, environmental regulation may be insufficient, and trade restrictions may be used as a second-best policy to correct the distortion at least partially. On the other hand, large countries can influence trade and, therefore, may be interested in using trade-policy measures. However, the question arises as to where the environmental motivation is to be sought in such cases.

Initially consider the case of insufficient environmental regulation or enforcement of environmental regulation. In the public debate on trade and the environment, this is often being viewed as a typical waste-importing country problem. The underlying argument is that waste-importing countries, particularly developing countries, accept hazardous waste not because they are endowed with superior storage facilities or because they have a lower preference for environmental quality but because they face regulatory and enforcement deficits in the field of environmental policy. As Chichilnisky (1994) has shown, this can explain North–South trade even in situations when endowments and tastes do not differ across countries. Most of the following analysis is devoted to the case of a waste importer. Figure 7.6 shows a situation where, as in figure 7.3, a small waste-importing country is insufficiently regulated. The SSP' curve represents the distorted supply of storage and processing services. In the free-trade situation, the domestic price in the hazardous-waste market equals the world market price, P^W. If a tax on waste imports is levied, the domestic price must be reduced. Foreign producers of toxic waste are still willing to pay no more than the world market price. This means that domestic suppliers of waste treatment and disposal have to offer their service at a lower

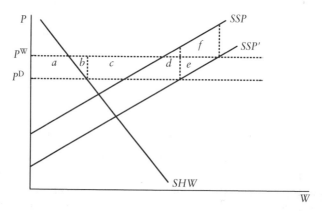

Figure 7.6. *Trade restrictions in the importing country.*

price. As a result, domestic producers of hazardous waste will experience a gain of a. The waste disposal sector will lose $a + b + c + d + e$. The government's tax revenue is $c + d$. Finally, the gain in environmental quality is $e + f$. The net welfare effect turns out to be $f - b$ and this is positive in this diagram. Note, however, that it is always possible to find an import tax such that the net welfare effect is positive. The smaller the distance between the *SSP* and *SSP'* curve, the smaller must this tax rate be. Of course, as already mentioned, the taxation of hazardous-waste imports is only a second-best policy. The first-best policy would be to eliminate the original distortion, that is, to internalize the environmental externality completely by moving the supply function of the storage and processing industry to *SSP*. An import quota would have the same effect as a tariff. Domestic supply of hazardous waste would be reduced and this would generate an excess supply of storage and processing services. The domestic price would be reduced and rents would be generated. Since the foreigners' willingness to pay is unchanged, this rent accrues to domestic citizens. However, there may be problems of rent seeking that reduce the rent.

In a waste-exporting country with insufficient regulation of the waste treatment sector, it is also advisable to reduce domestic disposal and processing of toxic substances. This is done by an export subsidy as a second-best policy, which moves the problem to the other side of the border. In the large-country case, trade interventions are desirable even in the case of perfect internalization of the environmental cost. This result is based on the well-known terms-of-trade argument of optimum-tariff theory. See also Schulze and Ursprung (this volume, Chapter 2). The waste importer benefits if the compensation payment made by the exporter is high, the exporter benefits from low payments. Thus, the importer is interested in increasing the scarcity of her services and the exporter is interested in reducing the supply of hazardous waste. Thus, it is optimal for the exporter to restrict her exports and for the importer to impose a restriction on imports. See Levinson (1999a,b) for similar results based on a tax–competition model. The empirical evidence derived by him from a data set on interstate waste movements in the US shows that waste-importing states indeed have an incentive to introduce a surcharge for imported waste. These import taxes deter interstate waste transport and lead to a decentralization of storage of hazardous waste in the US.

Of course, this standard optimum-tariff argument has nothing to do with ecological goals of trade policy. Green objectives enter the arena if there is transfrontier pollution. Then the optimum tariff has an antileakage component. Consider a waste-exporting country which fears that it will be negatively affected by hazardous waste being treated or disposed of on the other side of the border. An optimal trade policy takes this into account. An export restriction reduces the waste treated or disposed of in the neighboring country and, thus, the transfrontier externality. See figure 7.7, which depicts the international waste market from the point of view of the exporting country. *EHW* represents the desired exports of waste, *SSP** is the foreign country's

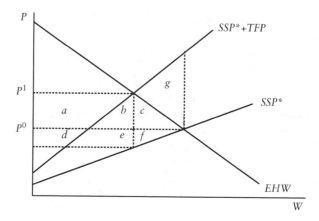

Figure 7.7. *Optimal tariff and transfrontier pollution.*

supply of storage and processing, and $SSP^* + TFP$ is the total cost of exporting if transfrontier pollution is taken into account. The free-trade situation is characterized by a relatively large volume of trade and a low price, P^0. The country can then restrict its exports of waste by taxing them, for example. The tax to be considered in this diagram is chosen such that the transfrontier externality is taken into account appropriately. The effect on the domestic waste generating and disposal sectors is $-a - b - c$. The tax revenue is $a + b + d + e$. Finally, the environmental dividend is $c + f + g$. The total welfare effect turns out to be $d + e + f + g$, which is unambiguously positive. Part of this is a terms-of-trade gain, the other part an increase in environmental quality.

The waste-importing country also has an incentive to use trade interventions in the case of leakage. Again, the objective is to reduce the storage of hazardous waste on the other side of the border. This can be done by a subsidy on toxic waste imports. The subsidy increases waste imports but it reduces the world market price of storage and processing of hazardous waste. Thus, less waste is stored on the other side of the border and the level of transfrontier pollution is reduced.

Summarizing the results derived thus far, we can state that trade restrictions may be explained by regulatory deficits in the downstream waste sector in the importing country, by transfrontier pollution and leakage in the exporting sector and by terms-of-trade objectives in both countries.

The first link probably explains much of the trade barriers raised by developing (but also by developed) countries against imports of hazardous substances. If the capabilities of dealing with hazardous waste are insufficient and these deficiencies cannot be repaired, for example, because of lacking access to new technologies or because of insufficient monitoring once the hazardous substances have entered the country, it is indeed advisable to control the waste at the border and refuse to accept major proportions of it. Nonetheless, it would be better to eliminate the source of the problem, that is, to introduce appropriate technologies or to improve the monitoring process.

The transfrontier pollution problem is of practical relevance only in border regions, and this is a case where bilateral negotiations are feasible and a cooperative solution can be implemented rather easily.

Finally, the real-world role of the terms-of-trade motive is probably rather limited and unimportant. So, what explains export restraints by industrialized countries? The first explanation is a political-economy argument. If one wants to explain an economic policy, particularly an inefficient one, one has to find out who gains from it. Two interest groups need to be considered here: the domestic waste disposal sector and the environmentalists. The waste industry benefits from trade restrictions since waste that cannot be exported must be treated or disposed of domestically. This increases the demand for the services this industry has to offer and, therefore, its producers' surplus. Another impediment to international trade in hazardous waste consists of strict requirements on the way in which the waste is treated or stored abroad. This raises rivals' costs – and domestic producer surpluses.

The environmentalists are interested in clean environment, often not only in their own backyards but also elsewhere in the world. For them, pollution in a distant Third-World country is not much different from pollution at home. Thus, they will exert political pressure in favor of export restraints or even bans. Moreover, they can be expected to lobby for stricter environmental standards abroad. The green view can also be visualized by means of figure 7.7. Now the $SSP^* + TFP$ curve does not represent the real cost of exporting toxic waste but the psychic cost due to the fact that the altruist suffers from environmental disruption abroad. Since an altruist is not interested in improving the terms of trade at the expense of the trading partner, the terms-of-trade deteriorating effect for the other country would be taken into account. Then the welfare gain from implementing the trade restriction is only g.[6] Thus, green lobbies and domestic producers have similar interests in restricting international trade in hazardous waste. A similar view has been expressed by Hillman and Ursprung (1994a). It should be noted, however, that there are also groups with opposing interests such as the domestic producers of hazardous waste. They want to dispose of their waste at lowest possible prices and, therefore, are interested in unrestricted trade.

Finally, one may argue that trade restrictions such as those contained in the Basle Convention originate from a paternalistic view on trade and the environment: in industrialized countries, many people think that lax environmental standards in developing countries are "wrong". Arguments to support this view range from wrong preferences regarding environmental scarcity and lack of knowledge about the environment to lax standards implemented by autocratic

---

[6] See Maestad (1998), for a formal elaboration of this argument in a slightly different model with trade and the environment.

governments that do not act in the interest of the people and lack of technologies or institutions to deal with hazardous substances appropriately. The "wrong preferences" argument is unsustainable. In the other cases, trade restrictions are only second-best. The first best solutions would be to provide information, to pave ways towards democracy, to provide technological assistance and to build up institutions that can effectively enforce the environmental standards (see Bhagwati and Srinivasan 1996). However, this is not always possible to the desired extent. In such cases, trade restrictions may be useful from the point of view of the waste-importing country. However, there seems to be no necessity that trade restrictions be implemented by waste-exporting countries.[7] A final argument in favor of trade restrictions is that they increase participation in the international environmental agreement. If nonsignatories are discriminated against, there are incentives to join the agreement. Credible threats can stabilize international environmental agreements and lead to better cooperation.

To summarize, there are some arguments in favor of trade interventions in the hazardous waste trade. However, with one exception, trade restrictions are never first-best. The exception is the transfrontier pollution case with leakage effects. However, this is usually a bilateral rather than a multilateral problem and cooperation is feasible. In all other cases, free trade is optimal given that environmental regulation takes the true cost of environmental regulation into account and provides correct signals to those who generate, store, and process hazardous waste. Free trade then improves the allocation of factors of production and allows countries to obtain welfare gains from a beneficial international division of labor. However, the world we are living in is not first-best in many respects and almost certainly not so in the international hazardous waste business. Particularly in developing countries, it may be easier to control and restrict waste imports at the border. Restrictions on exports of hazardous waste can be explained by the political influence of the recycling industry and green pressure groups as well as by a paternalistic approach to environmental policy.

7.9. CONCLUSIONS

This paper has dealt with the issue of trade in hazardous waste. It has been seen that standard economic theory and diagrammatical approaches can be utilized fruitfully to derive interesting results. From a purely economic-theory point of view, hazardous waste is not a particular commodity. Of course, waste movements offer the possibility of separating pollution geographically from the location where it has been generated but this does not add additional complications to the economic analysis.

---

[7] Even in a situation where an autocratic regime is in office and environmental standards are very lax, it is not known which standards would have been chosen if there had been a democratically legitimated government (see Bhagwati and Srinivasan 1996).

It has been seen that the possibility to trade does not guarantee that there will be positive gains from trade. Waste importers may lose if they have internalized the environmental cost of treatment and disposal only incompletely. This, however, is a result which has been established elsewhere in the trade-and-environment literature. Moreover, it is rather unlikely that there will be a race towards the bottom in the regulation of the waste industry. There are substantial environmental costs that accrue to the country becoming an importer of hazardous substances and there is no reason not to take them into account. However, it can be useful to soften environmental standards if leakage effects are substantial. Finally, trade restrictions are desirable in situations where other irremovable distortions are present. It is likely that this is true for the international market for treatment and disposal of hazardous waste. However, this paper argues that the Basle Convention goes too far by prohibiting trade with nonsignatory parties.

This paper was concerned with legal movements of hazardous waste only. The issue of illegal dumping has not been discussed. It is likely that increasing restrictiveness of regulations and agreements governing the trade in hazardous waste generate additional incentives to dump toxic substances illegally (see Copeland 1991, for instance). From a theoretical point of view, the interesting question is how environmental and trade policies are affected by this possibility. The practical problem is to cope with illegal international waste movements. The Basle Convention addresses this by the requirement to reimport hazardous waste under certain circumstances. This regulation shifts the risk of illegal wage shipments from the importer to the exporter. On the whole, the set of rules defined by the Basle Convention for the international trade in hazardous waste are rather restrictive. Particularly, the discrimination against nonsignatory countries can hardly be justified on purely economic arguments even if second-best considerations apply. It is still an open question as to how the WTO will deal with the foreign-trade implications of this international environmental agreement.

8

Environmental Taxation in Open Economies: Trade Policy Distortions and the Double Dividend

SJAK SMULDERS

8.1. INTRODUCTION

During the first half of the twentieth century, the world's economy experienced that protectionism is an evil that should be avoided. Considerable effort was taken to liberalize trade through international negotiations. In the form of GATT and WTO an institutional framework was developed to anticipate and prevent new forms of protectionism.

The world economy has not yet experienced a global environmental crisis. Yet, the developed countries in particular have become increasingly aware of the threats of global pollution problems. International negotiations and agreements are devoted to aligning domestic policies towards the environment and to reaching international targets for pollution abatement. Ideally, two sets of instruments are needed for these two sets of goals, internalizing environmental externalities and removing trade restrictions. Domestic environmental taxes and other instruments should ensure that private costs and benefits of domestic activity reflect social costs of pollution and benefits of environmental quality improvements. The removal of tariff and nontariff barriers to international trade should ensure that domestic activity fully exploits comparative advantages. Indeed, if an economy "gets prices right", liberalizing trade is always welfare improving (see Chapter 2 of this volume).

In practice, economies are riddled with externalities and distortions. Not only policy reforms concerning pollution externalities or trade distortions are proposed. We see attempts to reduce unemployment by removing labor market rigidities, to enhance competition and productivity by deregulating product markets, to improve mobility by addressing distortions in housing markets, and so on. It is practically impossible to address all distortions at the same time with

The author wishes to thank Lans Bovenberg, Partha Sen, Günther Schulze and an anonymous referee for useful comments on an earlier draft.

one grand reform of the entire tax and incentive structure. Hence, economic policy and policy reform always operate in a second-best world in which some distortions are likely to be left. As a result, the removal of one distortion might exacerbate other distortions.

Sometimes, however, opportunities arise to address several distortions at the same time. In international environmental agreements, trade liberalization may be linked to environmental policies. The threat of trade boycotts may support international cooperation on environmental issues. The growing awareness of environmental problems might pave the way for environmental taxes that provide new sources of revenues for governments, thus allowing them to cut other distortionary taxes and tariffs. The question then arises whether these new taxes are so powerful that they not only improve the environment but also provide the means to reduce other problems like unemployment or losses from restrained international trade. Policymakers that lack the time, resources, and political support to free society in one stroke from all distortions always like to kill two birds with one stone.

As of the early 1990s, a series of articles discussed the possibility, both theoretically and empirically, of reaping a *double dividend* by raising environmental taxes. The first dividend is the improvement of environmental quality. The second dividend is a welfare improvement due to reductions in distortions in other fields. The literature is concerned with an economy in which the government has to raise revenue by relying on distortionary taxation (i.e. lump-sum taxes are not available). Most of the literature focuses on the case that higher environmental taxes yield higher revenue that can be used to cut distortionary labor taxes. In this case, an increase in welfare through a rise in employment might arise as a second dividend. However, the insights from the double dividend debate can also be applied to the interrelation between trade liberalization and environmental policy.

The main results from the theoretical double dividend literature are rather pessimistic. The double dividend often fails to materialize since environmental taxes may exacerbate distortions from the rest of the tax system. However, general equilibrium effects sometimes work in the right direction, that is, environmental taxes mitigate existing distortions. Economists have been assisting policymakers by sorting out when general-equilibrium effects, which are complex in general, work in favor of the double dividend and when they do not.

A double dividend can only occur if initially, before the tax reform, the tax system was suboptimal from a nonenvironmental point of view. Hence, if the overall cost of tax distortions was minimized with the environmental distortion ignored, any tax reform that introduces environmental taxation must necessarily increase distortions in nonenvironmental fields. A double dividend then fails. This result applies to all distortions, including distortions in international trade. It applies to closed economies, as well as small or large open economies. However, as pointed out above, initial tax systems are often inefficient.

Political economy aspects and distributional concerns might explain why opportunities to improve the efficiency of a tax system remain unexploited. A change from low to high concern for the environment might change coalitions in favor of reforming the tax system.

This chapter focuses on interactions between international trade and environmental policy in a second-best world. Section 8.2 reviews the double dividend hypothesis for a small open economy that takes all world market prices as given. The result found that the double dividend hypothesis in general fails to materialize if environmental tax revenues are used to reduce distortionary labor taxes is elaborated upon. It is examined to what extent these results carry over to an economy distorted by trade tariffs. Section 8.3 considers an open economy that holds some market power in one of the markets. It is distorted by trade tariffs rather than labor taxes. Environmental tax reform may now yield a double dividend due to terms-of-trade improvements. Section 8.4 returns to environmental-for-labor tax swaps and examines the effects on international distribution of welfare and global welfare in an interdependent multiregion world.

8.2. LESSONS FROM THE DOUBLE DIVIDEND LITERATURE

Few authors have considered the simultaneous reform of tariffs and environmental taxes. However, there is a growing literature on tax reforms that aim at replacing labor taxes with environmental taxes (see Bovenberg 1995; Goulder 1995; de Mooij 2000, for excellent surveys). This literature is closely related to the central topic of this chapter since it addresses the question as to how environmentally motivated taxes can be used to serve not only the environmental goal, but also a second goal of reducing existing distortions.

In most of this literature on environmental and labor taxes, the labor market clears, labor supply is endogenous, which implies that unemployment is voluntary, but labor taxes distort the trade-off between leisure and employment. It is hoped that the tax reform yields a second dividend in the form of enhanced employment. Higher employment boosts welfare since, due to the labor tax, a wedge is present between costs and returns to labor, causing the social marginal returns to employment to exceed the private marginal cost of leisure. Other papers on the double dividend study imperfect labor markets in which search and hiring costs, efficiency wages, or union power result in involutary unemployment (e.g. Bovenberg and van der Ploeg 1998; Schneider 1997; Koskela and Schöb 1999; Holmlund and Kolm 1997). Also in this setting environmental tax reform affects labor markets: swapping environmental taxes for labor taxes changes net real wages and unemployment benefits, which may increase labor market search or reduce union power.

The literature focuses on gains from employment rather than gains from trade liberalization. However, some basic lessons can be learned from this

"Double Dividend" literature. In particular, the existing literature gives insight in what kind of welfare gains we can expect in general from using environmental tax revenues to reduce other distortions. These other distorions may include distortions in international markets and distortions caused by tariffs. In this section we summarize the main insights and relate them to our topic of trade distortions and environmental taxes.[1]

8.2.1. The Benefits of Recycling Environmental Tax Revenues

The first basic insight is that if environmental taxes are increased, it is better to use the associated increase in revenue to reduce distortionary taxes, than to return them in a lump-sum fashion to households. The welfare gains from the former scheme are higher, thanks to the clever *recycling* of tax revenues. This result is refered to as the "weak double dividend" (Goulder 1995: 159).

The intuition for this result is that, for efficiency reasons, we prefer lower distortionary taxes, that are by definition providing wrong incentives, over higher lump-sum transfers, which do not affect incentives. This result shows that there is, apart from the first dividend of improved environmental quality, a second dividend to be reaped *relative* to the case without reduction in distortionary taxes. Because it remains to be seen whether absolute gains can be realized, it is said that with a green-tax-for-distortionary-tax swap typically a *weak* double dividend arises. The weak double dividend claim can be directly applied to economies with trade policy distortions. It is welfare improving to recycle the revenues from environmental tax increases by reducing distortionary tariffs rather than return the revenues in a lump-sum fashion. Some qualifications are in place here.

First, a situation might exist where tariffs are nondistortionary, that is, tariffs might provide the right incentives. This happens if the country has some market power such that the optimal tariff is positive. Starting from a tariff that is equal or below the optimal tariff, a tariff is nondistortionary, and environmental tax revenues should obviously not be used to reduce tariffs.

Second, although reducing a particular distortionary tariff is better than returning tax revenues in a lump-sum fashion, there might be even better ways to use the tax revenues. If the economy suffers from many distortions, it is best to reduce the distortionary tax that has highest welfare cost. (This cost is usually measured by the "marginal cost of public finance" associated with each tax rate, defined as the change in welfare per unit of revenue raised by the tax, relative to the marginal utility of income, see for example Schöb 1996.)

---

[1] We will not discuss the entire range of outcomes and subtleties of the double dividend literature since we are only interested in the relation to the topic of trade and environment. For a more detailed discussion (without extensive attention to international trade aspects) see Bovenberg (1995), Goulder (1995), Pezzey and Park (1998), and de Mooij (2000). In particular, we will focus on taxation and ignore other types of regulation; see Fullerton and Metcalf (1997) for a discussion of command and control regulation and subsidies in the context of the double dividend.

Third, if distributional concern is taken into account, even the weak double dividend may fail. Although a reduction in distortionary taxes increases efficiency in the economy as a whole, it may benefit some groups more than (or at the cost of) other groups. For example, if the revenues from the environmental tax are used to cut labor taxes, workers benefit, but income groups relying on welfare payments do not. In contrast, if revenue was recycled to increase welfare payments, low income groups would gain relative to high income groups (workers). Indeed, if society cares about improving income distribution, the recycling of revenues in a lump-sum fashion might be more attractive (see Proost and van Regemorter 1995). In open economies with distorted trade policies, similar situations may arise. If high income groups spend a larger fraction on tradables relative to low income groups, a lump-sum subsidy might be preferred over a reduction in trade tariffs as it improves income distribution more. In general, different ways of recycling revenues have different distributional impacts which affects the welfare calculus as well as the likelihood of a tax reform from a political economy perspective. I will return to this below.

8.2.2. The Cost of Interaction between Distortionary Taxes

The second insight of the double dividend literature is that normally there is a positive (nonenvironmental) *cost* associated with higher environmental taxes rather than a (nonenvironmental, or second) dividend, despite the fact that this cost is reduced by recycling the revenues. In Goulder's (1995) terms, tax recycling allows a weak double dividend (see above) but normally not a *strong* double dividend. For example, higher environmental taxes that are revenue-neutrally combined with lower labor taxes result in an improved environment but lower employment.

A *strong* double dividend holds if the gains from distortionary tax cuts outweigh the costs of the environmental tax increase. If a strong double dividend holds, the tax reform is worth pursuing even if there are no environmental benefits since it improves the allocative efficiency of the tax system.

The theoretical literature based on general equilibrium modeling (starting with Bovenberg and de Mooij 1994) reveals that normally this free lunch does not arise. The basic model assumes a small open economy, a single factor of production, labor, which is immobile and supplied endogenously. One of the goods produced causes pollution as a byproduct. There are two taxes, on labor and on polluting goods. Tax revenue is used to finance a public good. Households care about private goods, leisure, public goods, and environmental quality. Many variations on this model structure exist, some of which will be discussed in some detail below. However, this structure suffices to point out the basic general equilibrium mechanism that may cause the double dividend to fail, and that can be labeled the *tax burden effect* (de Mooij 2000: 15).

A green tax swap (replacing other preexisting taxes by environmental taxes) is likely to increase the overall tax burden rather than decrease it, since environmental taxes tend to exacerbate rather than mitigate existing distortions. For example, introducing a tax on dirty inputs reduces input use, thus allowing for an improvement in environmental quality which provides the first dividend. However, it also lowers labor productivity, reduces labor demand and thus erodes the labor tax base. In this sense, environmental taxes and labor taxes interact: the increase in public revenue from the higher environmental tax is partly offset by a fall in revenue from labor taxation. This reduces the scope to cut labor taxes. In particular, public revenues from the environmental tax do not increase sufficiently to offset the decrease in labor demand by cutting labor taxes. Hence, we may distinguish between a negative *tax-interaction effect* (the erosion of the tax base) and a *revenue-recycling effect* (as explained above). If the revenue-recycling effect, which positively affects welfare, is more than offset by the tax-interaction effect, the strong double dividend fails (Goulder, 1995; Parry 1995).

More generally speaking, environmental taxes trigger abatement efforts which are costly in terms of nonenvironmental goals by reducing economic activity (per capita real output, income or consumption), thus eroding the tax base and making the tax system less efficient.

The tax-burden effect, which is the combined result of the tax-interaction and revenue-recycling effect, and which is responsible for the failure of a strong double dividend to materialize, is also at work in open economies distorted by tariffs. Environmental taxes and abatement costs for firms increase producer costs and harm competitiveness of domestic producers. Hence, they erode the tariff base for the government, tariff revenues fall, and a negative tax-interaction effect arises. If the economy relies on trade tariffs to raise revenue, thereby distorting international trade, the introduction of environmental taxes (or increases in existing environmental taxes) will not yield enough revenue to offset the abatement costs by cutting tariffs, in analogy to the case of labor taxes.

A similar effect arises if the environmental externality is associated with consumption rather than production and when environmental taxes are accordingly levied as a consumption tax on "dirty goods". Consumers will change their consumption mix by shifting from relatively high-taxed dirty goods, to relatively low-taxed clean goods. This will indeed reduce environmental harm, but also changes trade patterns which were already distorted due to tariffs.

The following example illustrates. Consider a small open economy which takes world prices as given, which produces and imports a good of which production causes pollution in the producer's country and which is levying an import tariff on this good. From an overall welfare point of view in a first-best world, the import tariff is a bad idea since it protects polluting producers. In a second-best world, a budget-neutral tax reform may try to replace this implicit subsidy on pollution by a tax on pollution, by simultanously reducing the import tariff and introducing an output tax (a tax on value added of the domestic producer). From a partial equilibrium perspective, the effects of the tariff and tax change can be separately studied. First, the reduction of the tariff results in the conventional gain from trade. Second, the output tax reduces producer surplus, but this loss is more than offset by the social gain of the improvement in the environment and the first dividend arises.

In order to examine whether there is a double dividend, we have to ignore the environmental gain. Nonenvironmental welfare falls on account of the loss of producer surplus, but this loss is (at least partly) offset by a gain from trade liberalization, thanks to the recycling of the tax revenues by reducing the tariff. Here we see the weak double dividend at work: recycling helps. Whether a strong dividend applies depends on, first of all, the relative size of the forces at work in the partial equilibrium story, and second, how the tariff and tax interact in general equilibrium.[2] The presence of the import tariff in the first place implies that consumers have shifted demand to other goods so that consumption of other goods is too high. For this reason an output tax is more distortionary if import tariffs are present. A higher output tax on the dirty good shifts consumption even further to clean goods of which consumption was already excessive. Hence, the gains from trade have to be accordingly higher in order to offset this adverse effect of the output tax. Sen and Smulders (2000) show that the overall nonenvironmental welfare effect is negative. Apparently,[3] as is often found in the double dividend literature, the tax-interaction effect outweighs the revenue-recycling effect so that on balance the tax reform is reducing nonenvironmental welfare.

8.2.3. Optimal Second-Best Taxes

If governments had no revenue requirement, or could raise revenue without relying on distortionary taxes, a first-best optimum is reached by implementing a Pigovian tax, that is, a tax on pollution that is equal to the marginal social damage of pollution. This tax balances the direct private marginal cost and social marginal value of pollution reduction. In the presence of distortionary taxes, however, the costs of environmental taxes are magnified due to the tax-interaction effect. Environmental taxes not only imply a direct (abatement) cost, but also a reduction in revenue from other taxes because of tax-base erosion. Therefore, the optimal environmental tax[4] that maximizes overall

---

[2] Note again that the partial equilibrium approach does not suffice to answer our questions concerning double dividends because it ignores the general-equilibrium or "tax-interaction" effect.

[3] Note that up to now, this particular trade-environment tax reform has not been formally analyzed by explicitly decomposing the welfare effects into a revenue recycling and a tax interaction effect.

[4] Here, an environmental tax is defined as a charge that comes on top of commodity taxes that might be in place to raise revenue. Suppose that the government has only pollution taxes and labor taxes at its disposal (and no commodity taxes). Then, the tax on dirty commodities is below the Pigovian tax. However, if the government levies commodity taxes to raise revenue rather than labor taxes, all commodities should be taxed according to the Ramsey-principle. In that case, dirty commodities are taxed at a higher rate to internalize pollution externalities, but the tax differential between clean and dirty taxes is less than the Pigovian tax because of the tax-interaction effect (see Fullerton 1997; Schöb 1997).

welfare (including utility derived from a clean environment) is lower than the Pigovian tax (see Bovenberg and van der Ploeg 1994; Bovenberg and Goulder 1996).[5]

The negative tax-interaction effect makes environmental taxes a costly way to raise revenue since each euro of revenue that is raised directly by environmental taxes is partly offset by lower revenues from other taxes. This amounts to saying that the marginal cost of public funds associated to environmental taxation is larger than one. The Pigovian tax that applies to an undistorted economy should be corrected by this marginal cost of public funds in order to find the optimal second-best environmental tax. A higher gap between private and social marginal damage from pollution calls for a higher pollution tax (according to the Pigovian principle), but the higher the marginal cost of public funds as raised by environmental taxes is,[6] the smaller is the fraction of this gap that should be bridged by the environmental tax.

Another way of interpreting this result is that the preexisting labor tax is an implicit tax on pollution, since labor taxes reduce consumption and production of polluting goods by reducing labor supply. Accordingly, environmental taxes in the second-best world are lower than without labor taxes.

8.2.4. Conditions for a Strong Double Dividend

Some economists (or policy makers, see Fullerton and Metcalf (1997) for a survey) interpret the double dividend literature less pessimistically than the previous subsection would suggest. In various articles, it is shown that under certain conditions there is scope for a double dividend.[7] In the light of the previous discussion, this requires that the tax-interaction effect is rather small as compared to the revenue recycling effect, or that the tax-interaction effect is positive rather than negative. In the latter case, the environmental tax *mitigates* other distortions. We can also say, that if a double dividend can be reaped, the tax reform improves the efficiency of the tax system in terms of revenue raising. To get a feeling for what gives rise to a double dividend, it is insightful to review some cases from the literature.

Several papers[8] show that a double dividend can arise if environmental taxes are raised and labor taxes cut. As discussed above, there is a burden associated with this tax reform due to increased abatement efforts. However, if it is not labor that bears this burden, the wedge between real wages and wage costs becomes smaller, employment expands and welfare is boosted because employment was too small initially. The tax reform yields a double dividend if

---

[5] Goulder, Parry and Burtraw (1997) consider the optimal form of other types of regulation in the presence of second-best taxes.

[6] To be more precise: the more it exceeds unity.

[7] From now on, I will use "double dividend" where Goulder's (1995) term "strong double dividend" applies.

[8] Including Nielsen *et al.* (1995), Felder and Schleiniger (1995), Koskela and Schöb (1999), Holmlund and Kolm (1997), Bovenberg and van der Ploeg (1998).

it shifts the burden of environmental policy from labor to factors of production that are less elastic in supply and demand. If fixed factors are relevant in production, it may be more efficient to tax these factors than labor. Taxing labor results in lower labor supply, while taxing the recipients of the rents to fixed factors will not change behavior, that is, is nondistortionary, by the definition of a fixed factor.

Consider the case that production employs energy, which is polluting, labor and a fixed factor (e.g. firm-specific capital). The tax reform introduces a tax on energy and reduces labor taxes. The energy tax leads to lower energy use, which is the abatement effort. The labor tax cut reduces labor costs, but this effect is at least partly offset by a reduction in labor productivity because of lower energy use. The energy tax acts as an implicit tax on labor.[9] However, the burden of lower energy inputs is shared by labor and the fixed factor so that the energy tax is also an implicit tax on the fixed factor. If the latter bears a large enough share of the burden, labor is better off on balance because the explicit labor tax cut dominates the implicit labor tax rise. If the fixed factor (capital) and energy are highly complementary (i.e. if substitutability between energy and labor is larger than between capital and energy), the tax on energy is an implicit tax mainly on the recipients of the fixed factor. The tax reform improves the efficiency of the tax system because the (explicit) distortionary labor tax is replaced by an implicit nondistortionary tax on the fixed factor.[10]

In an open economy context, fixed factors are less relevant because of capital mobility. If environmental taxes fall mainly on capital and capital is internationally mobile, capital flight might result. Lower capital inputs further decrease labor productivity. Indeed, with perfect capital mobility labor again bears the entire burden of dirty input taxation (Felder and Schleiniger (1995) show that capital mobility lowers the benefits (but also lowers the costs) that would arise without capital mobility). Of course, fixed factors need not be capital, but could take the form of firm-specific human capital and expertise which are likely to be internationally not very mobile.

More generally speaking, a double dividend arises if the tax reform shifts the overall burden of taxation from factors with high marginal excess burdens to factors with low marginal excess burdens (see Bovenberg and de Mooij 1996). This may happen in a closed and an open economy alike. To expand the term used in the previous subsections, we may say that not only a revenue-recycling and a tax-interaction effect are at work, but also a *tax-shifting* effect. The

---

[9] Hence, when labor and energy are the only factors, the double dividend fails because what labor gains in terms of a lower *explicit* labor tax is lost in terms of a higher *implicit* tax, and on top of that is a negative tax-interaction effect.

[10] Similarly, we could consider a case in which consumption, rather than intermediate input use, is the source of pollution. In this case, a tax on consumption of the dirty good should be introduced. This consumption tax is an implicit tax on labor and offsets the explicit labor tax cut. However, if there are nonworker consumers (pensioners, capital owners), labor will not bear the entire burden of the pollution tax and a double dividend may arise.

latter effect may mitigate a negative tax-interaction effect such that a double dividend arises.

One very effective way to realize a double dividend from environmental taxation in a trading economy is to make foreigners bear the burden of the environmental tax. This would happen if the domestic economy has market power in the international market. Environmental taxes then result in changes in international prices (an improvement in the terms of trade) such that foreigners bear part of the tax burden. I return to this in the next sections.

8.2.5. Distributional Effects of Environmental Tax Reforms

Since the double dividend requires a shift in effective tax burdens from one category of taxpayers to another, distributional effects will occur. For instance, the tax burden for fixed factor reward recipients or nonworking consumers increases as a result of dirty input and commodity taxes. This makes the double dividend less politically acceptable. If distributional considerations are important, the second dividend becomes a more complex concept. Looking for tax swaps that increase income may hide costs in terms of worsened income distribution. (Proost and Van Regemorter (1995) explicitly take into account income inequality to calculate welfare effects of tax reforms. See also de Mooij (2000), chapter 6).

Suppose that initially, before the green tax reform, the distribution of incomes and rents from existing regulation reflects a political equilibrium, that is political support is maximized (see Chapter 4, this volume) but environmentalists' interest are not organized. Once environmental concern rises and environmental interest becomes organized, a green tax reform may become viable. If an environmental tax is introduced, environmentalists will gain, but other will lose. Most likely, owners of firm-specific assets in polluting sectors will lose most. To restore the political equilibrium, these groups should be compensated by, for example, higher subsidies. This reasoning shows that in a political economy context it is perfectly possible that higher revenues from environmental taxes will be used to increase rather than decrease distortionary taxes. Similarly, a cut in labor taxes is probably not viable since it does not compensate specifically those who lose from the environmental tax increase.

The distributional effects may also explain why the government might prefer to use command and control rather than market-based instruments like taxes to implement environmental policy.[11] While the latter typically take away money from pollutors (since the government puts a price on pollution and

---

[11] Note that from efficiency point of view, revenue-raising instruments perform better than nonrevenue-raising instruments. The possibility of revenue-recycling makes environmental instruments that raise revenue, like taxes and auctioned pollution permits, more attractive than nonrevenue-raising instruments, like grandfathered pollution permits and quota (Goulder *et al.* 1997). While the former impose an explicit burden, the latter imply an implicit tax burden. Overall cost of the former is lower since revenue-recycling at least partly offsets tax-interaction effects.

becomes implicitly the owner of the right to pollute), command and control does not do so (pollutors are forced to cut back pollution, but keep a right to pollute which yields a scarcity rent). Hence, command and control involves smaller redistribution and may be more politically acceptable.

In an open economy, the result might be reversed. Bommer and Schulze (1999) show that an increase in environmental regulation may go hand in hand with a reduction in trade distortions. They consider a two-sector trade model with sector-specific factors and preexisting import tariffs that protect producers in the import competing sector. If the export sector is pollution intensive, this sector loses from a rise in environmental taxes. To compensate the producers in these sectors, tariffs can be lowered because this benefits export sector and hurts import sectors.

Rather than at the cost of some domestic sectors, a double dividend may arise at the cost of lower foreign income because of terms of trade shifts (see below). The domestic economy could ignore this distributional effect in deciding whether or not to undertake the tax reform. However, these international spillover effects of environmental tax reforms are likely to occur in large open economies that could expect retaliation. Hence also in a trade context distribution effects matter.

8.3. THE ROLE OF TERMS OF TRADE IMPROVEMENTS

Environmental taxes, whether used to reduce other distortionary taxes or not, impose a burden in the form of output reduction and/or abatement. If a tax reform shifts this burden to foreigners, the domestic economy may experience a double dividend. In a small open economy for which prices are determined in world markets, such a shift is not feasible: any change in domestic taxes leaves prices for foreign consumers unaffected. An economy with some market power in international markets, however, may be able to shift the burden through changes in international prices.

Consider for example a country (or group of countries, e.g. EU or OECD) that imports a dirty (intermediate or final) good and increases the domestic tax for this good.[12] If the country is a significant player in the world market and consequently has monopsony power, the world market price for the import good falls, the terms of trade improve for the domestic economy and foreign producers suffer lower prices. Hence, foreigners pay for part of the burden of environmental taxation. Similarly, if an exported good is taxed, export supply falls, the world market price rises so that domestic producers pass part of the costs to foreign consumers. The implication is that the domestic tax base is eroded accordingly less than without terms of trade improvement. The tax-interaction effect is mitigated by a tax-shifting effect in the form of a

---

[12] This example is also analyzed in de Mooij (2000: chapter 5), see next section.

terms-of-trade improvements, thus allowing revenue recycling to be powerful enough to offset the tax-base erosion.

Hence, if the terms of trade improve strongly enough, a double dividend arises. In that case, the environmentally motivated tax acts as an optimal tariff that exploits the monopoly or monopsony power of the country. While on the one hand the tax distorts consumption and/or production in the domestic economy and exacerbates existing distortions, it improves the terms of trade. For small initial taxes, the latter will dominate the former, and taxes improve domestic welfare irrespective of the environmental dividend.

The double dividend becomes more likely if a larger part of the burden can be shifted to foreign consumers. This is the case if nontradables sectors are less important, that is, if the economy is more open to international trade, if the exposed production sectors face less elastic demand, that is, if market power is stronger, and if the share of domestic consumption in total production of export goods is relatively small, that is, if export specialization is high. The model in Sen and Smulders (2000) illustrates this.

These authors explicitly analyze simultaneous trade liberalization and environmental taxation in an economy that has some power in a niche in the world market. In particular, the country produces two goods: an export good that is not consumed domestically and a nontraded good. The two production factors in both sectors are labor, which is internationally immobile and whose supply is fixed, and (nontraded) polluting inputs. While the world market price of the imported consumption good is given, the world market price of exports is affected by home supply. A tariff is levied on imports to generate public revenues. The government finances transfers to households from the tax revenues out of tariffs and pollution taxes.

An increase in the tax on polluting inputs reduces production of both nontradables and export goods. If the price elasticity of exports is high (as in the small economy case with given world prices), export revenue falls and imports have to fall to restore balance of payments equilibrium. Accordingly, private welfare is hurt on two accounts: not only nontradables consumption but also tradables consumption is lower. The lower volume of imports implies reduced tariff revenues for the government, thus making the double dividend infeasible.[13] The terms of trade improve but not sufficiently to offset the fall in output.

However, if foreign demand is inelastic, that is, if the home country has significant market power in its export market, higher pollution taxes increase not only export prices but also export revenues. This allows for more imports,

---

[13] Revenue-recycling (lowering tariffs) turns out to be infeasible since with elastic foreign demand the revenue from pollution taxes falls in the model. Production functions are of the Cobb-Douglas type so that the share of labor costs and pollution costs in (sectoral) output are constant. The latter share also represents the tax revenue per unit of output. Environmental policy reduces (physical) output by reducing polluting inputs. Hence tax revenue falls. Only if the value of output increases because of inelastic export demand, a constant share of taxes in output results in an increase in tax revenue.

thus providing the government with more tariff revenue. Moreover, environmental tax revenues can be used to reduce tariffs thus boosting imports even more. Nontradables production and consumption, however, fall due to lower inputs. On the one hand, consumers enjoy more tradables consumption but, on the other hand, they have to cut nontradables consumption. Hence, private welfare increases only if the weight of nontradables consumption in utility is not too large, that is, if the economy is sufficiently open to international markets. The lower the degree of openness as measured by this share of nontradables, the lower the price elasticity of exports has to be in order to allow a double dividend to arise.

The basic result that the double dividend requires inelastic export demand is robust to a number of generalizations of the model. Sen and Smulders (2000) first consider the role of substitution in production and consumption, and then introduce a third sector, producing import competing goods. The lower the elasticity of substitution in consumption between tradables and nontradables, the more likely it is that the double dividend arises. As explained above, the cut in pollution reduces output in both nontradables and exportables, and with inelastic export demand imports rise. With poor substitution in demand between importables and nontradables, the relative price of nontradables rises steeply in response to these supply changes. Thus labor and polluting inputs are reallocated from the exportables sector towards the nontradables sector, which reduces the supply of exports even more and which magnifies the terms-of-trade gains. The opposite holds for the elasticity of substitution in production between labor and polluting inputs. The better substitution is, the smaller the fall in output due to the pollution reduction, the lower the tax burden effect.

8.4. INTERNATIONAL FEEDBACK EFFECTS AND GLOBAL WELFARE

Once we consider large economies and changes in international prices due to unilateral environmental policies, international feedback effects become important. While the country (or group of countries) that initiates the environmental tax reform may gain a second dividend thanks to terms-of-trade improvements, other countries necessarily experience a worsening of the terms-of-trade and lose. International distributional effects matter and the question arises as to what happens to global welfare. Furthermore, environmental policies at home affect production and consumption abroad. If the rest of the world shifts to more polluting sectors, world pollution might increase. If transboundary pollution is significant, domestic environmental policy thus becomes less effective because of these feedback effects.

In his book on the double dividend hypothesis, de Mooij (2000) devotes a chapter to the analysis of trade and environmental taxation in a large economy with preexisting distortions. As explained in section 8.2, the bulk of the existing literature on environmental taxes in a second-best world abstracts from most trade issues by considering a closed economy or a small open economy in which the wage is the only price to be endogenously determined. In de Mooij's chapter, the home country initiating the environmental tax reform is assumed to be a significant player in the world market either for polluting inputs (which are imported by the home country) or for the single good it is producing and exporting to the rest of the world. Hence, not only real wages but also input prices or producers' prices are affected by environmental tax reforms, as is already discussed in section 8.3. In contrast to the previous section, trade is not distorted directly by tariffs, but labor taxes distort the domestic economy. This has special relevance in the context of trade agreements that prohibit the use of tariffs to raise revenue.

De Mooij shows that a double dividend is possible in this setup since pollution taxes act as an implicit optimal tariff to exploit monopoly or monopsony power and maximize the value of export revenues (see section 8.3). This has implications for the second-best optimal environmental tax. Without terms-of-trade effects, the second-best optimal environmental tax is lower than the Pigouvian tax since the tax exacerbates existing distortions. The ability of environmental taxes to act as an implicit optimal tariff and improve the terms-of-trade effects makes environmental taxes more desirable and raises their optimal level. In large open economies, the optimal tax may therefore be higher than the Pigouvian level. However, there is an opposing effect stemming from transboundary pollution effect. Domestic environmental taxes boost polluting activities in countries where environmental regulation is less tight. Foreign pollution affects domestic welfare if pollution is transboundary. Hence, because of this feedback (or leakage) effect, environmental taxes in a large open economy are less effective and optimal taxes are lower on this account.

Domestic, foreign and global welfare changes can be compared in de Mooij's three-region model. It is assumed that region 1 cares about pollution, has a distorted labor market, imports polluting inputs from region 2, and specializes in the production of intermediate goods that are exported to region 3. Region 1 is the region that initiates the unilateral environmental tax reform. Regions 2 and 3 are undistorted and do not care about the environment. Region 2 can be thought of as the OPEC region. It produces and exports polluting inputs (oil), and it imports final goods from region 3. Region 3, the rest of the world, imports oil and intermediate inputs and exports final consumption goods.

Suppose region 1 has monopoly power in its export market. If environmental taxes are swapped for labor taxes in region 1, prices of the intermediate goods this region produces increase and regions 2 and 3 face a terms-of-trade loss.[14] Country 1 gains because of not only a terms-of-trade gain but also reduced

---

[14] Alternatively, de Mooij assumes that the home country has monopsony power in the market for polluting inputs (oil). A tax on these inputs lowers the oil price, so that region 2 suffers a fall in its terms of trade, but region 3 benefits from the lower oil price.

labor market distortions. Hence, the gains in the home region are accompanied by losses in the other two regions. However, global welfare may increase, thus providing the possibility of a Pareto improvement if international transfers are introduced. The intuition behind the global welfare increase is that the tax reform redistributes income from undistorted economies to the distorted economy that has room for efficiency improvements. The international income redistribution provides the government revenue to reduce the labor market distortion that ultimately affects the entire world economy. It should be clear the incentives to undertake a tax reform in region 1 are stronger from a national perspective than from an international one. The reason is that region 1 captures a terms-of-trade gain at the expense of other countries. The optimal pollution tax from a global point of view is accordingly lower than the optimal pollution tax from a national point of view.

While de Mooij assumes that there is only one region with domestic distortions, Felder and Schleiniger (1998) study optimal second-best environmental taxes in a multiregion world in which domestic distortions differ across regions. Each region is small and takes the price of its single output as well as the price of imported energy inputs as given. Hence, the real wage is the only endogenous price (see section 8.2). Labor is the only factor of production. Labor taxes and energy taxes are available to finance government consumption in which the levels differ among regions. The authors show that in order to reduce pollution to a given target at minimum cost from a global point of view, regions with a high (low) labor tax should set a low (high) environmental tax. The intuition behind this result directly follows from the insights of section 8.2. In a small open economy without fixed factors of production, environmental taxes exacerbate preexisting distortions and there is no double dividend. Hence the burden of environmental taxation is highest in regions with high labor market distortions. Global welfare is maximized if the burden of environmental policy is shifted to countries in which the efficiency cost of environmental taxation is lowest. Note that this result is the opposite of de Mooij's result that environmental tax reform should be shifted to the distorted economy. The reason for this contrast is that in de Mooij's setting the distorted economy reaps a double dividend and is thus able to reduce distortions while in Felder and Schleiniger there is no double dividend so that distorted economies become more distorted.

The polluting input that plays such a large role in the double dividend literature is normally associated with energy, in particular fossil fuels causing CO_2 emissions. However, because of the static nature of the models, it is ignored that fossil fuels are a nonrenewable resource of which prices are set such that resource owners expect the entire stock to be exhausted before backstop technologies become economically viable.[15] Amundsen and Schöb

---

[15] Ulph and Ulph (1994) and Farzin (1996) consider optimal taxation (rather than tax reform, as is the topic of the double dividend literature) of polluting nonrenewable resources.

(1999) study tax coordination among oil-importing countries in a dynamic model. It is shown that it is optimal to tax energy use at a rate that exceeds the Pigovian tax. The additional tax component serves to extract rents from the oil-producing country. Since any coordinated energy tax increase in oil-imported countries reduces demand for oil, the oil-producing countries will react by setting a lower oil price to avoid that they will be left with unsold oil reserves in future. This implies a redistribution of income from oil producers to oil-importing countries. Hence, apart from internalizing environmental externalities, an energy tax appropriates rents from resource owners. This analysis has important implications for the double dividend debate. The price reaction by oil producers mitigates the tax burden effect of the environmental tax, while still the government raises revenue to reduce distortionary taxes or tariffs. Hence, the energy tax improves the efficiency of the national tax system.

8.5. CONCLUSIONS AND AREAS FOR FURTHER RESEARCH

In the early 1990s policymakers became excited about the idea of using environmental taxes, not only to provide incentives for a clean environment, but also to raise the revenue that allowed them to decrease distortionary taxes, thus reaping a double dividend of a clean environment and a less distorted economy. Economists reacted to this claim with a series of theoretical and empirical papers. While some papers support the double dividend claim and others firmly reject it, one general lesson emerges. Environmental taxes may exacerbate the distortions of the existing tax system through general-equilibrium effects. Hence, while intuition and partial thinking reveals the virtues of revenue-recycling, economists' analysis revealed that these benefits have to be balanced with tax-interaction effects. There is no general answer as to whether the former outweigh the latter. Theoretical work concludes that it is difficult to reach a double dividend, although it may be realized at the cost of worsening income distribution. In any case, the specific nature of preexisting distortions and the specific design of the tax reform determine whether environmental tax increases mitigate or exacerbate existing distortions, as is common in a second-best world.

In open economies, there is one clear way to ensure that higher environmental taxes *reduce* existing distortions, viz. the realization of terms of trade gains. Economies that exploit some market power in international markets may shift the burden of environmental taxes to foreigners, thus improving national welfare. However, two problems arise.

First, few countries are likely to be able to exert enough market power to mitigate the burden of higher environmental taxes. Developing countries, which are not very eager to clean up the environment for the international society if national nonenvironmental gains are absent, especially suffer from high price elasticities of their export demand. Even high income countries that have a strong position in some sectors of the world market could burden other sectors of their economy too much by applying stringent environmental standards.

Second, international redistribution could cause international tension and repercussions. Under some circumstances, tax reform increases global welfare so that it would, in principle, be possible to compensate the losers. However, such an international redistribution should be feasible in a nondistortionary (lump-sum) way, otherwise the international compensation scheme itself may cause additional distortions and introduce a new source of tax-interaction effects.

The international coordination of environmental tax reforms is an interesting area for future research.[16] On the one hand, this allows for studying the effects of cooperative abatement strategies. The cost of environmental policy of a group of cooperating countries is likely to be lower than that of an individual country because the group may exploit a larger degree of monopoly power. Tax-interaction effects therefore vary with the degree of cooperation which may seriously affect conclusions from the double dividend literature. On the other hand, it allows for the study of side-payment and compensation schemes which have major consequences on distribution and political viability of tax reforms. It also puts into perspective the linkages between global tariff and nontariff barrier reductions and international environmental policies.

The research so far ignores political and institutional constraints. If there is scope for a double dividend, why is it not reaped? In order to explain unexploited opportunities to gain from tax reforms, we should think about distributional problems, political constraints imposed by interest groups, and administrative costs and informational problems associated with the implementation of the tax reform (see Chapter 4, this volume). In all these areas much work remains to be done.

---

[16] See Chapter 10, this volume, on international environmental cooperation.

9

Sustainable Growth in Open Economies

LUCAS BRETSCHGER AND HANNES EGLI

9.1. THE DYNAMIC PERSPECTIVE: INTERNATIONAL SUSTAINABILITY

9.1.1. Introducing Dynamics

International environmental economics can be conducted both in a static and in a dynamic mode. This chapter introduces the dynamic approach view. We analyze the impact of international trade and the natural environment on long-term economic development. In the literature, there are hardly any contributions which cover environmental economics, trade, and growth at the same time. Most of the papers in this field concentrate either on the interaction between (i) international trade and the environment (see Chapters 2–4), (ii) economic growth and the environment or (iii) trade and economic growth. To obtain a comprehensive view of the intersection of all three issues, we proceed in the following manner. In the first section, we focus on economic growth and the natural environment by introducing the most important concept in this field: sustainability. Section 9.2 demonstrates how endogenous growth theory identifies sustainable development paths. We present the predictions of different growth models, which also portray changes in the natural environment. The international aspects are introduced by considering open economy growth models in section 9.3. Building on the insights of modern growth theory, we discuss the literature which analyzes the impact of trade relations, trade policy, and the international division of labor on the sustainability of growth in section 9.4. Empirical results are presented in 9.5. The concluding section summarizes the results and identifies important issues which emerge from the discussion and remain for future research.

The authors thank Sjak Smulders, Frank Hettich, and Günther Schulze for helpful comments.

9.1.2. The Sustainability Paradigm

The "Brundtland report" defines sustainable development as "development that meets the needs of the present generation without compromising the ability of future generations to meet their own needs" (World Commission 1987: 43). Development, thus, includes not only rising aggregate consumption or output, but also environmental quality, social factors, and the distribution of income. Taking such a broad view, economic theory can make substantial contributions to several focal points. The most important concept in economics is individual welfare, which is usually captured with the help of utility functions. Realistic utility functions include not only consumption but also environmental quality. In a dynamic setup, the final objective of any normative theory of economic growth is the long-term development of individual utility.

If the above definition of sustainability is interpreted in economic terms, a sustainable growth path can be characterized by nondeclining welfare between generations, where welfare is measured as average individual utility (see Pezzey 1989; Bretschger 1998a). By demanding constant or increasing welfare over the long run, sustainability is more than survivability, which only requires consumption to be kept above some subsistence minimum. Both fairness and efficiency criteria are crucial ingredients of sustainable growth paths. Efficiency considerations which underlie the normative view of environmental policy are, of course, at the heart of economic theory. Concerning fairness, that is, the intra- and intergenerational distribution of income, the main contribution of economics is to investigate the relation between ethical constraints and the development of welfare.

9.1.3. Fairness Criteria

One should carefully distinguish between efficient allocations of resources – the standard focus of economic theory – and socially desired allocations that may reflect intergenerational as well as intragenerational equity concerns. Due to physical limits and ethical constraints on resource use, a sustainable development path may not be the same as the efficient path predicted by standard economic theory (see Pezzey 1989; Toman 1994). For example, it might be optimal from today's perspective to use considerable amounts of natural resources. But this resource use might not be fair for future generations if it causes individual welfare to decrease in the far future (see Beltratti 1996: 98). Furthermore, it should be noted that incorporating environmental values in decision-making *per se* will not bring about sustainability unless each generation is committed to transferring to the next sufficient natural resources and capital assets to make development sustainable. Put another way, the problem of intergenerational equity must be viewed as an issue of ethics that is distinct from economic efficiency in the Pareto sense (see Howarth and Norgaard 1992). For sustainability, one has to take the distribution of welfare between present and future generations into account. One possible solution is to alter the welfare criterion, which is the guideline for optimization today. Chichilnisky (1977) proposes a welfare function which is a convex combination of a discounted sum of instantaneous utilities and a minimum condition for future utility levels. A more straightforward procedure is to include the additional element of intergenerational fairness in optimizing by excluding all efficient development paths which lead to a declining average utility in the future. To do so, the sustainability criterion is imposed as a prior constraint on the maximization of individual utility. Consequently, each successive generation ensures that the expected welfare of its children is no less than its own perceived well-being (see Howarth 1995).

9.1.4. Focusing on Capital Stocks

While individual welfare considerations are usually restricted to flow variables, some authors argue quite forcefully that capital stocks should be the primary focus in the debate. Emphasizing the quality of the natural environment, sustainability is interpreted as "requiring some constancy in the stock of natural environmental assets" (see Pearce *et al.* 1990: 23). In this context, "weak" and "strong" sustainability have become widely used terms. Weak sustainability means that any form of natural capital can be run down, provided that proceeds are reinvested in other forms of capital, for example, manmade capital. Strong sustainability, however, requires that the stock of natural capital should not decline (Pearce and Atkinson 1998). In this context, one distinguishes between the requirement to conserve every single natural resource and the requirement to conserve an aggregate natural capital stock which leaves room for certain substitution possibilities.

In order to put the theoretical principles of strong sustainability into practice, one can lay down two main rules for the use of renewable natural resources. First, the harvest rates should equal the regeneration rates. Second, the waste emission rates should be in line with the natural assimilative capacity of the ecosystem. For nonrenewable natural resources, the problem is quite different. A strong management proposal is formulated by Turner (1988). He argues that talking about sustainable use of a nonrenewable resource is useless because any positive rate of exploitation will lead to the exhaustion of the finite stock. However, it should be noted that the effect of the exhaustion of certain stocks on welfare is by no means obvious; at least, it is not directly given for all natural resources. It might be that utility remains constant even with a decreasing stock of certain resources or that several (very special) resources do not have an impact on utility at all. Nevertheless, the following arguments in favor of strong environmental sustainability are relevant (see Pearce and Atkinson 1998). First, there is uncertainty about the value of the elasticity of substitution between natural resources and manmade capital. Second, there is an asymmetry between the different types of capital with respect to reversibility: once certain critical natural capital stocks are lost, they cannot be reintroduced. Put differently, a constant stock of certain natural resources can be necessary because natural threshold and irreversibility effects may severely limit the trade-offs that can be allowed between different resources without threatening sustainability (see Pearce *et al.* 1988 and 1990). Third, the scale effects from the loss of critical natural capital are not known and, finally, consumers have an apparent "loss aversion" that arises when certain natural resources are depleted.

The arguments in favor of strong sustainability given above should make us more cautious about depleting natural capital, but the issues raised do not add up to a complete justification for implementing this criterion. Pezzey (1989), for example, stresses that trade-offs between natural and manmade resources can, in principle, be calculated if appropriate weights are used. A "quasi-sustainable" use of nonrenewable resources can be achieved by limiting their rate of depletion to the rate of creation of renewable substitutes (see Daly 1990). The concentration on natural capital stocks has also been challenged because this view does not consider intertemporal efficiency. According to Dasgupta (1995), any stock concept is a "category mistake" because it mixes up the determinants of human well-being and the constituents of well-being. This means that capital stocks may influence welfare directly or indirectly but they are not the final argument in the utility function.

9.1.5. Sustainability in International Perspective

The final goal expressed by the notion of sustainability is nondecreasing welfare between generations. Taking the special attributes of natural resources into account, it is appropriate to supplement this general goal with an intermediate target concerning certain requirements for the state of the natural environment. The higher the probability of irreversibilities and the larger the uncertainties about aggregate costs of damage, the safer the minimum standards for the respective natural capital stocks should be (see Bretschger 1998a).

Allowing for free trade between the economies and considering international externalities does not alter the basic concept of sustainability as a general guideline for policy. It seems to be straightforward to require sustainable development for all the regions of the world, which adds up to sustainable development for the world as a whole. What does change, however, is the way sustainability can be achieved. The international division of labor can decrease but also, under unfavorable circumstances, increase the difficulties in reaching sustainable development. On the one hand, free trade may allow natural resource use to be diminished in the economies where the associated costs are the lowest internationally. On the other hand, free trade might also intensify inefficient resource use in certain countries. If, for example, international prices of a free access natural resource are higher under free trade than under autarky, trade may cause an overuse of certain natural resources. A case in point is, for example, tropical timber (see Chapter 6 of this volume).

Simple maximization of the utility of present generations will generally not be sufficient to reach sustainable development paths. This is true both for a single country as well as for the decentralized world economy as a whole. Present value maximization incorporates neither positive and negative externalities nor the sustainability constraint. It is, therefore, useful to distinguish between three possible types of development paths: paths that are reached under free market conditions, paths that materialize when all external effects are internalized by economic policy ("optimal paths"), and paths that are sustainable, that is, exhibit nondeclining individual utility between generations. The free market and/or the optimal paths may be sustainable in certain cases, but not under all conditions. The following sections focus on the different links between growth, natural resource use and trade in detail. Important issues not dealt with in the existing literature are pointed out in the last section of the chapter.

9.2. NATURAL RESOURCES AND ECONOMIC GROWTH

9.2.1. Growth and Sustainability

In this section, we identify the relevant issues which arise when the concept of sustainability is introduced into several theories of economic growth. In particular, we aim at analyzing the interactions between the state of the environment and the achieved growth paths. In older theory of economic growth, the long-term growth rate is determined exogenously. Only the adjustment growth to the steady-state can be explained by this theory. The emphasis is on the accumulation of physical capital which is subject to diminishing returns. In recent models of endogenous growth, however, the long-term growth rate is explained endogenously, that is, it depends on the production technique, preferences, and fiscal policy. Economic decisions carried out under market conditions induce technical change and endogenously affect the stocks of all kinds of capital, such as physical, human, public, and knowledge capital. The mutual relation between economic growth and the environment can be analyzed as well. The relationship between the natural environment and long-term growth is important since there is only little gain in speculating about hypothetical development paths that have no factual basis. Similarly, any intuition about the consequences of discounting future well-being is elusive, unless feasible development paths can be identified (see Koopmans 1965). Thus, a profound study of sustainable growth is not possible without a sound theory of economic growth, in particular, of endogenous growth.

In the following, the different growth models which consider the effects of natural resources are categorized in figure 9.1. For each model, the relation between growth and the state of the environment will be analyzed. The figure shows that the final objective of the agents portrayed by any economic growth theory is (individual) welfare. There are, in principle, two channels which

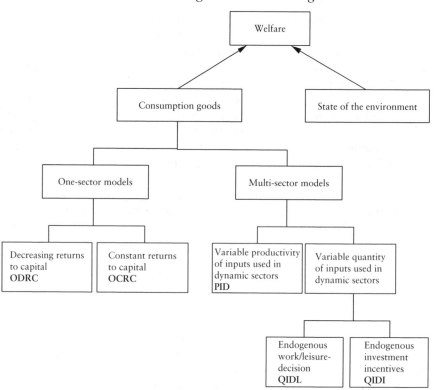

Figure 9.1. *Types of growth models with natural resources.*

transmit the influence of the environment on welfare. First, there is a direct impact of the state of the natural environment on personal well-being, which applies, for instance, to the cases of air quality or the amenities of the landscape. Here, it is obvious that *ceteris paribus* a better environment increases welfare. Second, environmental quality and natural resources affect the production process and by this, indirectly, future consumption possibilities. The influence of the environment on the production sector can be modeled in different manners which we will present in the following. We use the term "capital" for accumulable (manmade) capital and the terms "natural resources" and "pollution" for the influence of the environment on the production sector.

9.2.2. Applying Different Growth Models

Depending on the assumptions about the marginal productivity of capital, long-term growth is either an exogenous or an endogenous variable in the model. Postulating sufficiently decreasing returns to capital, the growth process peters out in the long term. The assumption of constant returns to capital, however, renders endogenous growth in the long term feasible; it keeps savings and investment incentives at a constant (positive) level. In addition, one can distinguish between one-sector models and multi-sector growth models. In one-sector models, the production technique is the same for consumption and investment goods. Multi-sector models usually have a dynamic sector, for example, research and development or education, and one or more static sectors for the production of consumer goods. In these models, the growth rate depends on the characteristics of the dynamic sector. Multi-sector models which include natural resource effects can further be divided into models where environmental quality directly alters the productivity of inputs in the dynamic sector, and models where the quantity of inputs used in the dynamic sector is affected. The latter is portrayed either by incorporating leisure into the model, which allows for more or less labor being used in the dynamic sector, or by varying investment incentives, for example, in research and development (R&D), which alters the quantity of inputs used in the dynamic sector.

In the following, we summarize the relevant literature and provide an intuition for the behavior of the archetype models. We show under which assumptions and conditions (positive) endogenous growth can be realized while the environmental quality does not deteriorate. Taking the limited pretension of this strand of literature into consideration, such a development can be called sustainable. In addition, the focus will be on the dynamic effects of environmental policy. After the general discussion, we will study the growth effects of a "green tax reform" in particular.

9.2.3. Model Type ODRC

A traditional approach to modeling the growth process is by way of a one-sector model with decreasing returns to capital (ODRC). The most prominent representative of this class is the neoclassical growth model. As a consequence of the decreasing returns to capital, growth can only be explained in the medium term, that is, only in the phase of adjustment to the long-term steady state. In the long term, the growth rate is determined by exogenous technical parameters (i.e. by exogenous labor-augmenting technical progress). This modeling approach in particular encompasses the standard "Solow"-model with a constant savings rate and the so-called "Ramsey-Cass-Koopmans"-model, which is based on intertemporal optimization. Using the "Ramsey-Cass-Koopmans"-approach, the state of the environment can be included in the utility function, so that abatement activities become desirable. A rise in the concern for the environment can be expressed by an increased weight of the environment in the utility function or by the introduction of taxes to internalize the negative externalities of pollution. According to the assumptions regarding the production function, more environmental care does not affect the growth rate in the long run in this setting. But environmental policy leads to a different steady-state. If the use of capital generates pollution, the environmental policy induced steady state is characterized by a lower capital intensity. This is due to an increase in the price of capital relative to the other production factors. This price increase is a consequence of the environmental policy, which is aimed at a less polluting production process, that is, a taxation of the polluting production factors, namely capital (see Gradus and Smulders 1993).

Now assume, alternatively, that the natural environment has an impact on capital formation, since a natural resource is used in the production process. With increasing exhaustion of this natural resource, its price is assumed to rise such that the quantity used in production decreases over time. This is the standard assumption given that the natural resource stems from a fixed stock, that is, an exhaustible (nonrenewable) resource. If, in the one-sector approach, all inputs, notably capital, natural resources, and labor, are treated symmetrically, growth is only sustainable if the elasticity of substitution between the inputs exceeds a critical value (see Bretschger 1998b). With a constant savings rate as in the "Solow"-model, the critical value is exactly equal to unity. For the corresponding Cobb-Douglas production function, Stiglitz (1974) shows that a necessary and sufficient condition for a constant level of consumption (with no technical change) is that the share of natural resources is less than the share of capital. Formulated in terms of the required savings, the consumption level can be sustained exactly, provided all returns on the natural resource are used for capital accumulation. This is the so-called "Hartwick-rule". However, if savings are realistically assumed to depend on interest rates as in the "Ramsey-Cass-Koopmans"-model, the critical value must be strictly larger than unity. This is necessary in order to obtain constant incentives for investments, which means constant returns to capital when the quantity of capital is increased over time. A value that is larger than unity leads to a decreasing share of natural inputs and an increasing share of capital in total income.

The prerequisite of elasticities of substitution between natural and other inputs being larger than unity has attracted severe criticism. Many authors have argued that physical laws limit the extent to which physical capital can substitute for the natural capital stock. On the other hand, not only physical capital but also human and knowledge capital should be considered. For the last two capital components, the restrictions of material balances do not apply (see Smulders 1995b). In addition, in a multi-sector model further substitution possibilities emerge, namely the substitution between sectors, which results in a new determination of the critical values for the elasticity of substitution (see the QIDI-models below).

9.2.4. Model Type OCRC

Up to now, we have seen that if we assume decreasing returns to capital one cannot explain long-term growth rates. From now on, we will therefore assume constant returns to capital to obtain an endogenous growth rate. In the one-sector approach (OCRC), this specification leads to the so-called

"Rebelo-model" of economic growth theory, which, in our case, is supplemented by the natural resources and pollution aspects. Again, the use of capital is assumed to be polluting and the state of the environment is included in the utility function, so that abatement activities are desirable. As a consequence of the one-sector structure, no resource-saving knowledge can be generated and therefore no transition to a less polluting production process is possible. There is just one production factor in the model, namely capital. The term "capital" is sometimes interpreted in a broad sense, so as to include physical, human, and knowledge capital. In the "Rebelo-model", however, capital is introduced as an aggregate variable so that we have no explicit substitution possibility for any form of input. Income can be used for consumption, capital investments or cleaning activities. Let us first assume that factor productivity is not affected by environmental quality. Under the specification of this model, a decrease in pollution due to a tighter environmental policy (or an exogenous rise in the disutility of pollution) gives rise to more cleaning activities and, therefore, to a crowding-out of consumption and investments. The crowding-out of investments reduces the growth rate; therefore, a negative relation between growth and environmental care emerges (see Gradus and Smulders 1993). However, welfare is higher in the new steady-state because of a better environmental quality.

The dismal consequences of this model will no longer materialize as soon as there are productivity effects of natural resources or substitution possibilities in production. If one gives up the assumption that the environment does not affect factor productivity, a different result can be obtained. As before, more concern for the environment leads to more abatement activities and, consequently, to a better environmental quality. But now, assume that a cleaner natural environment implies an increase in the productivity of capital. This positive effect boosts the growth rate. If the positive effect is high enough, the negative crowding-out of growth-generating investments brought about by abatement activities can be offset. Thus, an increasing growth rate and a cleaner environment can be achieved under these assumptions (see Smulders and Gradus 1996).

Aside from the productivity effect, the substitution effect between several input factors is another possible ingredient in the one-sector approach. Assume a production sector with capital, natural resources, and labor as inputs, where the assumption of constant returns to capital still holds *ceteris paribus*. (This is an extension of the "Rebelo-model".) The resulting production function has increasing returns to scale. But this property does not change the results on input substitution as compared to the case of decreasing returns to capital. The reason is that a decreasing input of the natural resource affects the return on capital negatively such that the *ceteris-paribus* condition does not hold in this case. Regarding the effects of capital accumulation which is accompanied by a decrease in the use of natural resources, the return on capital decreases over time. Thus, the behavior of the ODRC-model carries over to the OCRC-model in this case.

To conclude the discussion on one-sector models, one can summarize that a direct connection between the long-term growth rate and the state of the environment only exists if constant returns to capital are assumed. Whether a cleaner environment or a more moderate use of natural resources, respectively, is compatible with an increasing growth rate depends on whether one assumes factor productivity gains from a better quality of the environment.

9.2.5. Model Type PID

We now turn to multi-sector models. First, we analyze models in which the environment may alter the productivity of the inputs used in the dynamic sector (PID) where growth is generated. Many authors focus on the so-called "Uzawa–Lucas" model. This model has a production sector producing consumption and investment goods and an education sector, where skills (i.e. human capital) are generated. Growth is driven by human capital accumulation, that is, without advancements in the education sector, growth peters out. As before, assume the use of capital to be polluting. What is the effect of environmental policy, for example, of the internalization of pollution, in this case?

In this simple setting, environmental policy influences only the marginal value of physical capital but not the efficiency of human capital in the education sector. Therefore, the growth rate of the economy remains unaffected by environmental policy. However, the capital intensity decreases in the steady-state due to changed relative prices of capital components. As in the one-sector case, the price increase is a consequence of the taxation on the polluting production factors. The only direct way to stimulate growth is a rise in the productivity in the education sector. A further possibility (which is in fact valid in all growth models) would be a decline in time preferences. This applies here because lower time preferences lead to more education and therefore to an increased output in the future. But normally, the size of the discount rate is not explained by economic theory, that is, it remains an exogenous variable.

If, however, one extends the "Uzawa–Lucas"-model in the sense that pollution also affects learning abilities, that is, the marginal returns to education, the effects of environmental quality on the growth rate are different. Assume pollution to decrease learning abilities (which has been observed empirically in regions with heavy pollution of the air, for example in Mexico City). Consequently, investments in human capital become less efficient with higher pollution. Similarly, it can be assumed that the depreciation rate of human capital is increased by higher pollution. As above, more environmental care leads to more abatement activities and to a better environmental quality. But now, the output per unit of human capital rises as a consequence of environmental policy which stimulates human capital accumulation and, thereby, the growth rate of the whole economy. This positive effect can normally not be offset by the crowding-out of investments due to more cleaning activities. The result is a higher growth rate and less pollution (see Gradus and Smulders 1993).

Another version of the "Uzawa–Lucas"-model does not include physical capital at all. Moreover, in this pure human capital version, growth is driven by human capital accumulation. Pollution is assumed to have a negative impact on both the productivity in the final goods sector and on the acquisition of skills. In other words, the engine of growth is again impaired by pollution. Positively formulated, a cleaner environment due to higher pollution taxes or more concern for the environment leads to an increased productivity in the learning sector and, therefore, to a higher growth rate (see van Ewijk and van Wijnbergen 1995).

Now let us consider that Bovenberg and Smulders (1995) model in which the environment affects the production function of consumption goods in two different ways. Environmental quality is, on the one hand, modeled as an input factor. It is introduced as the stock of natural capital which provides productive services to economic activities. The stock of natural capital has a natural regeneration rate, which increases the stock. On the other hand, the stock is diminished by the use of natural resources in production (or by "pollution", respectively). As a further element of this approach, resource-saving technologies can substitute for the use of natural resources (or for "pollution"). These technologies are generated in the learning sector. Also the learning sector is using natural resources as an input.

Two different cases can be distinguished: one in which only the environment (meaning the natural capital stock) enters the utility function and another in which the environment enters the production function, too. In the first case, environmental concerns give rise to a declining productivity of both kinds of manmade capital, that is, physical capital and resource-saving knowledge. This is because natural resources and manmade capital are substitutes, so that a decline of one input decreases the productivity of the other. As a consequence, we have falling rates of return on investments and a declining growth rate. This effect varies positively with the influence of natural resources on production of consumption goods and negatively with the influence of physical capital on production in the learning sector. In the second case, lower resource use exerts a strong positive impact on factor productivity in final goods production in the long run. This growth effect varies positively with the production shares of natural and physical capital, and negatively with the impact of natural resource use on capital productivity.

An extension of this approach is to assume that natural resources are essential for the production of consumption and private capital goods as well as for the production of public knowledge. In this case, a tighter environmental policy results in a reallocation of economic activity away from the pollution-intensive private goods sector towards the production of knowledge, which is assumed to be a substitute for pollution. Furthermore, anticipating the future benefits of a better environment and an increased productivity of accumulated assets due to better public knowledge, consumers demand more final goods. These two effects result in a fall of private capital accumulation. In the long term, however, more and more environmentally friendly technologies are developed so that the private goods sector recovers gradually. Taken together, the tighter environmental policy leads to a higher rate of return and a higher growth rate which is sustained by higher investments in private capital, knowledge capital, and natural resources; for transitional effects see Bovenberg and Smulders (1996). In a special case of this model, the environment is assumed to be a pure investment good. Using this specification, the long-term behavior does not change, but growth also rises in the short term. Due to rapidly accruing productivity gains, the crowding-out of investments in man-made assets can be offset immediately (see Smulders 1995a).

9.2.6. Model Type QIDL

In the previous section we have looked at multi-sector models in which the influence of natural environment on the production sectors and, therefore, on the growth rate occurred through factor productivity changes. As already mentioned, there is another way to model the growth effects of the environment. We now assume that the production sector and the growth rate are influenced by environment-induced changes in the quantity of inputs used in the dynamic sector of the economy.

In many growth models, the available working time of the individuals is constant, that is, a fixed labor supply is assumed. For example, in the Uzawa–Lucas approach, time can only be spent either on the production of final goods or on studying, but leisure is disregarded in this model setup. Including leisure in the model (which leads us to the QIDL-model) means that the available time for productive activities is no longer constant, since leisure is not productive. However, leisure enters positively in the utility function.

The question now arises as to whether the introduction of leisure affects the time allocation. In particular, we have to ask if a tighter environmental policy, for example, the introduction of a pollution tax, increases the time spent in education. A higher pollution tax leads to more abatement activities which requires resources from the consumer sector. Decreasing consumption leads to higher marginal utility of consumption. To equate marginal utility of leisure and consumption, households choose to increase studying time and to decrease leisure. Put differently, individuals aim to counteract reduced consumption possibilities by reducing leisure time, while, indeed, spending more time studying, which increases human capital accumulation and the growth rate. Therefore, one conceivable result of this model setup is that a tighter environmental policy has a positive impact on economic growth (Hettich 1998).

9.2.7. Model Type QIDI

Another mechanism of varying quantities of inputs used in the dynamic sector works through investment incentives (QIDI). In multi-sector models manmade capital is not only an input into production but also an output of specific sectors in an economy. This fact has a large influence on the sustainability results. By modeling this double role of capital, one can show that it is the substitution between primary inputs (such as labor and natural resources) as well as the substitution between different sectors of the economy that matter for the long-term development. To derive this effect, it is assumed that the dynamic R&D-sector is relatively extensive in the use of natural resources.

Regarding the incentives to invest in the capital-producing sector, that is, in the R&D-sector which produces knowledge capital, there is a cost and a reward effect. Following an increase in the price of natural resources, costs in the dynamic sector decrease, provided that labor is a bad substitute for the natural resource (see Bretschger 1998b). So, viewed from the cost effect, a low elasticity of substitution between primary inputs does not prevent the economy from remaining on a sustainable growth path. In standard growth models, which use monopolistic competition to determine the rewards of the dynamic sector (e.g. R&D), rewards of investments remain unaffected by natural resources (see Bretschger 1998b). If, however, capital (e.g. knowledge capital) can substitute for natural resources, the rewards for investments increase with rising prices of natural resources. In this case, the positive incentive effect of rising resource prices on capital investments is strengthened.

Thus, it can be demonstrated that, while in a one-sector setup of the model, growth is only sustainable if the elasticity of substitution between natural and other inputs is larger than unity, in multi-sector models the elasticity may be smaller than unity in some sectors without making growth unsustainable. This suggests that the trade-off between long-term economic development and the protection of the environment is, under realistic assumptions, smaller than commonly postulated.

9.2.8. Green Tax Reform and Growth

A widely discussed issue today is the combination of an increase in environmental taxes and a parallel decrease in taxes on other factors, the so-called "green tax reform" (for an extensive survey of open economy models analyzing the effects of such a tax scheme, see Smulders, Chapter 8). In the contribution of Bovenberg and de Mooij (1997), there are two channels through which an environmental tax reform may yield not only an improvement in the environment but also a higher growth rate, that is, a so-called "dynamic double dividend". The first channel is effective due to a positive environmental externality on production; the second channel operates by a shift in the tax burden away from the return on capital accumulation towards profits. The second channel, contrary to the first, only works if the elasticity of substitution between pollution and other inputs is not too high, so that the base of the pollution tax is inelastic.

Considering several taxes simultaneously, Hettich (1998) and Hettich and Svane (1998) investigate the interaction between public finance, endogenous growth, and the environment. Whereas in both models human capital accumulation in the education sector is assumed to be the engine of growth, Hettich (1998) treats leisure as an endogenous choice variable and Hettich and Svane (1998) assume leisure to be given exogenously. Without governmental intervention in favor of natural resources the effect in both models is that pollution is too high, too few abatement activities are undertaken, and final goods production is too capital intensive.

In the endogenous leisure model, a pollution tax, contrary to a tax on consumption, capital or labor, leads to increased long-term economic growth as well as increased welfare as long as the tax is below the Pigovian level. One reason for the positive effect of the pollution tax on growth is that leisure is reduced in favor of studying time. Consequently, more human capital is accumulated and the growth rate increases. In the model with given leisure, factor income taxes reduce growth. A pollution tax, however, leads to a stimulation of the growth rate if the productive spillovers of a better environmental quality are high enough. Due to the increasing quality of the environment, welfare is improved as long as pollution exceeds the optimal level.

If the labor market is added to the model, one can analyze not only the interactions between environmental quality and long-term growth, but also employment effects (see the approach of Nielsen *et al.* 1995). These authors assume a further (rather special) ingredient, namely that the productivity of pollution abatement activities varies negatively with environmental quality. Two kinds of exogenous changes are considered. First, if there is a shift towards more concern for environmental quality, the optimal pollution tax rate increases, whereas the optimal level of abatement expenditures decreases due to the falling productivity of pollution abatement activities. The employment effect is ambiguous, that is, it depends upon the tax regime. Second, the tax regime may shift from a command-and-control tax regime, under which the firms can pollute the environment free of charge up to the limit given by the standard, towards a pollution tax regime, under which environmental property rights are assigned to the public and all firms are forced to pay for the services from the environment. This change implies an improved efficiency of environmental regulation. As a consequence, both the growth rate and consumer welfare increase, without any adverse effect on environmental quality. In addition, there are employment gains since the pollution tax revenues allow a reduction in the labor tax.

9.3. GROWTH IN OPEN ECONOMIES

9.3.1. Trade and Dynamics

Trade theory is mainly concerned with static models. If capital accumulation is introduced, a middle-term adjustment process to the steady-state with the help of the neoclassical growth model can be analyzed. Another possible direction in research is to look at the consequences of exogenous growth on the foreign trade position of a country. Here, how international trade is affected by economic growth abroad (increasing consumption demand from foreign economies) or by domestic growth (better production conditions in the domestic economy) can be studied. But as in the closed economy case, the prediction of sustainable development paths in open economies requires a theory of endogenous growth.

As has become clear in the last section, endogenous growth requires constant returns to capital. Constant returns are normally motivated by the existence of positive spillovers, for example, knowledge spillovers in the sense of learning-by-doing. Furthermore, spillovers are assumed to increase with the extent of economic activities. Taking the example of research, this means that current research is the more productive the more research has already been carried out in the past (see Romer 1990). Put differently, the greater the knowledge capital is, the greater the advantage of future additions becomes. The same is true for other capital components like human capital and public infrastructure. Therefore, it is decisive for the international division of labor in what way the economies of scale of a country can be shared by the other countries, and inversely, in what way each individual country can participate in the economies of scale of the other countries (see Grossman and Helpman 1991).

9.3.2. Scale and Reallocation Effects

Since the possibilities to exploit scale effects are broader in the international context, new growth prospects arise from the establishment of outward relationships. The international transmission of knowledge certainly provides such prospects. If knowledge arises as a side product of different kinds of investments, the size of the geographical spread of the knowledge spillovers is important for the growth of open economies (see Young 1991; Ben-David and Loewy 1998; Bretschger 1997). Positive spillovers are also effective in the accumulation of other factors such as public services and human capital. Typically, however, the effect of public services is limited to the geographical region which belongs to the political unit considered. Human capital is internationally mobile only if skilled labor migrates over the country boundaries. In reality, the share of internationally mobile skilled labor is small in comparison to the total amount of skilled labor. If economies of scale remain partly or totally limited to a country or a region, distinct specialization patterns arise in the interregional and international division of labor. A country or region can gain economies of scale in special industries which are not or barely existent in other places. By analyzing the history of clearly defined regions of economic specialization, one may be able to show how specialization, for example, in financial services, heavy industry locations, high-tech commodity production, etc., is generated via accumulated scale effects. The beginning of such a development can often be traced to rather accidental circumstances.

Another crucial point in the transition from autarky to free trade is the trade induced change in the intersectoral factor allocation (see Bretschger 1999).

After the opening of an economy to free trade, a country with comparative advantage in traditional production will specialize in this field. In contrast, countries with comparative advantage in research or in high-tech production will use more resources in these sectors (while becoming a net-importer of traditional goods). If the positive spillovers are of different intensity for the different sectors and, additionally, are not fully international in scope, resource reallocation caused by trade has an impact on growth. If the learning-intensive sectors which generate extensive spillovers increase in size, the growth rate rises, while in the opposite case, the growth rate decreases. Trade thus might, under unfavorable conditions, decrease the growth rate (see Grossman and Helpman 1991; Bretschger 1997).

To evaluate the consequences of international trade for sustainability, the environmental dimension needs to be added. In the following section we therefore introduce pollution and the use of natural resources into the open economy growth models.

9.4. TRADE, GROWTH, AND THE ENVIRONMENT

In this section, we look at models of open economies which take the environment and long-term growth into account. In particular, we are interested in the interaction between trade, economic growth, and the environment. Thus, the link between the environment and growth on the one hand and between trade and growth on the other hand need to be considered simultaneously. Obviously, this is a challenging subject and only some of the immanent issues have been dealt with so far. In this section, we will discuss the existing literature and consider the remaining issues in section 9.6.

9.4.1. Small Open Economies

A representative model of a small open economy is constructed as follows (see Elbasha and Roe 1996): there are two final output sectors and a research and development sector; output is produced with primary inputs and differentiated intermediate goods. The output of the R&D sector consists of patents which contain the knowledge to produce new intermediate goods. Pollution can either be assumed to be caused by final output or by differentiated intermediate goods. The small open economy grows at a rate which is different from the world growth rate because of somewhat specific assumptions, such as the assumption of no trade in intermediate goods. In the model, trade alters the relative prices of final goods and, thereby, the dynamic behavior of the whole economy.

In accordance with the R&D-models of endogenous growth of the closed economy, one finds that, in this setting, long-term growth increases with a country's endowments in primary factors. This is due to the fact that a larger resource base leads to a larger research sector and increased spillovers. Moreover, the growth rate increases with the degree of market power of patent holders since this means increasing incentives for R&D-investments. The effects of environmental policy on growth depend on the elasticity of inter-temporal substitution of consumption. If this elasticity is greater than unity, growth of output is reduced after a tighter environmental policy has been implemented. If this elasticity is less than unity, a tighter environmental policy promotes growth. Finally, if it is equal to unity (logarithmic utility function), growth remains unaffected. A high elasticity of substitution means that agents find it optimal to choose a low level of environmental quality (see Aghion and Howitt 1998: 152). According to the empirical literature, the elasticity of intertemporal substitution is close to unity. According to Elbasha and Roe (1996), it is less than unity, so that the relation between environmental quality and growth would be positive in this model.

The effects of trade on the environment and welfare are ambiguous. They depend on the price elasticities of the supply of traded goods, on the terms of trade-effects on growth, and on the pollution intensities of the different sectors. Numerical exercises suggest that trade worsens environmental quality but enhances welfare.

The added value of this model as compared to the static approach consists in including the dynamic dimension in the determination of welfare. Being able to treat, in a consistent framework, the effects of trade on the environment and on growth simultaneously, policymakers become better informed about the whole variety of consequences of different trade policy measures.

Whether an endogenously growing small open economy is able to implement an independent environmental policy crucially depends upon the tax system (see Hettich and Svane 1998). Under a residence-based income tax system, which discriminates between domestic-source and foreign-source income, it is possible to implement an independent environmental policy which has an impact on the domestic interest rate. The impact of a change in the interest rate can only be fully captured in a dynamic setting. While a static model will only provide information on the input mix at a certain point in time, a dynamic model is able to show the influence of the interest rate on intertemporal deci-sions determining the growth rate. By using a residence-based income tax, a country is thus able to determine its own growth rate. Under a source-based tax system, however, where the after-tax interest rate equals the world interest rate, the government can no longer implement its own first best environmental policy.

9.4.2. Comparative Advantage

Considering the supply of natural resources, labor, and capital in rich and poor countries, Anderson (1993) explains why exports of backward countries will first be concentrated in primary products. In a second stage of development, according to his contribution, the comparative advantage of lagging economies will gradually shift to manufactures and, eventually, to services. In this way, it is possible for later-industrializing countries to export their way out of poverty. This development process is, however, often hindered by government interference. Governments of poorer economies tend to discriminate against the primary sectors and in industrialized countries declining industries tend to be protected. In many cases, protectionist trade policy measures are claimed to address environmental problems (see also Chapter 4 in this volume). The agricultural and mining (coal) sectors are, however, policy fields in which good examples of trade liberalization is liable to improve the state of the environment and, thereby, the basis for international sustainable development (see Anderson 1993, and Chapter 5 in this volume).

9.4.3. North/South Relations

A different effect of environmental policy on R&D-driven growth, results if only certain countries are able to innovate, whereas others are not. This asymmetry underlies in the so-called dynamic "North/South" models, which divide the world in a dynamic and innovative region, called "North", and a less developed region, called "South". Bretschger (1998c) analyzes the effects of an environmental policy undertaken by the North on worldwide natural resource use which causes global pollution and on the growth rate of the world economy. To determine the economic activities in the South, two versions of displacement of production from the North to the South are considered. In the first version, the South imitates the product designs of the North ("imitation hypothesis"). In the second version, the North has the opportunity to shift production to the South ("production shift hypothesis").

Rather surprisingly, the results obtained are very similar for the two versions. Under realistic parameter constellations, a decreasing pollution in the North (achieved by environmental policy) is not offset by an increasing pollution in the South. Even in cases where pollution in the South rises as a consequence of an increased economic activity in the South, the worldwide positive effect of Northern environmental policy on nature is still guaranteed. The growth effect of the decrease in pollution in the North is ambiguous. It depends on the flexibility of the production process, which is measured by the elasticity of substitution between natural resources and labor (being the other primary production input). If this elasticity is small, the worldwide growth rate increases. This is due to the cost effect in the dynamic sector and corresponds to the QIDI-models (see above). If this elasticity is large, worldwide growth decreases. According to the results of this model, the slowing-down of growth does not, in most cases, mean that the growth rate becomes negative. Furthermore, the analysis shows that the imitation effect, compared to the production shift effect, has a larger impact on the environment but a weaker one on growth. Worldwide sustainability is a very likely outcome under all possible scenarios. Thus, free trade among different world regions does not, in general, appear to endanger the goal of worldwide sustainability.

9.4.4. International Cooperation

It is obvious that international agreements on environmental standards have growth effects (see van der Ploeg and Ligthart 1994). The noncooperative outcome of a differential game for a global economy is characterized by an excessive use of (renewable) natural resources due to the apparent international environmental externalities. International policy cooperation results in a reduced use of resources, lower growth, increased welfare, and an improved environmental quality, unless there are positive international knowledge spillovers in production and/or international spillovers in public spending. However, the opposite result concerning environmental quality and growth cannot be excluded. If there are international knowledge spillovers in production and if public spending in one country benefits productivity in other countries as well, optimal international policy coordination can harm environmental quality and boost the economic growth rate. These results are derived under the assumptions that (i) the period of commitment is equal to the planning horizon, (ii) only two countries are involved in negotiating and (iii) the countries are identical. For a detailed discussion of these assumptions see Schmidt (Chapter 10 of this volume).

9.4.5. Trade and Resource Growth

The impact of free trade on welfare in economies with open access to renewable natural resources is analyzed in Brander and Taylor (1997a) and (1997b). The authors establish that the pattern of trade and the structure of production depend on a simple ratio of the biological resource and the country's labor endowment growth rate. Aside from the resource good, manufactures are also consumed. For a broad range of parameter values, the resource exporting country will not fully specialize in producing the resource good. The steady-state utility levels fall in this country as a consequence of a move to free trade because the open access externalities are aggravated by trade, that is, by an improvement in the terms of trade. While the intuitive link from low resource-management standards to increased resource exports and lower welfare can be portrayed in this model, the authors show that this link does not emerge under various conditions. If we assume that harvesting of the resource becomes more difficult as the stock is depleted, productivity in harvesting rises with an increasing stock. A well-managed resource is then relatively cheap in the long run and a conservationist country may well be able to obtain a comparative advantage in the resource good. According to the authors, when introducing trade with a country that has a very poor resource management, the conservationist country exports the resource good and both countries experience an increase in welfare. If, on the other hand, the nonconservationist country overuses its resource to a lesser extent, the result is reversed and the conservationist country is not compensated for its resource management but experiences a welfare loss, as does the non-conservationist country.

9.5. EMPIRICAL EVIDENCE

Summarizing the behavior of the surveyed models, the impression of an ambiguous relation between the state of environment and economic growth emerges. The more complex models, which include elements such as different sectors, international trade, and abatement activities, suggest a negative relation between natural resource use and long-term growth. Under these circumstances it is paramount to look at the empirical relation between the income level and the state of the environment and, in addition, to compare the natural resource use and the growth performance of different countries.

9.5.1. Environmental Kuznets Curve

Looking at the correlation between per capita income and the pollution of the environment in an international comparison, pollution seems to follow a hump-shaped pattern, the so-called "environmental Kuznets curve" (see Grossman and Krueger 1995; Grossman 1995).[1] Structural changes in the composition of aggregate output, the replacement of old capital by new capital, and environmental policy are the most important reasons for decreasing pollution after a certain stage of development is reached (see McConnell 1997). The inverted U-shaped pattern especially applies to certain regional pollution effects, such as urban air quality and water quality in rivers, where abatement is relatively inexpensive. However, it does not apply to global pollution effects such as greenhouse emissions, commercial energy consumption, or municipal waste (see Moomaw and Unruh 1997; Rothman 1998). In addition, it has been shown that highly developed countries have been able to reduce their energy requirements by importing manufactured goods which used to be produced in the domestic economy at earlier stages of development (see Suri and Chapman 1998). The observed improvements in environmental quality might therefore well be a consequence, at least in part, at the increased ability of consumers in wealthy nations to distance themselves from polluting production.

9.5.2. Natural Resources and Growth

As has become clear from the theoretical analysis, economies with abundant natural resources do not need to grow faster than countries with only few natural resources. Sachs and Warner (1995) show for the period 1971–89 that economies with a high initial GDP share of natural resource exports tend to have low growth rates during the subsequent period. This negative relationship

---

[1] See also Chapter 3, section 3.4 of this volume.

holds even after controlling for initial per capita income, trade policy, government efficiency and investment rates, which are all considered to be important in explaining the growth rate. The authors mention motivation problems of individuals who get rich easy, increased rent-seeking behavior in resource-abundant economies, and the decrease of the manufacturing sector as possible explanations for their findings. The last argument refers to the motivation underlying industrialization policies in the 1940s and 1950s and the discussion of the so-called "Dutch Disease". The special attributes of industrialization are the high intensity of backward and forward linkages to the rest of the economy and the intensity of learning externalities, that is of positive spillovers. In some estimations, the authors find modest support for the Dutch Disease hypothesis.

Gylfason, Herbertsson and Zoega (1997) investigate the dynamic implications of natural resource endowments on per capita growth by approximating the supply of natural resources with the size of the primary and a secondary sector, in which primary goods (agriculture, fishing, forestry and mining) are produced by using an alternative technology. In a cross-section estimation with the explanatory variable initial GDP, initial share of the labor force employed in the primary sectors, external debt in proportion to GDP, real exchange rate volatility, initial primary and secondary school enrolments, and an Africa dummy, the authors find a statistically significant negative relationship between the size of the primary sector and the average rate of growth. The explanatory power of the education variable is reduced when primary sector employment or primary sector exports are used as additional explanatory variables. This seems to support the hypothesis that a preponderance of primary sector production tends to inhibit economic growth by discouraging investments in human capital or research and development. However, the measure for natural resource supply is very broad in this study. Moreover, the measure does not discriminate between natural resource problems and the well-known structural problems of farming and mining.

9.6. LESSONS AND OPEN ISSUES

9.6.1. Some Lessons

In the public debate, economic growth, the globalization of markets, and the increasing specialization of regions or countries are often viewed as major threats to the sustainability of long-term development. The surveyed literature reveals, however, that this negative view is not appropriate. Growth and free trade provide a variety of options for solving current environmental problems. This especially applies to global environmental problems. To be sure, the existence of different options does not mean that present generations automatically choose a sustainable development path. It should also be remembered that environmental problems are the consequence of market failures which are already present in a static representation of the closed economy. These inefficiencies may well be aggravated by free trade and economic growth, which has to be taken into account when formulating appropriate policy measures. Finally, for sustainability, the requirement of fairness regarding future generations has to be added explicitly to the general policy guidelines.

Most contributions to the literature in the field of environment, growth, and trade focus on the first two aspects and thus neglect the international dimension. What lessons for sustainability can be learned from this limited approach, and what results can be generalized for the general topic of sustainability in open economies? The most important lesson from the combination of growth theory and environmental economics is that economic growth and environmental care are compatible in principle. This statement is valid independent of the observation that certain natural resources are over-used in the present situation. Pollution is due to market inefficiencies and the current use of exhaustible resources hardly satisfies intergenerational fairness considerations. To better preserve the natural capital stock, it is therefore necessary to reduce the use of certain natural resources. On the other hand, sustainability requires that welfare does not decrease in the future. The decreasing amount of natural inputs therefore needs to be sufficiently compensated for by the accumulation of manmade inputs consisting of different forms of capital. The greater the saving effort of the present generation is, the easier the substitution of natural resources in production and consumption becomes. But saving means consumption renunciation and this renunciation is economically attractive only if the proceeds from saving and investment are sufficiently high. An adjustment of the relative prices under the title "removal of external costs" is already advisable in the name of present-day environmental protection. However, with internalization, we obtain optimal development paths but sustainability is not yet guaranteed. Even stronger measures and an additional acceleration of the substitution of natural resources in the production are necessary for future generations not to fall back to a lower utility level on optimal growth paths.

Traditional one-sector models conclude that conditions for the sustainability of long-term development are favorable, provided that the elasticity of substitution between natural resources and accumulated capital is high. Moreover, the existence of substantial positive spillovers is shown to be advantageous for sustainable growth. These preconditions may well be disputed. It is for this reason that an additional mechanism needs to be emphasized in this context. This mechanism, which is represented in multi-sector models only, is the continuous reallocation of resources between the different sectors of an economy. Learning-intensive sectors should increase their share of aggregate production while natural resource-intensive sectors should gradually shrink over the course of time. If one proceeds the realistic assumption that natural resource intensity and learning intensities are negatively correlated between the different economic sectors, intersectoral reallocation of resources becomes one of the most powerful instruments to achieve sustainability. In this way, sufficient new knowledge capital and human capital, which substitute for the natural resources in the long term, can be formed. New knowledge also allows one to realize a massive increase in efficiency when using natural resources.

The impact of trade on economic growth is determined by scale and resource reallocation effects. Analogous effects are working in the context of sustainable growth. The main difference is that economic growth often generates positive externalities while environmental problems generate negative externalities. Scale effects induced by international trade unambiguously support capital accumulation and thus the mechanics of economic growth. If the accumulation of capital harms the natural environment, it is necessary to tighten environmental policy according to the increased growth rate. On the other hand, it is possible that the additionally accumulated capital is a substitute for the use of natural resources, as is the case for human or knowledge capital. Then, a higher growth rate improves the conditions for sustainable development. One should note that the internationalization of the economy has, in the case of scale effects, only an indirect effect on the natural environment through the effect on growth. For the various resource reallocation effects caused by international trade, the impact on the environment and sustainability is more direct. The positive aspect of internationalization is that the principle of comparative advantage in international trade may improve worldwide conditions for knowledge production and for abatement of environmental damage. As relative costs of these activities are not the same across countries, sustainability can be achieved at the lowest economic cost, by means of international division of labor. On the negative side, there are cases where trade decreases the quality of the natural environment through the induced change in relative prices. As seen above, in small open economies and in the case of open-access renewable natural resources, this result can emerge under realistic scenarios. Nevertheless, in the case of renewable resources, if productivity positively depends on the natural capital stock, strong resource management rules increase a country's international competitiveness. Considering the global climate as a renewable natural resource, it becomes obvious that such a productivity incentive is not effective everywhere. For the greenhouse problem, the external costs are global but the effect of a single polluting country is small in comparison to the whole world. A comparable productivity mechanism cannot be assumed for non-renewable natural resources either, if the current use is found to be nonsustainable. The only way to decrease the worldwide use of these resources consists of international policy coordination (see Chapters 10 and 11 of this volume).

The lesson for environmental policy consists in the finding that appropriate tax instruments, usually summarized under the heading of "green tax reform", can improve the protection of the environment as well as produce additional economic growth. This double effect is made possible by a double market failure, consisting of negative environmental externalities and positive externalities (such as positive knowledge spillovers) in the growth sectors. Of course, direct instruments for internalizing the positive spillovers could be implemented as well. If this is the case, the growth stimulating effects of the green tax reform are radically diminished. If direct internalization is not under-taken, possibly for political reasons, the indirect way of correcting the growth deficit by increasing costs in growth-extensive sectors is effective and efficient. In this way, the government can provide benefits to spillover-intensive sectors without having to favor specific sectors or specific research projects. Thus, the allocation of resources within the dynamic sectors remains the decision of the firms (on the "double dividend" issue, see also Chapter 8 by Smulders, this volume).

One should also note here that even a globalized world does not require all sustainability policies to be implemented at the international level. If environmental externalities are purely local or regional in nature, free trade and economic growth cannot be used as arguments against environmental policy. Regional or local policy measures are efficient in open economies, if certain conditions are observed. For example, given the mobility of capital, the tax system needs to be especially designed in order to be effective. In particular, if capital taxes are to be used to reduce pollution via production (capital), a residence-based tax system is called for. Otherwise, the pollution tax has no effect on the environment.

The lesson emerging from the analysis of North/South models is that the shift of production from the North to the South as a consequence of a stricter environmental policy in the North is minimal under reasonable assumptions. The reason is that the production shift itself is not free. Rather, it requires resources of the South if Northern products are copied; it even includes Northern resources, when production is actively relocated by Northern firms. In addition, every shift from North to South sets Northern resources free which can, for example, be used for additional knowledge production. The conclusion is that environmental policy needs to be adjusted to the conditions pertaining to trade and growth. An appropriately adjusted environmental policy is still highly effective in bringing about sustainable development paths.

A different conclusion must be drawn for trade policy. Generally speaking, trade policy is not a good instrument to achieve global sustainable growth. Many existing restrictions on free trade, for example in agriculture and coal-mining, are not in favor of sustainable development. Lowering the barriers between countries, by intensifying the forces of comparative advantage, decreases natural resource use and increases knowledge accumulation. In general, free trade enhances the substitution of natural resources which is crucial to achieve sustainability. The reason that trade is blamed for certain problems regarding the long-term development is that trade can, under certain conditions, amplify problems that are already present in autarky. It is obvious that an adequate solution of these homemade problems serves to obtain both the benefits of a better environment as well as the gains from trade.

An important related issue is the promotion of international knowledge diffusion. The nonrivalry property of knowledge and the massive progress in communication technologies represent favorable preconditions for a successful international transmission of knowledge and thus for promoting worldwide growth. International policy coordination is appropriate whenever externalities cross national borders or the exhaustion of natural resources has worldwide consequences. Global environmental problems like the greenhouse effect require global policy coordination. Especially in the cases of rain forest protection (see also Chapter 6, this volume) and the protection of specific species (biodiversity), the losers from preservation should be compensated. Knowledge and capital transfers to poorer countries which are abundant in the supply of these natural resources are also good development policy instruments. If knowledge transfers are effective, Third World countries are in a better position to realize sustainable development paths.

9.6.2. Issues Remaining for Research

Sustainability is not only a worldwide political objective, environmental quality and economic growth are also both largely influenced by the economic relations between economies and world regions. We are thus led to conclude that combining elements of growth theory, international trade, and environmental economies represents a very promising field for further economic research. Many existing results of growth theory are only valid for closed economies. For open economies, one should try to confirm, reject, or refine the existing results. In any case, whether working with models of closed or open economies, in the future more effort should be spent on investigating which kinds of capital, under which circumstances, substitute for natural resources. Only with a solid grasp of this substitution process can we ever hope to obtain appropriate results for open economies in particular. The instances in which economic growth and environmental protection complement each other should be better identified by further research. Whenever pollution diminishes the productivity in the dynamic sectors, such as education and research, abatement measures reduce pollution and promote growth at the same time, which is an especially favorable constellation to implement environmental policy.

The findings applying to small open economies need to be generalized to capture large open economies with flexible prices and wages. It would be of great advantage to have a generally accepted model for international sustainability analogues to the traditional $2 \times 2 \times 2$ trade model. Such a framework could intensify and focus the discussion on certain crucial topics which proved to be very productive in other strands of economic theory. The different forms and impact of international knowhow transfers also require more careful study in the future. Moreover, a more subtle analysis of the implementation of environmental policy should be a focus of further research. For example, the joint implementation of environmental policies in the international community seems to be one of the most efficient ways to obtain worldwide sustainable development. To be successful in this area, the dynamic consequences of such agreements should be better understood.

A further issue is the treatment of risk. Since information on the impact of economic activities on the ecosystem is incomplete, the methods of decision-making under uncertainty should be better integrated in the theory of sustainable growth in open economies. For example, uncertainty and irreversibility may provide guidelines for the substitution of natural resources which greatly differ from the ones obtained under complete information. This applies especially to the sustainability objective of intergenerational fairness whose policy implications crucially depend on the available information with respect to the long-term development prospects.

Therefore, economic theory should aim at improving our understanding of the future consequences of environmental policies today.

10

Incentives for International Environmental Cooperation: Theoretical Models and Economic Instruments

CARSTEN SCHMIDT

10.1. INTRODUCTION

The 1990s were imbued by a growing concern for international environmental problems. This was reflected by political debates and activities at the international level such as the United Nations Conference on Environment and Development (UNCED) 1992, in Rio de Janeiro. At the same time, these issues were also intensively discussed in the academic sphere. The economic analysis of environmental problems has been investigated already in the 1960s and 1970s. Nevertheless, the increasing preoccupation with transboundary externalities led to a renaissance of research in environmental economics and to a new body of literature. The basic reason for a reformulation of the research program has been concisely summarized by Carraro and Siniscalco:

Standard solutions for [transboundary, C.S.] environmental externalities are therefore not available, and the protection of the international commons is left to voluntary agreements among sovereign countries. It is precisely this fact which requires a shift in our analyses, from a literature on government intervention to a literature on negotiation between nations and international policy coordination. (Carraro and Siniscalco 1992: 381)

Transboundary environmental externalities do not represent an exceptional class of environmental problems. They rather can be seen as the standard type of detrimental externalities between individuals sharing common natural

This chapter originated while the author was member of the *Sonderforschungsbereich* 178 "Internationalization of the economy" at the University of Konstanz. Financial support by the *Deutsche Forschungsgemeinschaft* is gratefully acknowledged. I would like to thank Michael Finus, Jörg Schmidt, Günther Schulze and Heinrich Ursprung for helpful comments and suggestions. All remaining shortcomings are my own responsibility.

resources. The environmental impact of economic activities does not stop at politically determined national borders, but depends on complex biological and physical regularities. Consequently, the spatial dimensions of environmental problems are often not congruent to the areas of political jurisdiction, and national governments are not in a position to pursue a centralized regulation. These problems must be addressed at the international or even global level by voluntary agreements among the countries concerned.

Many environmental problems share the features that a great number of countries are involved, which differ substantially with respect to economic, environmental, and other characteristics. These features make it difficult to coordinate environmental policies effectively. Although voluntary cooperation is by definition welfare improving, individual countries may, for two reasons, nevertheless lose by participating in an international environmental agreement. First, when countries are very asymmetric with respect to the benefits and costs of emission abatements, some of them may not profit from environmental cooperation at all although they contribute to pollution and therefore, should be part of a cooperative solution. This obviously holds for unidirectional externalities. In such cases, compensation payments to upstream countries are required in order to reach some kind of "Coasian" bargaining solution. In the case of more than one polluting country, the additional problem arises as to how to allocate the measures in a way that minimizes overall abatement costs. Again, the cost-effective allocation of abatement efforts may not be profitable for some countries without compensation. Second, even in the extreme case of identical countries, each single government has an incentive to abstain from an agreement as long as its own abstention does not cause a breakdown of environmental cooperation. The public-good character of environmental policy implies that outsider countries benefit from the measures taken by the cooperating countries without incurring any costs themselves. This incentive to free ride on the efforts of other countries is stronger, the more countries are involved, because the behavior of a single country then is only of minor importance for the cooperative outcome. In many situations it is thus difficult to create the necessary *participation incentives* so that all relevant countries join an environmental treaty.

An additional problem of cooperative approaches in international environmental policy is the lack of a supranational authority that can enforce formal agreements between sovereign states. This implies that national governments cannot credibly commit themselves easily to the obligations and actions stipulated in an environmental convention. The same free-rider incentives that lead countries not to participate in coordination of environmental policies also lead to an intrinsic instability of agreements. The stability problem is especially severe when many countries are involved, but it is not trivial even for a few countries. Therefore, one also has to provide appropriate *compliance incentives* which render cooperation incentive compatible once an environmental convention has come into force. From a global perspective, the basic economic problem thus is to maximize aggregate welfare gains from international environmental cooperation subject to the above incentive constraints. This translates into the question of how an agreement should be designed with respect to the international environmental standard, how the required policy measures should be assigned to the different contracting parties, and which additional treaty provisions and measures should be adopted to increase effectiveness and stability of the treaty.

In the following, I address the above questions by giving an up-to-date survey of the economic models which analyze the incentives for international environmental cooperation and by categorizing the different instruments proposed in the literature to support incentive compatibility. The survey does *not* claim to reflect the entire spectrum of the literature on transboundary environmental externalities.[1] For example, important aspects of international pollution control such as imperfect and/or incomplete information are reviewed only briefly. Moreover, all models discussed in this chapter assume countries to be "unit actors" whose governments maximize national welfare.[2] This survey focuses on the economic instruments that stimulate and sustain cooperation on the internalization of transboundary or even global environmental externalities.[3] It is thus focused on contributions which explore possibilities to compensate for the fundamental lack of institutional structure on the international level.

Instruments for promoting and stabilizing international environmental cooperation generally influence both the incentives to participate and to comply. Some provisions, however, even produce a conflict between cooperation incentives *ex ante* and *ex post*. There is thus no point for grouping conceivable strategies according to the type of incentives they affect. Instead, we classify the instruments according to how narrowly they are related to the environmental policies of national governments. Beginning with the choice and design of the internalization instrument itself, we gradually enlarge the strategy space of the players (the national governments) and increase the complexity of the underlying decision problem to discuss additional instruments. The strategy space is expanded by adding the time dimension to the decision problem (i.e. by assuming repeated decisions on pollution abatement) and by allowing for various forms of compensation as additional instrument variable. The underlying decision problem is extended in order to take utility interdependencies (i.e. additional arguments in utility/welfare functions) and interdependencies

---

[1] Mäler (1990) provides a comprehensive taxonomy of international environmental externalities and an excellent introductory survey on important theoretical aspects of their internalization through international cooperation. For a brief overview, see Carraro and Siniscalco (1992), and for a policy-oriented survey, Verbruggen and Jansen (1995).

[2] See Schulze and Ursprung (Chapter 4) for a survey on the political economy of international environmental policy. Congleton (Chapter 12) discusses political-economic aspects of international environmental agreements.

[3] We do not consider arguments that call for an international coordination of policies even in the case of purely national environmental problems (see e.g. Hoel 1997b; Kox and Tak 1996).

with other markets (i.e. general equilibrium effects) into account. Combinations of these extensions and of related instruments are certainly possible and realistic, but we discuss them separately to clarify the argumentation.

The structure of the survey is as follows.[4] In section 10.2 we discuss the influence of the choice and shaping of the internalization instrument on the incentives to engage in environmental cooperation. Section 10.3 expands the strategy set to multiple periods and considers treaty provisions that make future abatement efforts of observant parties dependent on the (potentially defecting) behavior of other countries. This way of warranting incentives for cooperation exploits the fact that environmental policy decisions are repeated and is referred to as *internal stabilization*. In section 10.4 the strategy space is expanded to compensations and sanctions of various forms (monetary side payments, issue linkage, trade sanctions etc.) to induce cooperation. They are labeled *external stabilization* instruments since they are not necessarily restricted to cooperation on environmental externalities. Section 10.5 investigates how unilateral and accompanying measures of single countries or sub-coalitions cohere with the incentives to participate in and comply with an environmental convention. These measures may be motivated by special preference structures (altruism, reputation, social norms etc.) or by general equilibrium effects of abatement activities. Finally, section 10.6 briefly addresses long-term measures in the form of flexible adjustments of an agreement to changing circumstances and of improved framework conditions for international negotiations.

10.2. THE CHOICE OF THE INTERNALIZATION INSTRUMENT

In general, one can distinguish between policy instruments for the internalization of environmental externalities that are market-based and those of the command-and-control type. The former comprise emission taxes or tradable emission permits, the latter emission quotas, technology standards or other forms of direct regulation. It is a central result of the theory of environmental policy that internalization by market-based instruments is the superior policy option since it guarantees the realization of a given environmental standard at lowest social costs. By contrast, uniform solutions like fixed quotas usually are (i) not cost-effective in a static sense, that is, the same environmental effect could be achieved at lower social costs through a different distribution of abatement efforts across polluters; (ii) they are inefficient in a dynamic sense because they do not create incentives to reduce emissions even further by investing in new and cleaner technologies. From the perspective of global efficiency, this holds for transboundary externalities as well.

---

[4] See also Heister *et al.* (1995) who distinguish internal and external stabilization instruments and flexible adjustments to an agreement.

The first-best instrument to provide sufficient incentives for national governments to sign and comply with international environmental treaties are lump-sum taxes and transfers that work as a reward (sanction) for (non-) cooperative behavior. In a second-best world where international lump-sum transfers and taxes are not or only to a limited extent possible, the question arises as to which type of internalization instrument is most compatible with the incentive compatibility requirement. In the following, we look first at explanations for the widespread use of uniform quota solutions in international environmental policy. Subsequently, we introduce the market-oriented instrument of *joint implementation*, which is proposed as a supplementary element of a quota agreement in climate policy. The section closes with a discussion on market-based instruments by comparing international emission taxes with tradable emission permits and by presenting some suggestions for increasing the incentives to cooperate if these instruments are chosen in an agreement.

Up to now national governments have been very reluctant to implement market-based instruments for the internalization of environmental externalities. International environmental negotiations in most cases lead to uniform or inflexible solutions, for example, in the form of equal percentage abatement obligations. If quotas are without any doubt an inefficient instrument, why are they nonetheless an element of so many negotiations on international pollution control in reality? One can think of several reasons for this phenomenon:[5] First, uniform solutions are apparently "fair".[6] Moreover, negotiating complex and differentiated solutions is associated with high transaction costs and manifold informational problems. Asymmetric information on the valuation of environmental quality and uncertainty about the working of the ecosystem make it often very difficult to determine economically efficient, differentiated strategies for internalizing international environmental externalities.[7] In addition, when several agreements are possible (i.e. in the presence of multiple equilibria), simple rules such as uniform quotas may serve as a "focal point" during negotiations.

Whenever countries are heterogeneous and compensation payments are ruled out, there is a close relationship between the choice of the internalization instrument, the agreed-upon international environmental standard, and the incentives to cooperate (Barrett 1992). If asymmetric countries are assumed to negotiate on equal (percentage) reductions and the outcome is determined by the median country of the coalition, some but not all countries involved in the pollution problem will cooperate in equilibrium (Hoel 1992). The stricter the

---

[5] In the context of purely national pollution, it is known from the political-economic analysis of command-and-control policies that quotas may create rents whereas taxes do not. This may be an important explanation also in an international context, but is disregarded here.

[6] The aspect of equity and the impact of different principles of burden sharing, especially rules of equal sacrifice, on the incentives to sign a global climate treaty are discussed by Welsch (1992).

[7] Due to lack of space, we do not discuss these problems in detail. See e.g. Larson and Tobey (1994) for the role of uncertainty in global climate policy.

chosen environmental standard and the stricter the requirements on incentive compatibility, the fewer countries will find it in their interest to participate in the agreement which, *ceteris paribus*, decreases the global level of pollution reduction.[8]

Partial cooperative solutions do not only emerge when implementing a quota instrument is an exogenously given constraint. Such solutions also ensure (see Finus and Rundshagen 1998) when the choice of the internalization instrument is endogenously determined, when participation as well as compliance incentives are taken into account, and when the preferences of the marginal signatory (instead of the median country of the coalition) are decisive for the outcome of the agreement. Moreover, the negotiating governments in most cases agree on a uniform quota and not an emission tax although the latter is (due to its cost-effectiveness) preferable from a global perspective. However, from the perspective of the country that is the "bottleneck" in the negotiations and decisive for the terms of the agreement – the country with the lowest environmental preferences – the quota is superior to the tax. In the quota regime, all countries carry the same abatement burden in relation to their perceived abatement benefits, whereas under the tax regime the relative burden varies negatively with the intensity of the preferences for environmental quality.[9] This statement is in line with an analysis of Kverndokk (1993) suggesting that the poorest countries in the world would have the highest costs of reducing emissions relative to GDP when carbon emissions are allocated in a cost-effective way. Hence, although uniform quotas restrict the number of participating countries, the use of taxes or emission permits is an even greater disincentive for many countries to sign an international environmental treaty. Especially when national governments have to compromise on "the smallest common denominator", the quota instrument is likely to be chosen in an environmental agreement. Of course, these conclusions are only valid as long as international side payments, an international redistribution of tax revenues, or an appropriate initial allocation of emission permits are not possible (see section 10.4).

Given the multidimensionality of international economic relations, an additional argument exists for not pursuing an environmental policy that would be optimal in a first-best world. This is shown by Mohr (1995) in a general equilibrium framework with overlapping generations where two countries are linked to each other, not only by environmental externalities, but also via international debt. Although the countries may agree to implement a

---

[8] Correspondingly, in the case of only two asymmetric countries, uniform emission reductions have to be set at suboptimal levels for an international environmental agreement to exist at all (Endres 1993).

[9] The relative burden is defined as national abatement benefits compared to their costs. Due to the functional specification of national abatement costs assumed by the authors, all countries will reduce their emissions to identical absolute quantities under the tax regime, irrespective of their abatement benefits.

tradable-permit scheme, they will not necessarily trade emission permits in such quantities that marginal abatement costs are equalized across countries. This is the case if the lender debt is constrained because countries are sovereign actors, that is, if the loans are not as high as they were with full international enforcement of debt treaties. Yet, if the debtor country is a net exporter of emission permits, selling permits functions as a substitute for the procurement of capital via international loans. Consequently, it may be in the interest of the creditor country to reduce its demand for emission permits below the cost-effective level, thereby ensuring that the supply of capital in the debtor country does not exceed the level where debts are not settled. This is a second-best argument: since another market (the capital market) is imperfect with respect to the enforcement of contracts on the international level, it is not necessarily optimal to select the first-best instrument for the environmental problem. Market-based internalization instruments may not only be inferior with respect to the incentives they provide for environmental cooperation, they may also increase imperfections in other markets such as the international capital market.[10]

The use of quotas in international environmental agreements may give rise to substantial cost inefficiencies since countries generally differ with respect to many characteristics, thereby giving rise to different marginal abatement costs. In order to avoid to some extent the cost-inefficiencies of uniform solutions, some governments have pushed for at least a limited possibility of trading emission rights internationally. The most prominent example for this attempt is the concept of "activities implemented jointly" that has been put down in the Framework Convention on Climate Change. It stipulates that two (or more) parties to the convention have the right to implement emission abatement measures jointly if they find it in their interest. According to this provision, contracting parties with relatively high marginal abatement costs can fulfill (part of) their abatement obligations by purchasing abatement activities in countries with low marginal abatement costs. The realized abatement quantities in the other country are at least partially credited to the donor country. Purchasing abatement implicitly introduces international transfer payments and is equivalent to trading emission rights: under ideal conditions, the combination of a quota agreement and *joint implementation* brings about international cost-effectiveness. It thus corresponds to the approach of Baumol and Oates (1971) of setting first a global environmental standard and then selecting a cost-minimizing instrument to achieve this target. Even if the scope for joint implementation projects is restricted and search and monitoring costs are high, some authors (see Bohm 1994) regard it as a first step towards a future system of tradable emission permits. Under ideal conditions with perfect foresight and no transaction costs, there is economically no difference between

---

[10] However, Mohr (1995) shows that the introduction of cross-default clauses allows for a full strategic stabilization of the permit scheme. See section 10.4 for a discussion of this strategy.

joint implementation and emission permit trading. From an economic point of view the issue of incentive compatibility is crucial in the context of joint implementation as well. It remains unclear why there should be incentives for joint implementation projects while it was not possible to sign a contract including a market-based internalization instrument in the first place. So far, this aspect has not been sufficiently addressed in the discussion on the superiority of joint implementation.

A comparison of emission taxes and tradable emission permits yields that both instruments possess similar characteristics with respect to cooperation incentives. The differences between market-based internalization instruments rather lie in institutional aspects.[11] Market based instruments, for example, do not differ substantially from each other in the degree of national sovereignty that has to be delegated to a supranational institution for implementation, monitoring, and administration. Moreover, the question of how to redistribute the tax revenue of a global carbon tax is equivalent to the question of how to allocate initial emission permits (and future assignments) to the individual countries. If there are no restrictions on how tax revenues or initial permits are allocated to individual countries, the allocation rule may be designed to support the participation of countries which would otherwise abstain from cooperation. This is possible even if the international assignment of emission permits has to be based on simple rules, such as status quo emissions (*grandfathering*), cumulated historic emissions, current GDP or national population size. The same holds for rules that determine the redistribution of tax revenues to the cooperating countries. Each of these rules distributes the gains from environmental cooperation in favor of specific countries. Uniform percentage reductions, the grandfathering of permits, and an assignment according to current GDP generally favor industrialized countries, whereas an assignment according to population size (equal per capita emission rights) or cumulated historic emissions (i.e. giving developing countries the right to "catch up") would benefit the developing countries, in particular because of the subsequent trade of permits between industrialized and developing countries.[12] An attempt to increase the broad acceptability of a tradable permits solution could be to mix different allocation rules by constructing a weighted average of different criteria, the weights being adjusted over time.[13] In order to reconcile the interests of industrialized and developing countries, Pearce (1990) proposes to start out with a grandfathering regime and to change emission entitlements

---

[11] See e.g. Hoel (1997b: 122–4), or Zhang and Folmer (1995) for a comparison of the institutional and implementation aspects of tradable carbon permits and an international carbon tax.

[12] A study by Kverndokk (1993) estimates payments to the developing countries of 6 percent of GDP from the USA and and 3 percent of GDP from the other OECD countries in the year 2000.

[13] See Zhang and Folmer (1995: 139), for a formal presentation. Additional instruments to provide cooperation incentives possible in a dynamic framework are discussed in detail in section 10.3.

over time in such a way that the number of permits allocated to developing countries is increased, but by a number less than the number by which the permits allocated to industrialized countries are reduced. Cline (1992) expects that an agreement which shifts the weights over time towards the population rule would have the best chance of broad and lasting support. All of the above proposals redistribute the gains from environmental cooperation to grant compensations to certain countries. In the preceding discussion on internalization instruments, this option had been ruled out since it constitutes an instrument which provides cooperation incentives of its own and is treated separately in section 10.4 in more detail.

The discussion of market-based and command-and-control instruments in international environmental agreements has shown that uniform solutions often generate greater incentives for international environmental cooperation. It has also transpired that the distributional effects of the internalization instruments become predominant when international lump-sum taxes and transfers are not (or only to a limited extent) possible. Without these first-best instruments, globally efficient cooperative solutions will not be accepted by countries which are worse off under cooperation. With distortionary international transfers, the optimal cooperative solution will be second-best. In general, suggestions with respect to the choice of the instrument of internalization have in common that they focus on stimulating the incentives to participate in an environmental agreement and neglect the incentives to comply with the obligations each party has committed to.[14] In order to cope with this time consistency problem, additional instruments are required. One road to enlarge the government strategy space is to extend the time horizon of environmental policy, that is, by assuming repeated decisions on pollution abatement. If breaching an environmental treaty can be sanctioned, for example, through lower cooperative abatement in future periods, this "shadow of the future" is likely to make countries comply with their obligations more carefully. This strategy is discussed in the subsequent section.

10.3. INTERNAL STABILIZATION

Countries willing to cooperate in an international environmental agreement can exploit the fact that environmental policy measures are repeatedly taken. They can thus agree to sanction unilateral noncompliance with less ambitious internalization efforts in future periods. This behavior of the contracting parties makes an opportunistic government weigh the gains from breaching the agreement against the future losses from being sanctioned. The purpose of such a strategy is to provide sufficient incentives for complying with a treaty in cases when a supranational enforcement authority does not exist and other

---

[14] See Laffont and Tirole (1996) for an analysis of the impact of spot and future markets for tradable pollution permits on the potential polluters' compliance and investment decisions.

stabilization instruments are not available. Since incentives for cooperation are provided exclusively in terms of abatement activities – which are the core element of an environmental treaty – we refer to this strategy as *internal stabilization* (Heister *et al.* 1995). Internal stabilization corresponds to the principle of "reciprocity" in international law. In order to represent an effective threat, the reactions of observant parties to a breach of the treaty have to be both predictable and credible.

Dynamic games of international pollution control can be divided into models in which identical decision-problems are repeated (supergames) and models in which not only current emissions matter, but also accumulated depositions (*stock externalities*). With stock pollutants, we are in the realm of differential game theory. In this type of dynamic games *open-loop* and *closed-loop* (or *feedback*) strategies are distinguished. With open-loop strategies countries establish an abatement policy that is pursued forever because governments expect no new information on the other countries' actions and the aggregate emission level. With closed-loop strategies countries expect to receive new information as time passes and reformulate their policy on the basis of current information. For both types of strategies it can be shown that the globally efficient (full cooperative) allocation of abatement activities can be implemented as a subgame perfect Nash equilibrium as long as future pay-offs are not discounted too strongly (Dockner and Long 1993; Mäler 1991). The strategies prescribe that a country – when observing emission levels that do not correspond to broad cooperation – terminates its own cooperation. They are thus very similar to trigger strategies in repeated (super)games. In the long run, the stage game emissions in the cooperative and in the open-loop noncooperative equilibrium are the same. However, the convergence towards the efficient level is faster with cooperative strategies, and they result in a lower stock of externalities.[15]

Decisions on the provision of an international environmental good often resemble a repeated *prisoner's dilemma* type of game.[16] Just as in differential games, global efficiency can be sustained as a cooperative equilibrium as long as the future is not discounted too strongly. This holds for the infinitely repeated prisoner's dilemma (Folk theorem), but under certain conditions and for a subset of rounds also for the finitely repeated prisoners' dilemma (see Kreps *et al.* 1982). In both cases cooperation is sustained by the threat to abort cooperation if one party does not stick to the cooperative strategy. One can distinguish different strategies according to how severe the sanctions are.

---

[15] The subsequent exposition concentrates on models with flow pollutants. For the analysis of transboundary stock pollutants, see e.g. van der Ploeg and de Zeeuw (1991, 1992), van der Ploeg and Lighthart (1994), and Kverndokk (1994).

[16] For a different view, see Heal (1994), who considers technological spillovers and fixed costs of abatement policies that have reinforcing effects on the formation of a *minimum critical coalition* in an international environmental agreement. The above assumptions imply a coordination problem in addition to the free-rider problem.

The most drastic form of internal stabilization is the *trigger* or *grim* strategy. It implies a return to noncooperation once and for all if one country unilaterally defects. This is a very strong punishment which effectively deters free-riding behavior as long as renegotiations are ruled out. By contrast, if countries can renegotiate a new agreement after the breakdown of the initial one, announcing a grim strategy is not a credible threat; the sanctioning countries obviously harm themselves when returning to noncooperation forever. In addition, it may be technically impossible or economically too costly to return to non-cooperative emission levels. Hence, to be of practical use, trigger strategies must be both effective and credible.

Credibility may require that countries agree to *reoptimize* their cooperative abatement efforts after a breach of the international environmental treaty. The basic mechanism of this strategy is illustrated by Barrett (1994a) both for a one-shot and an infinitely repeated game.[17] In his model, N identical countries suffer from a global environmental ills. A subset of cooperating countries is assumed to lead in the sense of Stackelberg and to maximize their joint net benefits of abatement. Since joint net benefits depend on the size of the coalition, cooperative abatement is readjusted when a country joins or leaves the coalition. A unilateral breach of the agreement by a single country induces a lower level of cooperative abatement and has a sanctioning effect on the disloyal country. This sanction is credible because it maximizes the coalition's welfare. Joining the coalition is individually profitable because a new member benefits from the additional abatement of the other cooperating countries. On the other hand, a new member increases the incentive to take a free ride on the cooperative abatement efforts of other coalition members and to leave the agreement. Consequently, the number of cooperating countries remains limited. Barrett shows that a coalition of more than three countries is only stable when marginal abatement benefits decrease with the extent of global abatement, implying nonorthogonal best response functions. In general, a stable coalition with many countries only emerges if the difference in global net benefits between full cooperation and the noncooperative Nash equilibrium is small, that is, if there is not much to gain from cooperation. If there are large potential gains from cooperation, only very small coalitions of at most three countries are stable, irrespective of the total number of countries involved in generating the externality.[18] Aside from the inability of the reoptimization strategy to support a full cooperative solution, the model cannot predict which of the N countries cooperate and which not.[19] Moreover, as a defection by one

[17] The one-shot game also mirrors a dynamic structure of the decision problem, but assumes, for the sake of simplicity, that actions are immediately followed by reactions.

[18] Coalition stability is defined with the help of the concept introduced by D'Aspremont and Gabszewicz (1986) and Donsimoni *et al.* (1986) to study cartel stability in an oligopoly. In the oligopoly literature, similar results are derived with respect to coalition size.

[19] This problem could be resolved by introducing asymmetries between the countries, e.g. asymmetries with respect to the countries' relative bargaining power (see e.g. Barrett 1997a; Schmidt 1997).

country would induce entry by another, an effective reoptimization strategy requires that the coalition refuses the entry of new members and credibly commits to a suboptimal size. The limitations of this stabilization strategy in securing gains from stable environmental cooperation also remains in the context of a supergame (i.e. infinite repetitions of the stage game) if renegotiation proofness is used to portray stability (Barrett 1994a). Although the full cooperative outcome can, for sufficiently small discount rates, be sustained as a subgame perfect equilibrium of the infinitely repeated game, the sanctions that guarantee incentive compatibility may be vulnerable to renegotiation.

A stabilization strategy that avoids the incentive to renegotiate is *modified tit-for-tat*: employing this strategy, countries cooperate until one of them defects. Then the remaining countries exclude the defecting country and readjust their emissions. In addition, they do not readmit the defecting country before it has paid a fine or has made a front end abatement concession. The latter serves to compensate the countries in the coalition for the losses they incur by executing the punishment, thereby making the threat of their execution credible. The defecting country will pass under the yoke if it can expect sufficiently high gains in the future after all countries have returned to environmental cooperation. This punishment strategy hence eliminates the gains from noncompliance without inducing the observant parties to renegotiate with the defecting government. This kind of "carrot-and-stick" strategy is used by Finus and Rundshagen (1998) in a supergame in which asymmetric countries form a coalition to cooperate on abatement efforts. It is shown that international environmental agreements stabilized in this way can reap only small aggregate gains from cooperation if the externality problem is severe, that is, if many countries suffer from transboundary pollution and if abatement is relatively costly as compared to the perceived environmental damages. In these cases, only small subcoalitions prove to be stable.

Black *et al.* (1993) analyze a *minimum ratification clause* as an instrument to create incentives for environmental cooperation. This clause prescribes that the environmental convention does not come into force until a specified number of countries has ratified it.[20] In this case a rational government needs to consider not only the effect of its participation decision on the terms of the agreement, but also the possibility that there may be no cooperation at all if it does not join the agreement. Due to the assumption of incomplete information with respect to the net benefits of environmental protection, it is risky to abstain from the ratification process. This risk of treaty failure must be balanced against the expected free-rider benefits. Although the minimum ratification clause provides participation incentives, these incentives hinge on the assumption that countries do not renegotiate after having failed to reach a

---

[20] Just as other forms of "internal stabilization", minimum ratification clauses make national abatement efforts contingent on the cooperative behavior of other countries. The distinctive feature of such clauses is that they represent a sanction for noncooperation even before the treaty has come into force.

minimum number of signatories. Moreover, by assuming that the signatories remain committed to their obligations after the convention has come into force, the issue of compliance incentives is disregarded. In fact, the more successful the minimum ratification clause is in making a large number of countries sign the agreement, the greater are the incentives to breach it afterwards. Thus, such clauses generate a conflict between the provision of participation and compliance incentives.

To summarize, internal stabilization strategies are suited to generate participation and compliance incentives for international environmental cooperation, but only to a limited extent. Although it is easier to provide cooperation incentives when environmental policy decisions are taken under the "shadow of the future", the requirements for successful sanction strategies are strict and often not satisfied under real world conditions. The weight that is put on future benefits – expressed by the discount rate – is crucial for all internal stabilization strategies. The higher the discount rate on national welfare in future periods, the less effective is stabilization via retaliation in terms of abatement efforts. Imperfect observability of the countries' real internalization efforts and time-lags in the implementation of sanctions are detrimental to internal stabilization as well. Finally, retaliation, by adjusting emissions to arbitrary levels, may be technically impossible or economically unfeasible. New and "greener" technologies that have been developed and adopted in the course of international environmental cooperation may not be easily reversed.[21] Hence, international environmental agreements have to rely on additional stabilization instruments. These are discussed in the next section.

10.4. EXTERNAL STABILIZATION

Instruments providing external stabilization of international environmental agreements modify the payoff of the players via incentives which are not related to the reduction of pollution. These instruments may be used to provide incentives for compliance as well as for participation and can be implemented in various ways. The two basic methods of external stabilization are transfers and sanctions which are stipulated in an environmental convention and executed according to its terms. Both instruments enhance the incentives for cooperation, but in different ways. While sanctions reduce the individual gains from breaching an agreement, transfers redistribute the gains from cooperation in a way that increases cooperation incentives for certain critical countries. In an incentive compatible agreement, transfers will always be executed, whereas sanctions never take place (provided that there are no unforeseen changes to exogenous circumstances). Hence (utility) transfers and sanctions are basically dual approaches to creating incentives for cooperation: an agreed-upon transfer not given to a country because of its noncooperation represents a sanction for defecting behavior.

---

[21] See e.g. Carraro and Siniscalco (1993), fn. 3, for this objection.

We start the discussion of external stabilization instruments with a survey of contributions that analyze the general profitability of compensation schemes (subsection 10.4.1). Subsequently, various forms of transfers and other external stabilization instruments are presented. As will become clear, a strict distinction is often not possible. Nevertheless, we will devote separate subsections to *issue linkage* (subsection 10.4.2) and trade sanctions (subsection 10.4.3) because these instruments are in the forefront of the discussion. The section concludes with an evaluation of the external stabilization instruments (subsection 10.4.4).

10.4.1. Transfers

International compensations are an important instrument in international environmental cooperation. Their basic purpose is to redistribute the gains and burdens from a cooperative solution in a way that makes it attractive to many – if not all – countries to join an environmental agreement. In an early contribution, Markusen (1975b) has shown that, in face of a transboundary environmental externality, international transfers are in general a necessary and sufficient condition for a cooperative solution that yields a Pareto-optimal allocation of world resources. In contrast, without transfer payments, international environmental cooperation will generally not result in global efficiency, even if one assumes that the countries can make binding commitments. The reason is that, in these cases, the cooperative solution depends on the characteristics of the countries such as the countries' initial endowments with resources. The only way to reach broad cooperation without compensations is to resort to a less ambitious treaty that is not fully efficient (see section 10.2). In these cases, the efficiency and distribution cannot be treated separately and the *burden sharing* between the cooperating countries becomes a crucial issue of negotiations.

The question of how to share the burden of an environmental treaty has extensively been analyzed in the framework of cooperative game theory, that is, under the assumption that national governments can make binding commitments. The focus of these contributions is how to attract – by appropriate transfer and burden sharing schemes – the participation of new members to an existing coalition in order to generate additional gains from environmental cooperation. The analyzed regimes often base the burden-sharing on the intensities of the countries' environmental preferences. In this respect, the proposed cooperative solutions are similar to the "Lindahl prices" of pure public goods in a Lindahl equilibrium.

Eyckmans (1997), for example, analyzes a *proportional cost sharing* mechanism that distributes the total costs of emission reductions in proportion to the participants' marginal willingness to pay for the international environmental good. The proposed mechanism is shown to have the following properties. First, it yields an efficient (i.e. cost-effective) allocation of abatement activities. Second, its proportionality is widely accepted as a form of fairness in international negotiations. It reflects the idea that countries which benefit more from environmental quality should bear a larger share of the burden. Third, proportional cost sharing can be implemented as a Nash equilibrium under complete information with the help of a simple tax/subsidy mechanism.[22]

The adoption of a burden-sharing scheme is a problematic issue when the cooperating countries are heterogeneous. As burden-sharing will be anticipated by governments that are considering joining a coalition, different cooperative solutions may emerge depending on the adopted scheme. In a numerical simulation analysis calibrated to a data set with five world regions, Botteon and Carraro (1997) compare the outcome of negotiations under burden-sharing based on the Nash bargaining concept with the one based on Shapley values.[23] According to the simulations, the latter concept seems to be preferable in an agreement that uses transfers to expand a coalition. The reason is that burden-sharing according to the Shapley value provides cooperating countries with a more even distribution of the gains from cooperation. This observation underlines the importance of distributional aspects in comparison to efficiency aspects in international environmental negotiations, as has already been emphasized in section 10.2. The policy implications of this analysis remain somewhat unclear, though, since the adopted burden-sharing rule depends on the relative bargaining power of the governments which only have their national welfare position in mind when negotiating an agreement.

The above contributions have in common that they apply cooperative game theory to the problem of coordinating environmental policies. Although the results derived in this framework certainly provide valuable insights into potential cooperative solutions, in a cooperative game, the players are taken to be able to engage in binding commitments, an assumption that does not correspond to the lack of enforcement on the international level. On the other hand, the nonexistence of a supranational enforcement authority does not imply the absence of any institutional framework at the international level. This, however, is implicitly assumed when modeling the strategic interactions of sovereign states as a noncooperative game. In the latter framework, agreements have to be fully self-enforcing and, cooperative solutions which cannot rely on transfers or other additional instruments consist of small subcoalitions that achieve only minor welfare gains.[24] In contrast to the extreme assumptions on the enforceability of noncooperative and cooperative game theory, the existence and widespread use of international environmental institutions calls for the consideration of limited forms of enforcement in theoretic models. One

---

[22] See also Chander and Tulkens (1995, 1997) and Germain *et al.* (1997) for transfer schemes in cooperative games of international pollution control.

[23] In an analysis of heterogeneous countries, Barrett (1997a) also employs the Shapley value.

[24] In most cases, coalitions involving more than three countries are not stable. This result is quite robust with regard to the specification of the countries' welfare functions (see e.g. Barrett 1994a; Carraro and Siniscalco 1993; Hoel 1991, 1992).

way of doing this is by assuming that binding commitments are possible for certain groups of countries (see Carraro and Siniscalco 1993; Hoel 1994), and to analyze the use of side payments within this framework.

Transfers are an important instrument for making countries not only sign, but also comply with an international environmental agreement. This is shown for the case of identical countries by Carraro and Siniscalco (1993) who consider a one-shot abatement game of complete information in which the cooperating countries induce the accession of additional countries by giving self-financing transfers, that is, side payments that are financed out of the former coalition's gains from increasing the number of cooperating countries. To sustain broader coalitions by means of transfers, it is necessary, however, to introduce a minimum degree of commitment. This means that at least some players cannot deviate from the cooperative strategy they have voluntarily agreed upon and it implies that the agreement is not fully self-enforcing.[25] Carraro and Siniscalco (1993) analyze different forms of commitments which – although limited and less demanding than full commitment by all governments – can, under certain conditions, even give rise to a stable "grand coalition" (i.e. cooperation of all countries). The achievable gains from cooperation crucially depend on the assumed type of commitment[26] and on additional assumptions concerning the costs and benefits of pollution abatements.

Self-financing transfers can reap even greater gains from stable cooperation if the countries involved in the coalition formation process are heterogeneous. In such cases, some countries may not benefit from free-riding, but experience a welfare loss from environmental cooperation. Yet, for a preexisting coalition, the entry of outsider is often profitable because it can help to reach a negotiated environmental standard at lower costs. Side payments are the only way to create cooperation incentives for these low cost countries and to generate additional gains with the help of internationally efficient cooperative abatement policies (see Hoel 1994; Kverndokk 1993).[27] Using a model of heterogeneous countries, Petrakis and Xepapadeas (1996) show that even the global first-best optimum can be implemented as a cooperative solution through appropriate self-financing transfers to initial "outsiders". This result corresponds to the analysis of identical countries in Carraro and Siniscalco (1993)

---

[25] In the scenario of identical countries, the enlarged coalition would not be stable since paying transfers reduces the interest of the donor countries in the agreement.

[26] If the group of countries that are precommitted to cooperation is, for example, endogenously determined, the anticipation of receiving transfers reduces the incentive to sign an IEA and to commit to cooperation. In a model of identical countries where social norms influence a national government's participation decision, Hoel and Schneider (1997) show that total emissions may be even higher with side payments than without.

[27] Kverndokk (1994) simulates the gains from expanding an existing subcoalition that is committed to a joint carbon emission abatement policy (analogous to the scenario of *internal commitment* in Carraro and Siniscalco 1993) by compensating joining countries for the losses they incur from reducing their emissions to the cooperative level. Even if cooperation is only partial, the simulations show that substantial gains are attainable.

and is feasible as long as (i) a subgroup of countries (the donor countries) is committed to cooperation, (ii) this group maximizes global welfare, and (iii) the marginal pollution damages in the recipient countries are not too high.[28]

Schmidt (1997) shows that substantial gains from environmental cooperation are possible even if the use of self-financing transfers is restricted to compensating countries with low marginal abatement costs for their incremental costs of increasing abatement to cost-effective levels. The greater the initial cost-inefficiencies in the noncooperative equilibrium are, the higher the transfers that are used as an instrument to enforce the agreement. The enforceable welfare gains from stable cooperation therefore increase with the heterogeneity of the countries. Again, a minimum degree of commitment is required to render the cooperative solution stable. Instead of assuming binding commitments for a subgroup of the players, a "third party" (e.g. an international agency) is introduced. The agency collects the side payments before they are given to the recipient countries. This setup allows to analyze how credible commitments of sovereign countries can be achieved and illustrates the usefulness of international institutions. The game structure solves the time consistency problem without resorting to a precommitment on the part of the donor countries. The results demonstrate that substantial gains from international environmental cooperation are enforceable if existing institutions are properly used.

A related instrument is the deposition of securities with a third party under the condition that the funds are lost if the depository country does not observe the treaty. The deposition of securities can be accomplished without recourse to a third party by exchanging "hostages" or "pledges" (see Williamson 1983). Hostages are of value only for the depositing country; pledges are also valuable for the country that can dispose of the securities in case the depositing country has breached the contract. The more valuable the securities are, the higher are the cooperation gains that can be secured. The exclusive purpose of deposits is to enable credible commitments. In contrast, transfers are paid to create positive cooperation incentives and may also be used to secure compliance. If an environmental agreement includes side payments, and a trustee is available, it may be easier to agree on depositing these transfers instead of additional securities. The deposition of securities or transfers at an international agency can generate additional compliance incentives if the agreement provides that the retained deposits of defaulting countries are used to compensate observant countries for their costs of sanctioning noncompliant countries, thereby making these additional sanctions credible (Heister *et al.* 1995: 38).

---

[28] Petrakis and Xepapadeas (1996) furthermore develop a mechanism which enforces a cooperative solution even if monitoring is difficult in the sense that information on global emissions is public, but information on national emission quantities is not. Using this mechanism, every country has an incentive to report its true emissions.

We now turn to the various ways in which welfare transfers between countries are conceivable.[29] The most straightforward form is monetary transfers flowing either directly from a donor to a recipient country or being granted by a common fund of the donor countries. Unfortunately, the fungible character of cash creates incentives for opportunistic behavior if a strict earmarking of compensation payments cannot be guaranteed. The risk that received monetary transfers are not used for the purpose they were intended is given also in the context of an agreement on internationally tradable carbon emission permits (see Mohr 1991, 1995). A breach of the contract in the case of an international permit market may occur if a (developing) country sells its excess permits and uses the revenue to boost its economic growth, thereby expanding CO_2 emissions as well. Once the country has sold its excess permits, it loses interest in complying with the agreement and may start to emit without possessing the corresponding permits. Such an opportunistic country may even decide to borrow against future income from (leased) permits and breach the contract later. A similar risk of defection is given in agreements, which provide side payments to certain countries in advance and grant them a grace period of abatement obligations afterwards, as it is the case in the Montreal Protocol on substances that deplete the ozone layer.

In order to circumvent potential time consistency problems arising from the fungible character of monetary transfers, national governments may resort to *in-kind* or earmarked transfers. These cannot be used for purposes other than the one they are granted for without incurring retrading costs. If these costs completely cover the value of the in-kind transfer, the incentive for opportunistic behavior is entirely eliminated. Hence, in-kind transfers are a superior instrument in situations where institutional arrangements for ruling out the abuse of side payments are not available (see Stähler 1992). The concept of "joint implementation" stipulated in the Framework Convention on Climate Change makes use of this concept, as it earmarks side payments for the replacement of "dirty" by "clean" energy technologies abroad. The same characteristic of an in-kind transfer possess compensations given for the development and use of environmentally friendly, irreversible technologies that exclude an increase in emissions after the new technology has been implemented. By paying for the introduction of a "clean" but capital intensive technology, for example, a switch back to an old and cheap "dirty" one will become prohibitively expensive. This ratchet effect secures compliant behavior of recipient countries. Stähler (1993) shows that transfers for "irreversible" abatement technologies provide commitment options which render these technologies superior even if they are more costly than alternative reversible technologies. This form of transfer is the more attractive, the less the recipient country takes the future impact of the irreversibility into account. A low

---

[29] The contributions discussed so far treat side payments as *utility* or *welfare* transfers and abstract from the way in which they are given.

valuation of the future favors the donor because it decreases the component of the transfer that compensates the recipient for being locked into an irreversible technology.

10.4.2. Issue Linkage

The linkage of different and otherwise independent issues in international negotiations is another way of providing international compensations. In a second-best world where monetary transfers between countries are excluded, issue linkage may work as a substitute and allow for cooperative solutions where isolated agreements would otherwise not emerge.[30] Concessions in other policy fields that are on the agenda in various international negotiations (e.g. negotiations on other international environmental problems, trade policy, international debt, development assistance or the membership in a military alliance) may alter the payoff structure of the countries in a way that makes the participation in an international environmental agreement profitable. Issue linkage can thus be regarded as an implicit transfer between countries. Accordingly, the withdrawal of existing international privileges can be used as a sanction against noncompliance.

Cesar and de Zeeuw (1996) analyze issue linkage involving two different reciprocal environmental externalities within a dynamic bi-matrix game. For each of the two environmental problems, both countries are simultaneously polluters and victims of pollution. National costs and benefits of abatement efforts, though, are distributed asymmetrically across the two countries so that one country is, without additional compensations, worse off under isolated "cooperation" in comparison to mutual noncooperation. Depending on the payoff structure, the initial situation can be characterized either as an asymmetric prisoners' dilemma or as a *suasion game*. In the former game both countries have an interest in the negotiated issue whereas, in the latter only one country is interested.[31] For both games, cooperation can be sustained if the games are infinitely repeated and the discount rates are small enough. However, without side payments, the cooperative equilibrium will not support global efficiency. Assuming a second game that represents the exact mirror image of the game described above, Cesar and de Zeeuw (1996) show that, by linking the two offsetting games, the social optimum can be sustained with renegotiation-proof trigger strategies where noncompliance of a country in one agreement can be credibly punished by suspension of cooperation on the other issue.

---

[30] For the merits of issue linkage from a global point of view and general conditions under which Pareto optimality emerges see Carraro and Siniscalco (1995).

[31] In the two-player suasion game, the payoff structure is such that noncooperative behavior is the dominant strategy for one player, whereas "cooperation" is dominant for the other. Thus, in contrast to the prisoners' dilemma, in the noncooperative Nash equilibrium of the suasion game, one player behaves cooperatively and the other does not.

Folmer *et al.* (1993) illustrate the linkage of an environmental issue with a nonenvironmental one with the help of an *interconnected game*. They consider an example with two repeated prisoners' dilemma games: a pollution game with a unidirectional transboundary externality and a trade game. It is shown that playing the two games independently results in cooperation only if (i) the games are infinitely repeated, (ii) the discount factor is not too low so that trigger strategies are successful and (iii) if one allows for (monetary) side payments. The resulting aggregate payoff is lower than in the case where the two games are strategically linked to each other. The model demonstrates that issue linkage is especially attractive when countries are strongly asymmetric with respect to their perceived damage from transboundary pollution and when monetary transfers are not available. This is most obvious for the case of a unilateral externality, a scenario that naturally calls for compensation payments. Issue linkage in this model is optimal even if international transfers are feasible. This superiority of issue linkage in comparison to a cooperative solution with financial transfers, however, hinges on the assumption that offering a transfer implies a loss in terms of being labeled as a "weak negotiator", that is, a loss in reputation.

Carraro and Siniscalco (1997) analyze the linkage of negotiations on an environmental agreement and on technological cooperation by identical countries. In their model, environmental coalitions are profitable but unstable, whereas coalitions that cooperate on R&D are both profitable and stable. Linking the two issues increases the number of countries that participate in a stable environmental coalition because the gains from R&D cooperation offset the environmental free-riding incentives. This result is established for a specific form of the payoff function. The decision process consists of three stages. In the first stage, individual countries decide whether to participate in the linked agreements and a stable coalition is formed. In the second stage, optimal abatement levels of cooperating and noncooperating countries are determined, and, in the last stage, the firms in all countries choose their profit-maximizing output levels and R&D expenditures. Technological spillovers are modeled as an excludable positive externality between firms located in different countries.[32] It is assumed that the innovation spillovers are always larger between countries belonging to the coalition than between outsiders. This assumption is debatable because countries may cooperate on R&D activities independently from environmental cooperation. It is not clear why the latter spillovers should be smaller, at least as long as research activities are not connected in some specific manner with environmental policy. For example, if there are economies of scale in R&D cooperation, these may be realized not only by a coalition that

---

[32] This approach resembles the idea pioneered by Olson (1965) who suggested that access to an excludable club good is made dependent on the individuals' contributions to the supply of a nonexcludable public good. For instance, in the case of labor unions, membership is rewarded with extra benefits that are excludable to nonmembers.

simultaneously cooperates on environmental protection, but also by a sufficiently large coalition of outsider countries. In general, issue linkage is of greater relevance when countries are asymmetric, in which case they have different interests in the various topics dealt within international negotiations. Issue linkage in most cases entails bargaining on concessions in different policy fields giving rise to some kind of "package deal".

Mohr and Thomas (1998) analyze the prospects of issue linkage between international debt contracts and environmental treaties in the presence of uncertainty.[33] They consider the simultaneous existence of an international environmental agreement between a state and a multilateral (or foreign) environmental agency and an international debt contract between the same state and a foreign lender, both contracts lacking any enforcement mechanism. The compliance problem in environmental agreements corresponds to the repayment risk for lenders in international loan contracts and the expropriation risk of foreign direct investments. Uncertainty is introduced via an exogenous random cost of violating any of the two contracts. It is shown that compliance with both contracts can be guaranteed with the help of a *cross-default* contract between the environmental agency and the lender stipulating that the government cannot discriminate in its compliance between the two treaties. Cross-default clauses are often used to stabilize international debt relations. Instead of being a contract between several creditors, the idea here is to pool risks between parties that have different relations with the sovereign. The pooling of sovereignty risks in addition creates incentives to engage in *debt-for-nature swaps* that would otherwise not exist.[34] Induced swaps provide additional gains for the agency and the lender so that both may accept some loss in terms of higher risk from the cross-default contract. However, several limitations exist with respect to the pooling of sovereignty risks.[35] First, both the creditors and the environmental agency must have an incentive to agree on pooling compliance risks. This is given in some but not all constellations where the respective compliance risks are not too different and not too high. Second, the effect of the cross-default contract on the country's welfare position is indeterminate. Thus, the environmental treaty may have to include a clause which allows the pooling of sovereignty risks even without the consent of the sovereign. Finally, successful risk pooling requires the execution of the cross-default clause itself to be incentive compatible *ex post*. If the compliance risk was the same for all kinds of contracts, the risk pooling strategy would not be effective. Yet, in many cases it may be possible to protect cross-default clauses from the compliance problem if the parties contract under their national law, which is fully enforceable. The advantage of stabilizing environmental treaties

---

[33] See Mohr (1995) for the strategic linkage of international debt and pollution permit trade in the absence of uncertainties.

[34] In a debt-for-nature swap a country's debt is canceled in exchange for undertaking additional environmental policy measures.

[35] See also Kirchgässner and Mohr (1996), section 4.2, on problems of cross-default clauses.

in the above manner is that the long-term compliance problem which governments face is delegated to the private sector. In contrast to national governments, private firms or institutions are able to engage in binding commitments because they are subject to national law enforcement.[36]

10.4.3. Trade Sanctions

The discussion of cross-default clauses has illustrated that the threat to withdraw some preferential treatment or to terminate cooperation in another policy field is a special form of issue linkage. As compared to the previous approaches these actions represent sanctions, not transfers. The most prominent and widely discussed forms of sanctions are trade restrictions levied against noncooperative countries. Just as cross-default clauses, trade sanctions are often stipulated in an environmental treaty so that they can be properly anticipated.[37]

The potential of trade sanctions as a measure to stabilize environmental cooperation is analyzed by Barrett (1997b) in a partial equilibrium model with homogeneous countries and intra-industry trade. Barrett assumes that imperfectly competitive firms produce an identical output but segment their markets. Trade sanctions are used to enforce the cooperative supply of a global public environmental good. The threat to exclude countries from trade that do not cooperate on international pollution control is credible because the sanctioning countries gain from executing the sanction via increased firm profits. The decisions of the firms and the governments are modeled as a game with several stages. First, governments decide whether the environmental treaty should employ trade sanctions and, if so, under what conditions. Then, the countries simultaneously choose to become a signatory or a nonsignatory. In the third and fourth stage, first the signatories and than the nonsignatories determine their respective abatement standards. In the final stage, firms choose their segmented outputs according to Cournot–Nash behavior. Governments that do not cooperate take the abatement standards of other countries as given, whereby cooperating countries are assumed to maximize their collective welfare.[38] With the help of numerical simulations for agreements with and without trade sanctions, it is shown that in many (but not all) cases an agreement with trade sanctions is preferred and that the social optimum can be sustained with such agreements. Optimality requires the introduction of a minimum participation level to secure coordination on the full cooperative solution.

---

[36] Of course, this requires the existence of an international agency or other independent party which engages in cross-default contracts.

[37] The general influence of uncertainty and reputation on the effectiveness of sanctions as a stabilization device is discussed by Heister *et al.* (1995).

[38] This is not problematic since countries are assumed to be identical and the question of how to distribute the gains from cooperation does not arise.

Unfortunately, it is not intuitively clear why the execution of trade sanctions is also incentive compatible *ex post*, that is, after an agreement has been violated, by an individual country. Moreover, the results of the analysis need to be taken with a grain of salt, because the setup is very specific.

Global environmental policy will typically be linked to international trade policy also when stabilization is not an issue.[39] One reason for using the trade policy link is the "leakage" phenomenon, because leakage can be curbed by imposing appropriate tax and tariff adjustments.[40] Another reason refers to the problems of implementing regulatory measures in cooperating countries in which pollution takes place. These can be administrative difficulties or lack of enforcement on the national level, for example, in developing countries. Trade policy measures then can principally serve as a second-best instrument for international pollution control. Of course, trade restrictions often also serve protectionist purposes. Moreover, the causes for noncompliance sometimes do not lie in deliberate, but rather in endemic and erratic decisions. Multilateral trading rules should therefore not permit the parties to an (environmental) agreement to impose arbitrary trade restrictions. On the other hand, it can be argued that the political support for trade restrictions due to national protectionist interests can turn into an advantage because it increases the credibility of this instrument (see Kirchgässner and Mohr 1996: 210–13).

10.4.4. Evaluation

Summing up, the above discussion shows that issue linkage can substantially contribute to achieving mutual gains from environmental cooperation. This is especially the case when the welfare-improving removal of trade restrictions is used as a carrot to make countries participate in and comply with international environmental agreements. The general advantage of linking different policy issues consists of the fact that linkage mechanisms provide participation and compliance incentives for package deals in cases where isolated agreements would not be signed or would not be stable. On the other hand, linking previously unrelated topics in international negotiations may cause substantial negotiation and transaction costs. For example, a complication arises when the issues to be linked in an international environmental agreement are already regulated by international law as in the case of trade policy (rules of the WTO). In addition, substituting many isolated compliance risks by a small number of bundled risks may create the danger of a deep crisis in international politics, once a contract violation, for whatever (accidental) reason, has occurred.

The crucial precondition for the applicability of sanction schemes intended to sustain international environmental cooperation is that they are credible.

---

[39] See Kirchgässer and Mohr (1996) for a general discussion of the effectiveness, efficiency and credibility of trade restrictions to promote international environmental policy.

[40] See section 10.5 for a sketch of this problem.

After a unilateral breach of an environmental convention it must not be rational for the observant parties to continue cooperation with the defecting country. Given that, in most cases, sanctions will be costly for the punishing countries as well, it is not an easy task to satisfy the credibility requirement. This is one major aspect in which transfers differ from sanctions. Self-financing transfers are credible by definition.[41] The basic advantage of external stabilization by transfers is that it allows for a separation of national abatement efforts from related economic burdens. Thereby, larger and stable coalitions with a more efficient distribution of abatement activities across countries are attainable, resulting in a higher degree of internalization.

Even if feasible, transfers also have a number of limitations for the stabilization of international environmental agreements. Side payments may give rise to inefficiencies in as far as they are given by downstream countries to bribe polluting countries to internalize these externalities. In these cases, transfers imply the application of the *victim pays principle* rather than the *polluter pays principle*.[42] Although the distribution of property rights makes no difference for global efficiency in a static framework, the polluter pays principle is preferable from a dynamic perspective because it creates appropriate incentives for innovations in abatement technologies (see e.g. Mäler 1990: 82). It is true that the polluter pays principle has been adopted by various agreements for domestic implementation by national environmental policy, but its application on the international level is unrealistic given the status quo in which each sovereign state claims a right to pollute. Moreover, countries that are less concerned with environmental quality and thus represent potential recipients of side payments may reduce their own abatement efforts below the non-cooperative level if the anticipated future compensation is sufficiently high and outweigh the present cost of pollution. In such cases, strategic behavior will lead to a crowding out of cooperative and noncooperative abatement efforts (see e.g. Mäler 1990: 99). Especially when the choice of abatement technologies is political, it is difficult to determine the hypothetical scenario that would have been realized without any agreement. In many cases the anticipation of agreements creates incentives for prenegotiation behavior that is detrimental to environmental protection. For example, it may be a rational strategy for a national government which anticipates an international environmental agreement to deliberately chose a "dirty" production technology with high per unit costs of emission reduction, although a cleaner one with lower per unit costs is available (Buchholz and Konrad 1994).[43]

---

[41] Of course, this requires that donor countries can credibly commit themselves on the *execution* of transfers, an aspect often neglected in the literature.

[42] Issue linkage represents an implicit transfer from the polluted to the polluting countries and thus implies the application of the victim pays principle, the payment being made in these cases not in cash but in kind.

[43] Similar problems arise in the presence of asymmetric information between the donor and the recipient countries. Under these circumstances countries have an incentive to report private

To sum up, instruments for the external stabilization of international environmental agreements substantially contribute to making such agreements incentive compatible. This holds for the incentives to sign international treaties as well as for the incentives to comply with their obligations. Moreover, as the strategies discussed in this section are "external" to the environmental objective of an international treaty, they may, in principle, be applied in order to stabilize international cooperation in other policy fields as well. It must be emphasized, though, that in many cases even the combined recourse to all of the instruments that have been presented so far will not suffice to attain globally efficient cooperative solutions in international environmental policy. This is not only a consequence of the second-best character of the discussed strategies in comparison to the enforcement of treaties by a supranational authority, but follows mainly because the maximization of joint welfare is generally not compatible with the pursuit of the individual interests of single countries. Given the diverging interests with respect to national pollution control, the governments of some countries may therefore consider taking additional measures which go beyond what can be achieved by international negotiations. The effects of such unilateral and accompanying policy measures on the incentives to participate in and comply with environmental agreements are discussed in the following section.

10.5. UNILATERAL MEASURES

Cooperation with other countries is only one option of coping with international environmental problems, each government is free to undertake unilateral measures at any time. This may happen in situations without any agreement but also in different stages of negotiating and implementing a treaty if countries take up the role of pioneer. Two questions arise if unilateral policies are chosen to internalize transboundary environmental spillovers: (i) What are the motives of an individual country to invest in international environmental improvement? (ii) Which accompanying measures can an individual country or a subcoalition of countries take to protect their unilateral policies against attempts to dilute the intended environmental effect?

The first question refers to motivations to engage in international environmental policy which are not captured by those presumed in the previous sections. For example, *social norms* may play an important role not only for the explanation of individual behavior, but also for the decisions of national governments in the diplomatic arena (see Hoel and Schneider 1997).[44] Governments may fear to be blamed as opportunistic and noncooperative

information in a distorted manner. By doing so, recipient countries try to receive higher transfers, whereas donor countries try to free ride on the side payments of other donor countries.

[44] For the role of social norms, intrinsic motivations and altruistic behavior see e.g. Elster (1989), Holländer (1990), or Sugden (1984).

and therefore sign and stick to an environmental convention, even if free-riding pays in pure economic terms. Especially in the political debate, it is often proposed that certain countries should take the lead and adopt measures for the protection of the global commons independent of the behavior of the other countries. By "setting a good example", they may act as a catalyst and initiate a similar behavior from other countries because the laggards would lose reputation if they did not cooperate or simply because they will feel morally obliged to. Unilaterally acting countries may also win recognition ("moral leadership") in the international arena.[45] In addition, by implementing unilateral policies (e.g. through pilot projects or by promoting new and "greener" technologies), forerunner countries may provide valuable information which facilitate negotiations. Of course, the provision of this information causes a public-good dilemma and is subject to free rider incentives as well. Nevertheless, the marginal environmental benefit of one dollar invested in providing such information may be greater than the marginal benefits of investing in domestic abatement activities.[46]

Under the traditional assumptions regarding the actors' preferences, unilateral actions often are detrimental to the welfare position of the country undertaking these actions. Hoel (1991), for example, derives negative effects of unilateral abatement activities on a country's own bargaining position in a cooperative Nash bargaining game. Individual countries suffer welfare losses if they unilaterally engage in abatement activities before or during international environmental negotiations. Detrimental effects of prenegotiation behavior on environmental protection are also analyzed by Buchholz and Konrad (1994). In their model, the governments anticipate negotiations on the internalization of transboundary externalities. In order to strengthen their bargaining position, they choose an inefficient technology with high per unit costs of abatement before the negotiations start. The technology choice is also inefficient from a national perspective, but works as a commitment device for this country and therefore pays off. It is assumed that the choice of the (abatement) technology is irreversible and that the outcome of the negotiations is determined by the Nash bargaining solution. A country with relatively high marginal abatement costs will then have an advantage because it is globally efficient to assign a relatively small share of the abatement burden to this country.

Unilateral abatement measures may not only weaken the bargaining position during negotiations, they may even have a detrimental effect on environmental quality. This perverse result is established by Hoel (1991) in a Nash bargaining model with sufficiently elastic reaction functions and a marginal abatement

---

[45] Taking into account such reputational effects often amounts to expanding the set of arguments in the utility functions of the governments by including immaterial values.

[46] A reason for conducting a unilateral environmental policy that is not motivated by environmental concerns could be to obtain a competitive advantage in the development of new (and environmental friendly) technologies. See Porter and van der Linde (1995) for this argument and Palmer *et al.* (1995) for a critique.

cost function of the unilaterally acting country, which is sufficiently steep. The model does, however, not provide a rationale for a single country to engage in unilateral abatement activities. On the one hand, it is assumed that a unilaterally acting country does not behave according to its "true" best reply function. On the other hand, the "true" net benefits of the abatement activities are used in deriving the cooperative solution. If intrinsic motivations are present in the case of unilateral action, these motivations should also underlie the cooperative solution. If, however, the unilateral action is undertaken to manipulate the outcome of the negotiations, the true net benefits should be used in deriving the unilateral action. Finally, it is hard to justify why one does not assume the same behavior for (at least some of) the other players involved. Of course, in that case the perverse result would break down.

A different approach is taken by Endres and Finus (1998) who analyze the effects of increased environmental awareness (i.e. a change in preferences) on the incentives for international environmental cooperation and environmental quality. In their framework, unilateral measures are compatible with optimizing behavior of national governments. In one part of their analysis, the authors assume that the cooperative solution is characterized by uniform emission reduction quotas – which corresponds to the outcome of many real world negotiations on environmental agreements. They analyze the ecological effects of unilateral actions taken before the negotiations start, as well as the effects of over-fulfilling the agreed upon targets after the agreement has been signed. As in Hoel (1991), the result emerges that, in many cases, global emissions will *increase* as a consequence of unilateral environmental policy measures. However, the deterioration of environmental quality in this model is caused by strategically adjusted proposals and reduced abatement activities of countries that observe or anticipate unilateral measures by others. It is shown that the results depend on the stage at which the commitment of over-fulfillment becomes known to the parties involved and how they react to this knowledge. Hoel (1991) as well as Endres and Finus (1998) assume non-orthogonal reaction functions. Although this assumption is plausible, no supporting empirical evidence is available. The orthogonality assumption implies that the countries' abatement activities are interdependent even in the non-cooperative equilibrium.[47]

The second question raised at the beginning of this section refers to the problem of protecting individual or joint efforts to reduce transboundary pollution against detrimental reactions from outsiders. This is relevant for the provision of cooperation incentives for subcoalitions of countries because the more the environmental impact of a cooperative strategy is eroded by offsetting adjustment processes of outsider countries, the less attractive it is to participate in an international environmental agreement. Such leakage effects can work

---

[47] For a critical discussion regarding this scenario see e.g. Carraro and Siniscalco (1993: 323–5).

through two channels: direct leakage occurs if marginal abatement benefits decrease with decreasing pollution, indirect leakage occurs if, due to a changed price vector, general equilibrium effects lead to an international reallocation of polluting industries or to a change in (international) demand and supply of polluting goods.[48] In the context of global warming, the latter effect is called *carbon leakage*: Joint efforts of a subcoalition of countries to reduce their consumption of fossil fuels would reduce world market fuel prices and thereby increase fuel consumption in nonsignatory countries. In addition, an increased demand for imported goods whose production is fuel-intensive would increase the demand for fuel in the nonsignature countries even further.

One can think of different ways to counteract offsetting emissions (leakages) in countries with weak preferences for environmental quality. In principle, it does not make any difference whether such accompanying measures are taken by a single country or by a subcoalition of countries which coordinate their environmental policies. A crowding-out strategy put forward by Bohm (1993) is to reduce the fuel supply to nonsignatories by having signatories buy or lease sufficiently large fossil fuel deposits from the producer countries. If the unilaterally acting countries are suppliers of fossil fuels themselves, an alternative option is to reduce the international supply by domestic policies such as a tax on production and/or consumption of fossil fuels. Hoel (1994) shows that it is often impossible to determine whether a demand or a supply policy is superior for the subcoalition without additional information on the shape of the demand and supply functions in the "carbon market". In general, some combination of production and consumption taxes will be better than a single instrument policy.[49] However, to the extent that a single "large" country or a group of cooperating countries exerts monopolistic power on the world carbon market, the imposed policy is not likely to pursue only environmental goals. The policy will also be used to alter the terms of trade in a favorable manner. From the point of view of this optimal tax argument, the optimal policy mix will be to tax consumption and subsidize production or vice versa, depending on whether the coalition is a net importer or exporter of carbon in the equilibrium. Hence, the optimal carbon policy from the perspective of the unilateral actor depends on the relative importance the actor assigns to the environmental damages (climate change) and the gains from manipulating the terms of trade.[50] An alternative strategy would be to induce the cooperation of other countries in order to influence demand and/or supply abroad. This requires instruments which have already been discussed in the previous sections, in particular the

---

[48] See e.g. Felder and Rutherford (1993); Merrifield (1988).

[49] Golombek *et al.* (1994) investigate under which circumstances such a subcoalition should differentiate the imposed carbon tax rate across *sectors*.

[50] Similar results are obtained by Killinger (1996) in a general equilibrium framework in which one "large" country uses its market power to *indirectly internalize* transboundary externalities from abroad. See Böhringer and Rutherford (1997) for an argument against a unilateral German carbon tax that includes exemptions for energy- and export-intensive industries.

compensation of the free-rider gains which these countries forego (see Bohm 1993; Hoel 1994).

It remains to be noted that, in contrast to the legal meaning of the term "unilateral", it does not make any difference whether the measures discussed above are taken by a single country or whether they are the outcome of a coordination between a subgroup of countries. With respect to the external effects of the agreed upon measures, a coalition faces the same pattern as an individual country. Therefore, no strict separation can be made between cooperative and unilateral internalization strategies and the above considerations apply also for subcoalitions.

10.6. FLEXIBILITY AND FRAMEWORK PROVISIONS

The last category of instruments which create incentives for stable international environmental cooperation consists of measures which are effective in the long run. Such provisions, which accommodate new and previously unknown circumstances, increase the flexibility of an agreement. Provisions may also aim at ameliorating the general framework of international negotiations. At the time of negotiating and signing an environmental treaty, it often is impossible to foresee the future development of all relevant factors. If at some point in the future, important parameters such as income and technology change in a substantial and not foreseeable way, compliance with an existing agreement may no longer be optimal for some countries. In order not to endanger the whole agreement, treaty provisions may be included which allow a flexible adjustment in future periods without undermining the substance of the treaty. Examples of such flexibility clauses include the indexation of national obligations to central economic variables such as national (per capita) income or population size and the concession of *escape clauses* for special circumstances (see Heister *et al.* 1995). More far-reaching adjustments could be arranged by renegotiating the treaty. Renegotiation, however, may have a destabilizing effect and may be counterproductive if it does not take place because of unforeseen changes, but is rather triggered by defecting behavior. Hence, there might be a trade-off between the flexibility and the effectiveness of an international environmental agreement (see Kerr 1995).

In the long run, incentives for international environmental cooperation can also be affected by improving the fundamental framework conditions of international negotiations. One crucial factor for the success of negotiations is the information available to the governments involved, especially the information about the physical and biological regularities and the economic consequences of global environmental externalities. An important instrument for facilitating environmental cooperation thus consists of improving the relevant information for negotiations and making it accessible to all parties. This is a traditional task of international organizations and research facilities. Another long-run strategy is to pursue a general policy of global integration in order to strengthen the political and economic interdependencies between national jurisdictions; the rationale being that dependent countries are less likely to behave in an opportunistic manner. The literature on issue linkage suggests that, in general, the number of available punishing strategies varies positively with the degree of integration.

10.7. CONCLUSIONS AND OUTLOOK

The purpose of this chapter was to survey recent contributions to the rapidly growing theoretical literature on the incentives for international environmental cooperation. We proposed a taxonomy of the instruments which are used to create incentives for the participation in and compliance with international environmental agreements. The conceivable strategies for promoting a successful coordination of environmental policies have been grouped into (i) the choice of the internalization instrument itself; (ii) carrot-stick strategies which make cooperative abatement efforts dependent on the past behavior of the other countries (internal stabilization); (iii) transfers and sanctions of various forms (external stabilization); (iv) unilateral and accompanying measures by single countries or subcoalitions; and (v) long-term provisions to increase the flexibility of agreements and to improve the framework conditions of international negotiations.

The surveyed body of literature has to some extent replaced the traditional economic theory of environmental policy because this literature focuses on incentive compatibility and time consistency problems, that is, issues which the traditional theory neglects. The principal merit of the surveyed research is that it systematically addresses the fundamental institutional restrictions which apply to the management of international environmental resources. The second-best character of international environmental policy is mirrored by the common feature of numerous models that international environmental cooperation is incomplete. Moreover, since first-best lump-sum transfers are not available, the focus often is not on mere efficiency aspects but on the distributional implications of international environmental policy. Despite these restrictions the discussed contributions show that, in theory, a large variety of instruments exist which provide incentives for stable environmental cooperation. In practice, though, the preconditions for these strategies to be successful (e.g. with respect to intertemporal discount rates) are often not satisfied.

The limitations of the discussed body of literature must not be overlooked. Although game theory definitely is the appropriate theoretical framework to analyze the strategic interactions of national governments, this method nevertheless abstracts from many factors which are important for the outcome of international environmental negotiations in practice. One striking simplification in game-theoretic models of transboundary pollution concerns the enforceability of international treaties. It is true that the surveyed literature considers different scenarios regarding the ability of sovereign countries to credibly commit themselves. Recent contributions do not only look at international environmental agreements which are completely self-enforcing, but also analyze cases of limited enforceability where some but not all countries can make binding commitments. Both approaches, however, neglect the institutional dimension of environmental cooperation. The lack of a supranational authority does not imply the absence of any institutional framework on the international level. This, however, is implicitly assumed when we analyze international environmental agreements which are fully self-enforcing. On the other hand, a limited degree of commitment (and enforceability) is usually simply assumed without explaining how this is achieved. It would be more appropriate to explicitly model the institutional structure on the international level. Such an institutional approach would allow to analyze how credible commitments of sovereign countries can be achieved and it would illustrate the usefulness of particular international institutions.

Another shortcoming of most models on international environmental cooperation is that they work with quite strong assumptions with respect to the available information, that is, these models do not sufficiently take into account crucial aspects of real-world decision-making. Informational imperfections and asymmetries certainly play an important role in environmental decision-making, especially when dealing with global environmental problems such as the anthropogenic greenhouse effect and the depletion of the ozone layer. Nevertheless it is usually assumed that national governments have perfect information both on their own costs and benefits of environmental policy and on the costs and benefits of all other countries affected by transboundary pollution. In the real world, national governments neither know exactly the characteristics of all other countries, nor are they able to precisely identify their own costs and benefits of abatement programs. Since countries may have an incentive as well as the opportunity to conceal their true characteristics, incomplete information expands the strategy space and generates new and potentially inefficient outcomes. The question therefore arises whether and how mechanisms and bargaining processes can be designed in order to make fruitful reporting incentive compatible and to achieve the utmost mutual gains from international environmental cooperation.

Furthermore, the central paradigm underlying all of the contributions discussed is that of *rational opportunism* of national governments which act on behalf of their principals. Outside the realm of economics, this rather narrowly defined notion of rationality is challenged as an appropriate assumption to analyzing the incentives for cooperation between sovereign states. From the point of view of regime theory (a branch of political science analyzing international relations), negotiating an agreement is part of the formation of a regime where countries "alter their behavior, their relationships, and their expectations of one another over time in accordance with its terms" (Chayes and Chayes 1993: 176). This perspective emphasizes the communicative and informative character of the whole international negotiation process, which itself modifies the structure of the decision problem, but is treated as exogenous in traditional economic investigations (see for example Young 1989).[51]

Most importantly, the central assumption underlying the majority of models on international environmental cooperation is that governments are assumed to act in a benevolent manner on behalf of their national population which is portrayed as a homogeneous entity (the so-called *unit actor* assumption). Contributions analyzing environmental decision-making from a political-economic perspective are still limited in number and are very often restricted to local externalities.[52] One of the most promising directions of future research consists in mending this shortcoming and to develop models that focus more on the *positive* aspect of international environmental agreements. In this context economists may also draw inspiration from already existing approaches in the political science literature, such as the analysis of "two-level-games".

---

[51] Congleton (this volume, Chapter 11) focuses on the negotiation *process* as well.

[52] See Schulze and Ursprung (this volume, Chapter 4) and Congleton (this volume, Chapter 11) for a discussion of political economic aspects of international environmental policy.

11

Governing the Global Environmental Commons: The Political Economy of International Environmental Treaties and Institutions

ROGER CONGLETON

11.1. INTRODUCTION

The literature reviewed in this chapter analyzes political aspects of negotiating and implementing environmental treaties, and develops empirical evidence on the kinds of treaties most likely to be signed, the types of countries that are most likely to sign them, and the extent to which environmental treaties tend to affect environmental quality. Although the literature that simultaneously addresses political and economic aspects of international environmental agreements is not very large, the problems addressed are very important, complex and multifaceted. The aim of the present chapter is to provide the reader with a sense of the main issues addressed, the strengths and weaknesses of the analyses, and the problems that remain.

A good deal of the economic work on international environmental problems assumes that the central problem is specifying policy instruments to address well-understood environmental externality problems. Because of this, international environmental policy is generally approached as a fairly direct extension of work done on domestic policy. Many useful analytical points have been developed from such assumptions as evidenced in the present volume, especially by Schmidt in Chapter 10. And, perhaps surprisingly, there is a sense in which international environmental solutions will necessarily resemble domestic policies. National sovereignty implies that any pattern of environmental policies agreed to in treaty documents have to be implemented via domestic legislation because no international body can impose laws on a sovereign government.

However, besides restricting the range of policy options, national sovereignty also implies that international environmental policy makers address political and institutional problems that can be ignored in ordinary domestic environmental legislation or regulation. The literature reviewed in this chapter suggests that these problems largely determine the content and character of environmental treaties.

The point of departure for the literature reviewed in this chapter is that international environmental problems differ from domestic environmental problems because the *nature of the externality problem* confronted differs. Most domestic environmental legislation addresses problems that arise because self-interested firms and consumers have little reason to fully account for the broad environmental consequences of their production and consumption decisions. Most international environmental problems arise because *governments* have little reason to fully account for the broad international consequences of their domestic environmental regulations. International environmental treaties, thus, attempt to correct instances of *government failure* rather than market failure.

Remedies for international environmental problems attempt to coordinate the regulatory policies of independent governments, rather than to coordinate the economic behavior of private decision makers within a particular polity. The independence, or sovereignty, of the parties involved in international environmental problems implies that any policies adopted must be in the interest of each party involved.[1] It is for this reason that international environmental solutions necessarily resemble long term contracts – Coasian contracts between governments. This contrasts with domestic environmental policies where only agreement by pivotal members of a ruling coalition is required.

However, if solutions to international environmental problems are necessarily voluntary contracts, they are contracts that are very difficult to negotiate and implement. The same public goods aspects of domestic regulations that give rise to international externality problems in the first place largely remain after the agreements are negotiated insofar as no independent authority is available to assure that signatories own up to their contractual duties (see Hoel 1991; Sandler 1996). Moreover, accommodating the domestic political interests of a multitude of national governments clearly has a large effect on the feasible content of the environmental treaties.

---

[1] See Black (1958: 152) for an early public choice perspective on international treaties. Tollison and Willet (1979) model international agreements as a method of realizing mutual gains from exchange (or treaty terms) in a setting where national governments act as perfect agents for their respective citizenry. It would be more accurate to say that decisive members of national governments necessarily *expect to gain* from any treaties consummated. Balwin (1989) discusses the GATT treaty as a device for internalizing externalities. Sykes (1991) provides a public choice based contractual explanation of the Article XIX of the GATT, the so-called escape clause. Buchanan and Tullock (1975), and Maloney and McCormick (1982) develop the first rational choice based analyses of the politics of environmental regulation. Congleton (1996) includes a nice cross section of empirical and theoretical papers that analyze both international and domestic political aspects of environmental regulation.

The literature reviewed in this chapter attempts to explain the observed pattern of environmental treaties. The chapter is organized as follows. Section 11.2 provides a brief history of environmental treaties signed during the last half of the twentieth century. Section 11.3 analyzes the political and economic content of the treaties that we should expect to see negotiated. It focuses on institutional and temporal aspects of negotiation which suggests that the content of treaties will be more institutionally oriented than economic analysis alone would imply. Section 11.4 analyzes the pattern of signatories that we should expect to observe on those treaties negotiated, and reviews empirical evidence on the pattern of treaty signatures and emissions reductions to date. Section 11.5 summarizes the chapter and suggests extensions for future work.

11.2. SOME BACKGROUND FACTS: A DIGRESSION ON THE RECENT HISTORY OF ENVIRONMENTAL TREATIES

The historical record of environmental regulation is very rich and extends well into antiquity. No society can long prosper if it ignores the importance of clean water and fresh air to public health. Rules to control access to common property resources, to regulate waste disposal and assure potable supplies of water have been adopted by all enduring societies. Indeed, appreciation of the environment's role in health and aesthetics evidently coincides with the dawn of recorded history.[2] A relatively modern example, written well after sunrise, is Aristotle's (330 BC/ 1969: 278) description, in passing, of optimal policies concerning water and air quality in his characterization of the ideal community.

I mention air situation and water supply in particular because air and water, being just those things that we make most frequent and constant use of, have the greatest effect on our bodily condition. Hence in a state which has [the] welfare [of its citizens] at heart, water for human consumption should be separated from water for all other purposes.

Evidence of the environmental regulations adopted by the great and not so great civilizations of antiquity include, for example, aqueducts, centralized waste disposal sites and burial grounds. (It is perhaps striking that these ancient environmental policies provide one of the most important windows through which modern archeologists attempt to induce the greater sweep of long forgotten civilizations.)

---

[2] Natural phenomena are of course central to the high cultures of both hunting and agricultural based societies. Such societies very often rely upon nature-based metaphors to make sense of the world at large, to address both practical and metaphysical questions. Even today, various forms of nature worship or pantheism are among the most common world religions, and increasingly central to western ideas about intergenerational duties and responsibilities. Environmental prerequisites to a comfortable and healthy life have long been a practical matter at the core of personal health and economic prosperity.

The history of environmental treaties appears to be much shorter. I have not been able to find many instances that predate the twentieth century – although one imagines that there must have been formal and informal agreements about water rights along waterways shared by ancient cities.[3] The lack of a clear record of international agreements predating modern times may reflect the subtlety of many international environmental problems, the difficulty of negotiating and implementing agreements with neighboring empires, nations, or tribes – which are often engaged in various territorial disputes – or simply insufficient research on my part.

The modern record of environmental treaties begins late in the nineteenth century. The *United Nations Treaty Series* and the *US Treaties and Other International Agreements* series catalogue international agreements on a variety of topics. International environmental and pollution treaties are separately indexed and document the evolution of treaty arrangements. According to these reference treaty series, most of the environmental treaties in force have been signed in the second half of the twentieth century.[4] Table 11.1 lists all of the treaties on environmental matters ratified between 1969 and 1985 that are presently included in these two treaty series. Table 11.2 lists more recent multilateral environmental treaties. Table 11.2 does not attempt to include every treaty modification or minor extensions negotiated, but rather lists major treaties and protocols. The two tables report signatories, the date at which the treaties were completed (opened for signing), their focus, and the principal institutional and substantive action taken.[5] The pace of international environmental negotiations has greatly accelerated in the past 30 years.

---

[3] Within polities, formal water rationing schemes and rules regarding effluents were evidently common in areas where irrigation was important for agriculture. Insofar as irrigation networks extended across the domains of independent sovereignties, or were fully within the autonomous powers of local governments, agreements similar to early international agreements presumably were worked out between the relevent decision making units.

[4] For example, the 1909 Boundary Waters treaty between Great Britain (Canada) and the United States establishes an International Joint Commission of the United States and Canada which "shall have jurisdiction over and shall pass upon all cases involving the use or obstruction or diversion of the waters" within the described area in the treaty (Articles VII and VII). "No use shall be permitted which tends materially to conflict with or restrain any other use which is given preference over it." Water for domestic and sanitary purposes is given the highest precedence followed by uses for navigation, power and irrigation.

[5] The data for the multilateral treaties are from treaty documents, *World Resources 1994–95* and the United Nation's web site: http://www.unfccc.de/. The modern historical record is consistent with the discussion of "treaties as contracts" perspective developed in the introduction.

There have also been several multilateral agreements negotiated over this time period, not included in the treaty volumes available at this time. For example, the European Economic Community also promulgates environmental directives from time to time. For a discussion of the coordinating efforts of the EEC, see Smith and Kromarek (1989) or Ashworth and Papps (1991). Generally, the policies require member nations to "set up programs for handling, storing and eliminating waste in all forms (Smith and Kromarek: 113)." Smith and Kromarek also note that these directives "are differently implemented by member countries (p. 113)."

Table 11.1. *International environmental treaties*

Signatory	Signatory	Year	Focus	Action	Responsibility
UK	W. Germany	1969	Oil slicks	Coordination	Inform each other of existing or potential oil spills
France	Switzerland	1971	Lake Geneva	Commission formed	Recommend policies and monitor water pollution
USA	USSR	1972	General	Commission formed	Exchange of scientific information, joint conferences
Italy	Switzerland	1972	Border Lakes	Commission formed	Recommend policies and investigate pollution sources
USA	Canada	1972	St Johns River	Commission	Monitor water quality and coordinate policies
USA	Canada	1974	Oil spills	Contingency planning	Development of a marine contingency plan
USA	W. Germany	1974	General	Cooperation	May harmonize policies and share information
Poland	Czechoslovakia	1975	Air Pollution	Commission (Plenipotentiaries)	Coordinate monitoring and exchange information
Denmark	Sweden	1975	Oresund Sound	Commission	Recommend policies and coordinate research
USA	Canada	1978	Great Lakes	Commission	Recommend policies and report on treaty programs
USA	Mexico	1980	Maritime Boundaries	Contingency plan	To coordinate a joint response to hazardous substance spills
USA	Canada	1980	Air pollution	Commission	Recommend policies and coordinate and share research
USA	Mexico	1983	Border area pollution	Commission (2 coordinators)	Coordinate policies and meet at least once a year
USA	Canada	1984	St Johns River	Continuation of 1972 agreement	Monitor water quality and recommend targets
USA	Mexico	1985	Hazardous Substances	Contingency plan	Coordinate responses to accidents along the border

Table 11.2. *Major international treaties on air and water pollution*

Treaty name and focus	[a]CPs (1998)	Year	Institutional	Substantive goal/obligations
Stockholm action plan for the human environment	UN	1972	Recommended UNEP	109 general and nonbinding recommendations
Convention on prevention of marine pollution	57	1972		
MARPOL: ship pollution	63	1978		
Geneva convention on long-range transboundary air pollution		1979	Representative executive body	Exchange of information, consultation, research and monitoring, develop policies
Helsinki protocol concerning the reduction of sulfur emissions	(Europe)	1985		Agree to reduce sulfur emissions by 30% of 1980 levels by 1993
Vienna convention on protection of the ozone layer	100	1985		Promote research and monitor the ozone layer
Montreal protocol on substances that deplete the ozone layer	92	1987		Requires nations to cut consumption of 8 substances to 50% of 1986 levels
Sofia protocol on the control of nitrogen emissions	(Europe)	1988		Require reductions in NOx emissions to 1987 levels by 1994
Basel convention on the control of movements of hazardous wastes	34	1989	Secretariat and conference of the Parties	Requires notification by waste exporting countries and consent by waste importing countries
Rio: framework convention on climate change	176	1992	Secretariat and conference of the Parties COP	Technology and information sharing, aim to reduce relevant emissions levels to 1990 levels
Kyoto protocol to the convention on climate change	2	1998	Secretariat and conference of the Parties	Reduce emissions of greenhouse gases (generally to about 8% below 1990 levels by 2012)

Note: [a]Number of contracting parties.

All of the treaties prominently mention the anticipated mutuality of treaty benefits. However, most of the treaties devote relatively little space to articulating the terms of regulatory exchange or to specifying specific targets or regulations for addressing environmental hazards. It would doubtless surprise environmental economists to observe how few of the treaties mention effluent targets or specify timetables for addressing environmental concerns. Even those treaties that do directly address such issues devote relatively little of their text to characterizing policy solutions. (It certainly surprised me when I first began reading environmental treaties many years ago.)

The literature surveyed in this chapter attempts to explain why many of these features of international environmental treaties, contrary to the expectations engendered by much of the purely economic literature, should have been anticipated.

11.3. WHAT KINDS OF TREATIES DO NATIONS NEGOTIATE?

11.3.1. Four Phases of International Environmental Agreements

The process of addressing an international environmental problem begins with finding common interests in new environmental policies. It ends with the joint implementation of those policies agreed to. This often lengthy process normally passes through four stages of development: (i) The first stage entails the recognition of the possibility of mutual advantage. Without agreement that mutual gains can be realized there is no point to further negotiation. (ii) The second stage attempts to establish procedures by which alternative policy targets may be evaluated and chosen from. Without some process of collective decision making – especially in multilateral treaties – it will be difficult if not impossible to proceed to the third stage. (iii) In the third stage, negotiators attempt to agree on specific environmental targets that can solve or at least ameliorate the environmental problem of interest. (iv) Finally, after negotiators have agreed to effluent targets or specific regulations, each country must pass and enforce new domestic environmental legislation to meet its treaty obligations. It is only after this last entirely *domestic* stage that environmental improvements may be actually realized.

Congleton (1995) notes that after each of the first three stages of negotiation there is an environmental treaty that can be signed. Treaties negotiated during the first stage may be categorized as symbolic treaties. *Symbolic* treaties do not characterize environmental regulations, targets, or even procedures by which such substantive matters might be explored. They simply express sentiments about the prospects for better environmental policy. Agreements negotiated in the second stage may be categorized as procedural treaties. *Procedural* treaties develop institutions, often fairly rudimentary institutions, by which substantive matters regarding environmental targets or regulations may eventually be explored or developed. Such treaties build international institutions for collective decision making on specific environmental matters but do not explicitly proscribe environmental targets or regulations. (The actual text of procedural treaties often deals fairly extensively with institutional development, and nearly always includes text on matters very similar to those of symbolic treaties.)

Agreements negotiated in the third stage allow what might called substantive treaties to be signed. *Substantive* environmental treaties specify environmental targets or regulations to be implemented via new domestic legislation by all signatory nations. (Substantive treaties normally reflect their history, and contain lengthy symbolic and procedural sections as well.)

– In order for environmental treaty negotiation to be initiated, the policy makers of at least two countries must believe that participation in the negotiation process yields net political advantages *for themselves*. It does not necessarily imply that implementation of a properly drafted and coordinated set of environmental regulations will be beneficial for all of the governments participating, although in many cases it may. Participation, itself, often generates domestic and international political advantage.

Incentives for governments to sign the three kinds of treaties can be more readily discussed if a bit of notation is introduced at this point. Let $P(\text{Proc}\,|\,\text{Symb})$ denote the subjective probability of consumating a procedural treaty given that a symbolic treaty is signed, and $P(\text{Subst}\,|\,\text{Proc})$ denote the subjective probability of consumating a substantive treaty given a procedural one has been signed. Let $C(\text{Symb})$ and $B(\text{Symb})$ denote the direct cost and benefit realized by signing a symbolic treaty, and so forth. In this case, the expected net environmental benefits of *signing and implementing* each of the three kinds of treaties can be written as:

$$\text{(i) Net(Symb)} = B(\text{Symb}) - C(\text{Symb}) + P(\text{Proc}\,|\,\text{Symb})[B(\text{Proc}) - C(\text{Proc})] \\ + P(\text{Proc}\,|\,\text{Symb})P(\text{Subst}\,|\,\text{Proc})[B(\text{Subst}) - C(\text{Subst})] \quad \text{(1a)}$$

$$\text{(ii) Net(Proc)} = B(\text{Proc}) - C(\text{Proc}) + P(\text{Subst}\,|\,\text{Proc})[B(\text{Subst}) - C(\text{Subst})] \\ \text{(1b)}$$

$$\text{(iii) Net(Subst)} = B(\text{Subst}) - C(\text{Subst}). \quad \text{(1c)}$$

Symbolic treaties are signed because of direct net benefits associated with them and/or because they are believed to increase the likelihood that procedural and substantive treaties will be negotiated in the near future. Procedural treaties are signed because of direct net benefits associated with signing them and/or because they are believed to increase prospects that a substantive treaty will be signed. Substantive treaties are signed because of direct benefits from participation and/or anticipated environmental (net) benefits from mutual implementation.

This highly simplified, but quite general, characterization of the negotiation process has several implications for the kinds of environmental treaties that we should observe.[6] First, it is clear that the greater is the expected flow of environmental improvements, the more likely such negotiations will be undertaken and yield substantive treaties. $B(\text{Subst}) - C(\text{Subst})$ is an argument in the sufficient conditions for signing each kind of treaty. In the case most analyzed by economists, where the principal benefit of participating in the negotiation processes arises from *environmental* benefits, $B(\text{Symb}) = 0$ and $B(\text{Proc}) = 0$, and *every* negotiation process that begins will eventually arrive at a substantive treaty. Such an environmental treaty game can be said to be subgame perfect at the level of negotiations. On the other hand, not every potentially beneficial treaty will be negotiated because anticipated negotiation costs may exceed anticipated environmental benefits.

A somewhat weaker conclusion follows in cases where political or other benefits are realized by governments that merely participate in negotiations, for example, when $B(\text{Symb}) > 0$ and $B(\text{Proc}) > 0$. In this case, at least some countries may sign symbolic and procedural treaties even if their anticipated net environmental benefits from future implementation are negative. All that is required in these cases is that the policy maker's *direct* domestic and international political benefits from participating in symbolic and procedural treaties be larger than the direct cost of participation. (They can avoid most of the costs of substantive treaties by not signing or implementing them.) Such countries would never implement the final substantive treaties in the absence of side payments or other nonenvironmental benefits. Implementation is not subgame perfect in this negotiation process.

A third implication, and the most relevant for the present discussion, is that symbolic and procedural treaties will be more commonplace than substantive treaties. This follows because sufficient conditions for symbolic and procedural treaties *are always reached en route* to a substantive agreement. At any moment in time, many environmental negotiations are underway, and are likely to be at different phases of negotiation. Both existing substantive agreements and those not yet consummated can be preceded by symbolic and procedural treaties. Consequently, there necessarily will have been more opportunities to have signed symbolic and procedural treaties than substantive treaties.

The relative frequency of symbolic and procedural treaties is likely to be greater than implied by the natural order of negotiation if the number of agreements being negotiated is increasing through time, or if there are significant *direct political benefits from participating* in the negotiations and signing procedural and symbolic treaties. In the latter circumstances, some

---

[6] See Congleton (1995) or Hoel (1991) elsewhere in this volume for more detailed analyses of the negotiation process.

governments will begin negotiations for substantive environmental treaties that they have no intention of signing or implementing. (That is to say, some environmental treaties may be signed as a form of environmental cheap talk.) It is also possible that later treaties may fail to be adopted as the power to make environmental policies is assigned to different persons or parties by national elections and/or intraparty politics.

11.3.2. Some Evidence from the Treaty Record

Evidence of the kinds of treaties negotiated to this point may be taken from tables 11.1 and 11.2 above. Only two of the bilateral treaties listed in table 11.1 are substantive treaties insofar as they explicitly list effluents, targets or establish an independent regulatory commission empowered to implement such regulations. The two substantive treaties are the *Oresund Sound Treaty* and the *1978 Great Lakes Water Quality Treaty* which clearly specify which effluents are to be controlled. However, only the Great Lakes Treaty mentions specific target levels for effluents and hazardous materials (from acetaldehyde to zirconium tetrachloride), although even it does not include specific time-tables for meeting the targets.

Consistent with the above analysis, these two substantive agreements are the result of negotiation efforts begun many years earlier which have gene-rated a series of increasingly substantive treaties. The 1974 Oresund Sound Treaty between Denmark and Sweden replaced nonbinding protocols signed in 1960 (*United Nations Treaties Series* 1975, 13823). The 1978 Great Lakes Water Quality Treaty superseded and expanded a similar treaty nego-tiated in 1972 with distant roots in the Boundary Waters Treaty of 1909. In both of these cases, successive treaties led to more rigorous monitoring of the common pool resource of interest and to more extensive environmental obligations.

The history and substance of multilateral treaties parallels that of bilateral treaties. Multilateral substantive environmental treaties are generally preceded by a series of symbolic and procedural treaties. For example, a series of suc-cessively more stringent treaties have been negotiated concerning maritime pollution over the last seventy years. The treaty series began in 1926 when an international conference of major oceanic nations was held in Washington DC. Seven maritime nations accepted a 50 mile discharge prohibition zone for nontankers in coastal waters near major sea ports. (See M'Gonigle and Zacher 1978: 81–3.) In 1948, the Convention on the Intergovernmental Marine Consultative Organization was negotiated, and ratified in 1958 by 21 states, by which time it had been delegated "bureau powers" by the 1954 Oil Pollution Conference. In 1983, a more stringent treaty took effect, the Convention on the Prevention of Pollution from Ships (MARPOL) (See Caldwell 1990: 84).

By 1994, 63 countries had ratified the convention (*World Resources 1994–95*: Table 24.1).[7]

Several major multilateral agreements on chloro-fluorocarbons (CFC) emissions and greenhouse gases have been recently negotiated under United Nations auspices. The Vienna Convention (1985) was a largely procedural agreement that established a process by which future substantive agreements could be achieved. The Vienna Convention did not call for specific targets for CFC emissions. Rather, Article 6 of the Vienna Convention established a

"Conference of the Parties" which "shall keep under continuous review the implementation of this convention, and, in addition, shall ... consider and undertake any additional action that may be required for the achievement of the purposes of this convention."

The substantive Montreal Protocol (1987) was an outcome of the process established.[8] Similarly, the 1992 Climate Control Convention negotiated in New York and Rio de Janeiro established a process by which future more substantive regulatory targets might be negotiated with the aim of reducing anticipated increases in global temperatures. The procedural framework institutionalized the "Conference of the Parties" and the use of the UN Secretariat which reduced the institutional requirements of subsequent negotiations. A series of meetings in New York, Berlin, Geneva and elsewhere eventually yielded a substantive treaty, the Kyoto Protocol of 1998, which specifies targets for reductions in greenhouse gas emissions by nation to be implemented by the year 2012. (As of January 1999, 71 countries have signed the Kyoto protocol, although only two have ratified it.)

---

[7] The evolution of environmental treaty obligations is often fairly complex. The roots of MARPOL may be traced back to an unsuccessful conference sponsored by the US in 1926 dealing with dumping waste oil in the ocean by ships. See M'Gonigle and Zacher (1979, chapter 4). Shortly after the conference, the British government appealed to the International Shipping Conference to adopt a 50 mile discharge prohibition zone. The ship owners of seven countries agreed to implement this prohibition. During the 1930s the League of Nations promoted an accord on oil pollution control. After the Second World War, the 1948 Convention of the Intergovernmental Maritime Consultative Organization was negotiated under United Nations auspices in 1948. This convention did not itself mention pollution or environmental matters but the organization founded by it was assigned bureau powers for the conventions negotiated at the 1954 Oil Pollution Conference. These conventions gave it the responsibility to monitor international agreements regarding intentional oil spills (previously a normal part of the process of ship maintenance). This authority was extended to unintentional spills after the Torrey Canyon spill in 1967. In 1973 a separate "Convention on the Prevention of Pollution from Ships" (MARPOL) was negotiated under United Nations Auspices which was subsequently revised in 1978. As of 1994, the 63 contracting parties to MARPOL, as negotiated in 1978, include all major maritime countries: the major Western industrialized nations, Korea, China and the USSR.

[8] Murdoch and Sandler (1994) argue that although the Montreal Protocol is a substantive treaty, it has had minimal impact on signatory trajectories of CFC emissions. They argue that the observed pattern of curtailed emissions are consistent with a Nash model of pollution abatement.

It bears noting that the analysis above allows the possibility that all three kinds of treaties may be forms of what game theorists call cheap talk, here *"environmental cheap talk."* Symbolic and procedural treaties can be signed and implemented at minor cost by any interested party but do not, themselves, affect national or international environmental quality. Substantive treaties can also be signed at low cost insofar as the principle economic cost of substantive treaties arise only after they are implemented by adopting new domestic legislation. Indeed, substantive treaties may be signed by nations with no intention (or political feasibility) of implementing the policies agreed to.

However, although the content of many of the above treaties seems consistent with a cheap talk interpretation, game theorists would be surprised at how few nations sign treaty documents. No environmental treaty has yet achieved the "unanimous support" that a cheap talk interpretation seems to imply. If signing is free, and "backing a good cause" is politically useful to policy makers, all nations should be expected to sign environmental treaties. This fact remains something of a puzzle. Evidently, indirect reputational effects discourage such insincere signing from taking place, at least to some extent. A government's reputation as a reliable agent or negotiation partner evidently has value in other policy areas in which international negotiations take place. For example, a politician or party's reputation for faithfully delivering on its promises may help assure its success in future elections. Similarly, a negotiator's reputation in environmental and other policy forums – as a political agent, as a trading partner, as a facilitator of a military alliance, or as a potential member of other international treaty organizations – may have value in that it reduces transactions costs.

Such reputational effects may partly account for the reluctance of pivotal policy makers to sign treaties that *have no provisions to punish* signatory nations that fail to comply with treaty mandates. Just as the main private benefits that policy makers realize by implementing environmental treaties tend to be reputational rather than environmental, so apparently are the main costs of not implementing such treaties. In other cases, as within the European Union, modifying domestic environmental policies may be seen as a necessary step for advancing other policy interests in related multilateral forums.

On the other hand, if the predictions of a cheap talk equilibrium do not seem to hold, neither does the other extreme of perfect implementation. As noted above, the substantive part of those treaties that actually specify targets or regulations is generally a relatively small part of the total text of environmental treaties – which is generally more focused on assuring mutual advantage, and specifying procedural arrangements. Moreover, the fully implemented environmental treaties that attract so much attention in theoretical analyses of environmental treaties are rare. For example, very few signatory nations have taken any steps to implement the recent series of treaties concerned with global warming.

11.4. WHO SIGNS ENVIRONMENTAL TREATIES?

At every stage of negotiation, participants are forward looking. Taken at face value, symbolic treaties increase the prospects for procedural and substantive treaties. Procedural treaties affect expectations by increasing the likelihood that a substantive treaty will be signed. Substantive environmental treaties affect expectations by committing signatory governments to enacting and enforcing more stringent domestic environmental regulations. In every case, the interests of policy makers must be advanced if treaties are to be signed, and must advance those of a majority of the national legislature or ruling junta if they are to be implemented.[9] The question arises as to whether there are any systematic economic or institutional factors that affect the net benefits that national policy makers realize from environmental treaties, and thereby whether there is a systematic component to the pattern of signatures observed on the treaties negotiated.

11.4.1. Utility Maximizing Pollution Standards for Democracies and Dictatorships

Congleton (1992) first modeled and estimated the propensities of nations to adopt domestic environmental legislation and to sign international environmental treaties. The model examined how differences in political institutions can affect national propensities to enact domestic environmental regulation and sign international environmental treaties. More recently, Murdoch and Sandler (1997) have demonstrated how public goods aspects of adopting domestic environmental regulation may affect propensities to sign and implement environmental treaties.

These two approaches complement each other and may be combined to form the basis of an analysis that captures both public goods and institutional aspects of propensities to engage in domestic and international environmental regulation. Again, a few equations may clarify the discussion. Suppose that all individuals, whether dictators or ordinary citizens, maximize a two-dimensional utility function defined over measured real income, C (consumption), as per GNP accounting practices and environmental quality, E, that is: $U_i = u(C_i, E)$. These two areas of choice, while technologically linked, are disjoint since environmental quality is not included in the most widely used measures of national output. Environmental quality may be thought of as an index of the concentrations of undesired effluents in the untreated air and water supply. Real national income or gross domestic product can be considered to be a pure private good, an index of manufactured private goods and services purchased in markets. *Personal* incomes can be approximated as a monotone increasing function of national income.

---

[9] See Mueller (1989) for an extensive overview of rational choice based models of government decision making.

Assume that every individual, i, receives a constant fraction, α_i, of national product, Y, as personal income. This fraction, α_i, clearly varies among individuals. For example, a dictator receives a much higher fraction of national income than a typical citizen or the median voter. Given national income and individual i's income share, his personal income or nonenvironmental consumption is $C_i = \alpha_i Y$. Environmental regulations affect personal income through effects on national output. The discussion above suggests that national income is affected by its resource base, N, and domestic environmental regulations, $Y = y(R, N)$. The relationship between environmental regulations and national income is nonmonotonic. Over an initial range, more stringent domestic environmental standards increase national output by improving the health and productivity of labor, and/or by freeing resources that would otherwise have been used by individuals to reduce exposure to the local environment – air filters, water purifiers, and the like – for more valuable uses. Beyond this initial range, more stringent environmental standards reduce nonenvironmental output as less productive technologies are mandated and inputs are diverted from ordinary economic production to environmental improvement without offsetting productivity gains.

In this last economically relevant range, a trade-off between personal income and environmental quality exists. More stringent regulations improve average environmental quality but increase the cost of consumer goods relative to income, reducing measured national income and thereby the nonenvironmental consumption of a typical individual. This tradeoff is the private marginal cost of environmental regulation. In other words, the private marginal cost of environmental quality generated by more stringent regulation is not reflected by changes in ordinary tax burden, but rather by indirect effects that more stringent environmental standards have on personal income. In this respect, environmental standards (and many other regulations) are unlike ordinary government expenditure programs where tax revenues are used to finance the provision of a public service.

Not only does environmental quality affect national and personal income, but national income affects environmental quality. The link between environmental quality, environmental standards and national income perceived by policy makers is stochastic. This reflects random elements of the underlying natural processes involved and the unpredictable nature of scientific progress in understanding the physical and social mechanisms involved. In the case of interest here, the probability F of high domestic environmental quality levels increases as domestic and world environmental standards, R and R^w, become more stringent and falls as national and world output, Y and Y^w, increase, $F = e(E \mid R, Y, R^w, Y^w)$. Standards adopted by other nations affect the base quality confronted and the extent to which domestic regulations may improve local environmental quality.

Domestic environmental regulations serve as a form of social insurance which reduces downside environmental risk. Each individual prefers the environmental standard that maximizes life-time expected utility given various personal constraints. This characterization of the environmental regulation at issue together with a finite time horizon, T, implies that an individual's preferred regulation or environmental standard can be characterized as that which maximizes:

$$U^e = \int_0^T \int_{-\infty}^{\infty} u(E, \alpha_i y(R)) e(E \mid R, y(R), R^w, Y^w)\, \delta E\, \delta t. \tag{2}$$

Assuming that the expected utility function is strictly concave, we can characterize the individual's ideal environmental policy with the first order condition:

$$U_R^e = \int_0^T \int_{-\infty}^{\infty} [U_C \alpha_i Y_R e(E \mid R, y(R), R^w, Y^w)$$
$$+ u(E, \alpha_i y(R))(e_R + e_Y Y_R)]\, \delta E\, \delta t. \tag{3}$$

(Subscripted variables represent partial derivatives with respect to the variable subscripted.) Equation 3 demonstrates that the effect of environmental regulation on personal welfare occurs through its effects on personal income and on the probability distribution of environmental quality. Each individual prefers the environmental standard that sets the expected present value of his subjective marginal cost for more stringent environmental regulation (in terms of reduced measured income or consumption) equal to the present discounted value of the time stream of marginal utility from greater environmental quality. The geometry of a typical person's preferred environmental standard is illustrated in figure 11.1.

The relationship between economic output and environmental quality determines the shape of the regulatory opportunity set faced. It represents the steady-state relationship between nonenvironmental (or pecuniary) income and environmental quality. The indifference curves represent constant levels of expected lifetime utility over the time horizon of interest. The relevant part of the constraint for policy makers is from R^y to $R^{\max}$. A policy maker who values income but cares nothing for the environment, would prefer the standard which maximizes national output, R^y. A policy maker that values both manufactured consumption goods and environmental goods tends to prefer an intermediate level of regulation such as R^* between R^y and $R^{\max}$.

The implicit function theorem applied to eqn. (3) implies that each person's ideal level of environmental quality can be written as a function of their domestic national income share individual time horizons, the national resource base, and the regulatory regimes and output of the rest of the world, for example,

$$R_i^* = r_i(\alpha_i, T_i, N_i, R^w, Y^w). \tag{4}$$

These five parameters of the individual's optimization problem all affect the pivotal policy maker's preferred domestic environmental regulation. The qualitative effects of changes in these parameters on preferred policies can be

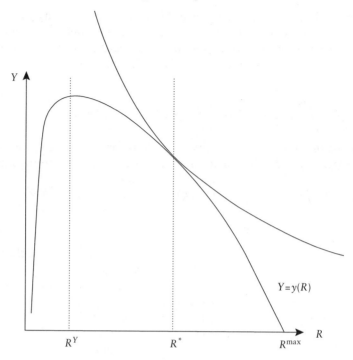

Figure 11.1. *The geometry of a typical person's preferred environmental standard.*

characterized by differentiating eqn. (4) with respect to each of the parameters of the individual's choice problem.

For example, an increase in the share of national income received by an individual has two effects.[10] First, there is an income effect which tends to increase consumption of all goods, including environmental quality. Such effects may include reductions in personal income based on changes in the costs of consumer goods or reductions in national income generated by reduced

---

[10] For example, differentiating with respect to α_i allows the effect of income share on a person's preferred environmental policy to be characterized as:

$$R^*_{i\alpha} = U^e_{R\alpha} / -U^e_{RR}. \qquad (5)$$

The denominator is less than zero since this is simply the second order condition of the original optimization problem, $U^e_{RR} < 0$. Consequently the numerator determines the sign of $R^*_{i\alpha}$. The numerator is:

$$U^e_{R\alpha} = \int_0^T \int_{-\infty}^{\infty} \{[U_{C_iC_i}Y\alpha_i + U_{C_i}]Y_Re(E_\cdot|R, y(R)) + U_{C_i}Y(e_R + e_YY_R)\}\,\delta E\,\delta t$$

or

$$U^e_{R\alpha} = \int_0^T \int_{-\infty}^{\infty} \{\alpha_i U_{C_iC_i}Y_RYe + U_{C_i}[Ye_R + Y_R(e + Ye_Y)]\}\,\delta E\,\delta t. \qquad (6)$$

exports (Leidy and Hoekman 1996). Second, there is a relative price effect which increases the marginal cost of environmental quality, and reduces the preferred degree of environmental protection. In the case where the relative price effect dominates the income effect, a policy maker that receives a higher share of national income will prefer less restrictive environmental standards than one that receives a smaller portion of national income as personal income. (Increases in national resource base, on the other hand, have an income effect but no relative price effect and would increase the demand for both ordinary consumption and environmental protection.)[11]

The model implies that utility maximizing policy makers within and among nations will disagree about the optimal environmental standard, even if they agree about the "proper" tradeoff between environmental quality and material consumption, and also agree about the underlying environmental science.[12] For example, national policy makers with similar incomes and time horizons would tend to disagree about ideal environmental regulations if domestic environmental policies have different effects on expected national environmental quality. As emphasized by Murdoch *et al.* (1997) and implied by Boadway and Hayashi (1999), this would be the case if domestic emissions fell mainly on neighbors or fell mainly within their own national boundaries. Similarly, policy preferences would vary if nations face different environmental spillovers from their neighbors.[13]

---

[11] Note that the model sketched out above may be extended to characterizes the overall pattern of world regulation and income as a Nash equilibrium. At the Nash equilibrium to the domestic regulation game, the world pattern of regulation, R^w and the world income, Y^w, would reflect equilibrium decisions by policy makers in every country in the world. Because the Nash equilibrium is unlikely to be Pareto optimal, potential gains to trade in environmental regulations exist. Negotiating substantive international environmental treaties is, of course, one method by which those potential gains to trade may be realized.

[12] Current scientific disagreements and the evolving state of scientific knowledge clearly make the political and economic benefits and tradeoffs of environmental policy more difficult to calculate. The economic costs and benefits of long term environmental policies also tend to change through time as innovation and business cycles take place.

Such problems are often greater for international environmental problems than domestic ones. The long-term nature of regional and global environmental problems and the cumulative nature of the processes involved often allow wider scope for reasonable scientists and policy makers to disagree about the consequences of alternative environmental policies. Because of all these political, scientific, and valuation problems, the net benefits that a particular government may secure by signing a particular treaty often remain highly uncertain.

[13] National interests in domestic and international environmental regulation differ for many reasons even for polities with similar forms of governance. Patterns of wind or water often generate quite different environmental consequences for upstream and downstream nations. The thinning of the ozone layer is uneven, and has larger effects on countries near the polar regions than near the equator. Global warming is likely to improve the economies of countries near the polar regions while those near to the equator will be economically disadvantaged. Even if similar losses from environmental hazards are expected, national demands for environmental regulations may differ because incomes, tastes, lifestyles, political institutions, or time horizon differ among nations. Even similar national governments facing similar environmental losses imposed on economically and culturally similar citizenries may disagree about ideal environmental policies if they face different costs for implementing environmental regulations.

Political institutions also affect the environmental policies that will be adopted by a given country insofar as different regimes imply different pivotal policy makers with systematically different incomes, interests, and time horizons as noted by Congleton (1992). This effect is clearest for the extremes of political organization: democracies and dictatorships. Recall that the pivotal decision maker within a democratic country can be approximated as its *median voter*. The median voter is approximately the voter with the median income share and time horizon. The median voter, as such, can not be an "outlier." On the other hand, the pivotal decision maker within an authoritarian regime is the ruler (or ruling council) who tends to be an outlier in many dimensions. Authoritarians have far greater than a median share of national income, and tend to have a shorter than average time horizon and greater risk tolerance than the median voter given the high turnover of authoritarian regimes (see Bienen and van de Walle 1989). Under the restricted circumstances previously discussed, a larger national income share and shorter time horizon tends to reduce the stringency of the pivotal policy maker's desired environmental standard. Consequently, authoritarians tend to prefer lower environmental standards than the median voter of an otherwise similar democracy.

Democracies will be more inclined to sign and implement environmental treaties than dictatorships, other things being equal. In such cases, authoritarian regimes will require additional inducements – direct cash or in kind transfers – to persuade them to sign an environmental treaty and adopt its more restrictive domestic environmental regulations. The following section will show that the empirical evidence is largely in accord with these conclusions.

11.4.2. Evidence on the Pattern of Treaty Signatures

Several papers have used statistical methods to determine whether the pattern of signatures on various multinational international environmental treaties can be explained as functions of such variables as national resource endowments, income, market structure, and political institutions as indicated above.

Murdoch and Sandler (1997) estimate reductions in CFC using 1989 data. The data set, thus, is after the conclusion of the Montreal protocol (1986), but before the date at which signatories were obligated to reduce emissions (1993). They found that national reductions in CFC emissions are larger in high income states than in low income states, and that reductions in CFC emissions are greater in free countries (democracies) than in nonfree (dictatorships). The latter results are consistent with the Congleton (1992) estimates of propensities of dictatorships and democracies to engage in domestic regulation as proxied by their ratification of the international treaties regarding the control of CFC emissions.

However, Murdoch and Sandler note that the CFC treaties, themselves, appear to have done little to reduce CFC emissions. Only 38 of the 61 countries that reduced CFC emissions between 1986 and 1989 had ratified the Montreal

Protocol. Nonratifiers were essentially as likely to have reduced emissions as ratifiers. Rather, they argue, that observed reductions in CFC emissions were simply the voluntary provision of a public good rather than evidence of cooperative behavior. The CFC treaties appear to have ratified reductions that national policy makers were already prepared to make on the basis of their own independent self-interest.

Fredriksson and Gaston (1999) analyze the time that it takes nations to sign and ratify environmental treaties. They focus on the United Nations Framework Convention on Climate Change (FCCC) negotiated in New York and Rio De Janeiro in 1992. The average time from signing the convention to ratifying it was 810 days. Fredrissson and Gaston find that the time a particular nation takes to ratify the treaty (once signed) can be explained with many of the same political and economic variables used in the Congleton and Murdoch and Sandler studies. They found that nations with greater civil liberties and smaller CO_2 emissions had a more rapid ratification of the FCCC than those with low civil liberties and high CO_2 emissions in all their model specifications. The estimated effects of other variables used to characterize national preferences for environmental policies were less robust. National area and population, interpreted as proxies for national resource endowments, were found to be significant in several of their estimates.

The effect of international spillovers ("spill ins" and "spill outs") on propensities to sign international treaties and to adopt domestic legislation is examined in Murdoch *et al.* (1997). They analyze the impact of two substantive protocols to the Long Range Transboundary Air Pollution convention negotiated in 1979. The Helsinki protocol was negotiated in 1985 and required sulfur emissions to be reduced to 70 percent of 1987 emission rates by 1993. The Sofia protocol was negotiated in 1988 and required reduction in nitrogen oxide emissions to 1987 rates by 1994. Murdoch *et al.* estimate the effects that various national parameters have on emission rates for the relevant effluents before and after the treaties were in effect. Generally, they find greater reductions in national effluent emissions in countries where relatively more of domestic emissions fall within a nation's boundaries. Reductions are smaller if national air quality is caused by the emissions of upwind countries. They interpret these empirical results as evidence of strategic (Nash-like) behavior on the part of domestic policy makers.[14] Civil liberties again appear to affect the stringency of the regulations adopted.

Perhaps the most interesting of their many empirical results is that the countries that signed the Helsinki and Sofia protocols are inclined to make

---

[14] The effect of wind direction on the propensies of nations to negotiate international agreements is also well illustrated by a case along the German French border analyzed by Feld *et al.* (1996). In that case, money was raised in Kleinbittersdorf, the downwind city, to upgrade a new incinerator in Grosbliederstroff, the upwind French town. Here wind direction not only determined the incentives for international negotiation but also the direction of monetary flows in the Coasian contract negotiated.

larger reductions in domestic emissions than those that do not. They interpret this pattern as evidence of a *screening effect* rather than of a treaty effect because the Sofia protocol *had not yet entered into force* at the time of their study. That is to say, nations with smaller emissions are more inclined to sign international environmental treaties than those with larger emissions.

In general, the results to date have affirmed the conclusion that political institutions, national income, and public goods aspects of environmental regulation affect the propensities of countries to adopt domestic environmental regulations, and to sign international environmental treaties. The degree to which the treaties have been implemented has not been extensively studied although the Murdoch *et al.* papers suggest that treaties have had only a modest effect on the level of national emissions of targeted effluents.

11.5. OVERVIEW AND CONCLUSION

In principle, environmental treaties attempt to coordinate domestic and international environmental regulations in order to secure long term mutual advantage for the signatories. The regulatory aim of environmental treaties suggests that environmental treaties will resemble domestic environmental regulations that deal specifically with emissions targets and policy instruments. The voluntary nature of environmental treaties suggests that environmental treaties will resemble ordinary long term contracts in as much as they are based upon the mutual agreement of all participating parties, and are only partly motivated by immediately observable benefits for the parties involved.

However, the treaties that we observe differ substantially from ordinary contracts and from ordinary domestic environmental regulations. The specification of regulatory terms of trade takes up surprisingly little of the text of most environmental treaties. Much more space is devoted to listing potential gains to trade and to establishing institutions and procedures by which further negotiations may take place. Unlike ordinary contracts, the actual parties to international environmental agreements are often not fully known for many years after a treaties is negotiated and opened for signing. Nor do all signatories of treaties ultimately implement the regulatory changes negotiated. The delay between opening negotiations, signing, ratifying and implementing is often substantial. Evidently, environmental "contracts" between sovereign nations are not ordinary contracts, nor domestic environmental regulations writ large.

This chapter has summarized efforts to understand the nature of the international treaties negotiated and the pattern of signatures observed. The literature reviewed argues that treaty content cannot be accounted for by economic factors alone, nor can agreements be taken as necessarily indicating policy coordination. From a political vantage point, the treaties negotiated are signals about the direction that domestic environmental policies in signatory nations may take rather than environmental policies, *per se*. Consequently, the political advantage from signing symbolic, procedural, and substantive treaties

are very similar in the short term, and all may be expected to attract the signatures of governments with environmental constituencies. Each form of treaty
signals future environmental benefits without imposing immediate costs on
signatory nations.[15] Sovereignty implies that any new environmental policies
agreed to will be implemented via domestic legislation. Thus international
environmental agreements must yield wide spread political advantage within
all signatory governments if they are to be fully implemented.

The pattern of signatures that we observe suggests that political institutions
and public goods aspects of the environmental regulations at issue play a
large role in determining whether or not environmental treaties are signed.
Democracies are more likely to sign than dictatorships. Countries that experience large spillovers from other countries are more likely to sign than others.
The latter suggests that environmental treaties address genuine environmental
problems, and are not entirely politically expedient cheap talk. Moreover,
efforts to address institutional and political concerns that make substantive
treaties feasible are very much in evidence in treaty documents.

All but the most symbolic treaties establish or augment standing international institutional arrangements. For example, international environmental
treaties generally establish specialized commissions with representation from
all signatory countries. The environmental commissions established generally
use unanimous agreement as their decision rule, which assures mutual gains
from treaty terms. However, relatively little explicit authority is granted to the
commissions created. Final approval and implementation of environmental
policies resides with the ruling legislatures and councils of signatory countries.
The commissions established are generally delegated the power to make proposals to signatory nations and to file periodic reports. In substantive treaties
the commissions are explicitly given responsibility for monitoring, implementing treaty obligations, and coordinating information flows.[16]

The institutions established have not to this point fully addressed the final
implementation problems analyzed in Chapter 10. Rather substantive treaties
have implicitly relied upon the good faith of signatories or implicit incentives
for compliance. The absence of explicit treaty provisions for penalizing

[15] Relative price effects generated by environmental regulations may be politically as important
as the regulations themselves. I focus on environmental quality throughout this chapter largely to
simplify the analysis. Political agency issues are only indirectly analyzed in this chapter insofar as
decisions are cast in terms of the interests of pivotal government decision makers. These may not
assure domestic Pareto optimality.

[16] Environmental treaties also depart from the Coasian perspective in that the contracting parties
are governments, rather than individuals, who may or may not promote the interests of their
citizenry by aiming for Pareto optimal domestic and international policies. See McGee (1989) or
Vaubel and Willet (1991) for examinations of government incentives in the negotiation of international trade arrangements.

Environmental treaties have not been subject to similar scrutiny. A preliminary look at incentives
faced by dictatorships and well-functioning democracies in signing environmental treaties is
developed in Congleton (1992).

noncompliance together with the absence of credible international enforcement agencies empowered to impose costs on nonperforming countries implies that incentives to implement treaty obligations are results of reputational effects or various international forms of *continuous dealings*.

The reputational "performance bond" can be significant for nations that continually deal with each other in a variety of economic and political policy areas, as among members of the European Union. Renegade signatory countries lose their reputations as reliable trading partners which places at risk benefits from future dealings with the other signatory nations. Such implicit performance bonds require no external agency to enforce them, and can be effective if reduced future transactions pose a credible threat to potential violators.[17]

Although treaties are a more cumbersome method of solving externality problems than other supranational regulatory solutions that might be imagined, environmental treaties are likely to remain the principal vehicle by which international environmental problems are addressed. The contractual nature of treaties reduces domestic political risks by guaranteeing that *all* participating governments benefit from the regulations finally adopted.[18] Even if greater policy making authority were delegated to international commissions, or incentive compatible treaty language were agreed to, sovereignty implies that domestic legislation would remain the method by which such international environmental policies are implemented.

The literature reviewed in this chapter indicates that international contractual solutions to environmental problems are clearly more challenging to

---

[17] For such a threat to be creditable, the parties must believe that nonviolators are better off ending a treaty and/or other future relationship with a nonperforming "partner" than continuing them. This is often the case in treaties where only two parties are involved and environmental concerns are transparent, simultaneous, and reciprocal. It can also be the case in multilateral treaties where the parties deal with each other on a wide range of policy areas. In such cases, implementing environmental treaties can be sub-game perfect in cooperation. The logic for bilateral environmental treaties is straight forward. Since each party in a bilateral treaty only has an interest in observing the treaty if the other party adheres to treaty terms, return to the pretreaty state is a credible threat. Thus, expected net gains from breach are negative for each party at every instant. See Telser (1980) or Schmidt (this volume, Chapter 10) for an overview of the theory of incentive compatible contracts.

[18] Analysis of the internal operation of the various international organizations established by both procedural and substantive environmental treaties is left for future analysis. Analysis of other international organizations (see Vaubel and Willet 1991) suggests that principal-agent problems at the government commission level of analysis are likely to occur. Even without policy making powers, environmental commissions may have a substantial impact on the agendas of domestic governments through their ability to make policy recommendations and by their superior knowledge of environmental detail in their area of responsibility. The delegation of relative little authority to the institutions established implies that governments are well aware of political agency problems. Such policies reduce the likelihood of "capture" whereby interest groups unduly influence regulatory agencies to promote their own narrow ends. Legislative oversight of treaty implementation does not rule out such influence, but does make extreme outcomes less likely inasmuch as legislatures cannot freely ignore the electorate's welfare.

achieve than solutions to domestic environmental problems because the political and institutional problems that have to be overcome are more complex. Those challenges seem likely to remain as long as nations are sovereign. Consequently, analysis of the complex interplay between the political and economic determinants of international agreements is likely to remain a fruitful area of research for the foreseeable future.

References

Abel, M., Daft, L., and Early, J. (1994). *Large Scale Land Idling Has Retarded the Growth of US Agriculture*. Alexandria, Va.: National Grain and Feed Association.

Abler, D. G., and Shortle, J. S. (1992). "Environmental and Farm Commodity Policy Linkages in the US and EC", *European Review of Agricultural Economics*, 19: 197–217.

Adams, M., and Johnson, S. (1998). *Turmoil in Asian Markets*. ITTO Secretariat, Yokohama, Japan.

Aghion, P., and Howitt, P. (1998). *Endogenous Growth Theory*. Cambridge Mass.: MIT Press.

Aidt, T. (1998). "Political Internalization of Economic Externalities and Environmental Policy", *Journal of Public Economics*, 69: 1–16.

Albrecht, J. (1998). *Environmental Regulation, Comparative Advantage and the Porter Hypothesis*, FEEM NOTE di LAVORO59.98, Milan: Fondazione Eni Enrico Matthei.

Amelung, T., and Diehl, M. (1992). *Deforestation of Tropical Rain Forests: Economic Causes and Impact on Development*. Kieler Studien 242, Institut für Weltwirtschaft an der Universität Kiel, Tübingen, J.C.B. Mohr.

Amundsen, E. S., and Schöb, R. (1999). "Environmental Taxes on Exhaustible Resources", *European Journal of Political Economy*, 15/2: 311–29.

Anderson, K. (1992). "The Standard Welfare Economics of Policies Affecting Trade and the Environment", in Anderson, K. and Blackhurst, R. (eds.), *The Greening of World Trade Issues*. Ann Arbor: University of Michigan Press, 25–48.

—— (1993). "Economic Growth, Environmental Issues and Trade", in Bergsten, F. and Noland, M. (eds.), *Pacific Dynamism and the International Economic System*. Washington, DC: Institute for International Economics in association with the Australian National University, Pacific Trade and Development Conference Secretariat, 341–63.

Anderson, K., and Blackhurst, R. (1992). *The Greening of World Trade Issues*. New York: Harvester-Wheatsheaf.

Anderson, K., and Strutt, A. (1996). "On Measuring the Environmental Impact of Agricultural Trade Liberalization", in Bredahl, M. E., Ballenger, N., Dunmore, J. C., and Roe, T. L. (eds.), *Agriculture, Trade, and the Environment: Discovering and Measuring the Critical Linkages*. Boulder, Colo.: Westview Press, 151–72.

Antle, J. M., and Just, R. E. (1991). "Effects of Commodity Program Structure on Resources and the Environment", in Bockstael, N. and Just, R. E. (eds.), *Commodity and Resource Policies in Agricultural Systems*. New York: Springer-Verlag.

Antle, J. M., Crissman, C. C., Wagenet, R. J., and Huston, J. L. (1996). "Empirical Foundations for Environment-Trade Linkages: Implications of an Andean Study", in Bredahl, M. E., Ballenger, N., Dunmore, J. C., and Roe, T. L. (eds.), *Agriculture, Trade, and the Environment: Discovering and Measuring the Critical Linkages*. Boulder, Colo.: Westview Press, 173–97.

Antweiler, W., Copeland, B., and Taylor, M. S. (1998). *Is Free Trade Good for the Environment?* National Bureau of Economic Research working paper no. 6707. Cambridge Mass.: NBER.

Asako, K. (1979). "Environmental Pollution in an Open Economy", *Economic Record*, 55: 359–67.

Asante-Duah, D. K., and Nagy, I. V. (1998). *International Trade in Hazardous Waste*. London: Spon.

Ashworth, J., and Papps, I. (1991). "Equity in European Community Pollution Control", *Journal of Environmental Economics and Management*, 20: 46–54.

Baharuddin, Hj. G. (1995). "Timber Certification: An Overview", *Unasylva*, 46/183: 18–24.

Baharuddin, Hj. G., and Simula, M. (1994). *Certification Schemes for All Timber and Timber Products*. Report for the International Tropical Timber Organization, Yokohoma.

——(1996). *Study of the Development in the Formulation and Implementation of Certification Schemes for all Internationally Traded and Timber Products*. Report to the International Tropical Timber Organization, Yokohoma.

——(1997). *Timber Certification: Progress and Issues*. Report to the 23rd Session of the International Tropical Timber Council, 1–6 December, Yokohama.

Baldwin, R. E. (1989). "The Political Economy of Trade Policy", *Journal of Economic Perspectives*, 3/4: 119–35.

Ballenger, N., and Krissoff, B. (1996). "Environmental Side Agreements: Will They Take Center Stage?", in Bredahl, M. E., Ballenger, N., Dunmore, J. C., and Roe, T. L. (eds.), *Agriculture, Trade and the Environment: Discovering and Measuring the Critical Linkages*. Boulder, Colo.: Westview Press, 59–78.

Barbier, E. B. (1996). *Impact of the Uruguay Round on International Trade in Forest Products*. Rome: FAO.

——(1999). "The Effects of the Uruguay Round Tariff Reductions on the Forest Product Trade: A Partial Equilibrium Analysis." *World Economy*, 22: 87–115.

Barbier, E. B., Bockstael, N., Burgess, J. C., and Strand, I. (1995). "The Linkages Between the Timber Trade and Tropcial Deforestation: Indonesia", *World Economy*, 18/3: 411–42.

Barbier, E. B., Burgess, J. C., Bishop, J. T., and Aylward, B. A. (1994). *The Economics of the Tropical Timber Trade*. London: Earthscan Publications.

Barrett, S. (1992). "International Environmental Agreements as Games", in Pethig, R. (ed.), *Conflicts and Cooperation in Managing Environmental Resources*. Berlin: Springer, 18–33.

——(1994a). "Self-Enforcing International Environmental Agreements", *Oxford Economic Papers*, 46: 878–94.

——(1994b). "Strategic Environmental Policy and International Trade", *Journal of Public Economics*, 54: 325–38.

——(1997a). "Heterogeneous International Environmental Agreements", in Carraro, C. (ed.): *International Environmental Negotiations: Strategic Policy Issues*. Cheltenham: Edward Elgar, ch. 2, 9–25.

——(1997b). "The Strategy of Trade Sanctions in International Environmental Agreements", *Resource and Energy Economics*, 19/4: 345–61.

Bartik, T. (1988). "The Effects of Environmental Regulation on Business Location in the United States", *Growth Change*, 19: 22–44.

Bartik, T. (1989). "Small Business Start-Ups in the United States: Estimates of the Effects of Characteristics of States", *Southern Economic Journal*, 55: 1004–18.

Basle Convention on the Control of Transboundary Movements of Hazardous Wastes and Their Disposal, 1992, Canada: Treaty Series 1991/92, Ottawa: Queen's Printer for Canada.

Baumol, W. (1971). *Environmental Protection, International Spillovers and Trade.* Stockholm: Almqvist and Wiksell.

Baumol, W., and Oates, W. E. (1971). "The Use of Standards and Prices for Protection of the Environment", *Swedish Journal of Economics*, 73/1: 42–54.

——(1998). *The Theory of Environmental Policy*. Cambridge: Cambridge University Press.

Beltratti, A. (1996). *Models of Economic Growth with Environmental Assets.* Dordrecht: Kluwer Academic Publishers.

Ben-David, D., and Loewy, M. B. (1998). "Free Trade, Growth, and Convergence", *Journal of Economic Growth*, 3/2: 143–10.

Berger, N. (1998). *North-South Trade in Recyclable Waste: Economic Consequences of the Basel Convention.* Adelaide: Centre of International Economic Studies Seminar Paper 98,03.

Bevans, C. I. (1974). *Treaties and Other International Agreements of the United States of America*. Department of State Publication, 8761.

Bhagwati, J. (1993). "The Case for Free Trade", *Scientific American*, 269/5: 42–9.

Bhagwati, J., Srinivasan, T. N. (1996). "Trade and the Environment: Does Environmental Diversity Detract from the Case of Free Trade?", in Bhagwati, J. and Hudec, R. (eds.), *Fair Trade and Harmonization: Prerequisites for Free Trade?* Vol. I Economic Analysis, Cambridge Mass.: MIT Press, 159–223.

Bienen, H., and van de Walle, N. (1989). "Time and Power in Africa", *American Political Science Review*, 83: 19–34.

Birdsall, N., and Wheeler, D. (1992). "Trade Policy and Industrial Pollution in Latin America: Where are the Pollution Havens?", in Low, P. (ed.), *International Trade and the Environment*. World Bank Discussion Paper 159. Washington, DC: World Bank, 159–68.

Black, D. (1987/1958). *The Theory of Committees and Elections.* Boston: Kluwer Academic Publishers.

Black, J., Levi, M. D., and de Meza, D. (1993). "Creating a Good Atmosphere: Minimum Participation for Tackling the 'Greenhouse Effect'", *Economica*, 60: 281–93.

Boadway, R., and Hayashi, M. (1999). "Country Size and the Voluntary Provision of International Public Goods", *European Journal of Political Economy*, 15: 619–38.

Bohm, P. (1993). "Incomplete International Cooperation to Reduce CO_2 Emissions: Alternative Policies", *Journal of Environmental Economics and Management*, 24/3: 258–71.

——(1994). "Making Carbon Emission Quota Agreements More Efficient: Joint Implementation versus Quota Tradability", in Klaassen, G. and Forsund, F. R. (eds.), *Economic Instruments for Air Pollution Control*. Dordrecht: Kluwer, ch. 9, 187–208.

Böhringer, C., and Rutherford, T. F. (1997). "Carbon Taxes with Exemptions in an Open Economy: A General Equilibrium Analysis of the German Tax Initiative", *Journal of Environmental Economics and Management*, 32: 189–203.

Bommer, R. (1996). "Environmental Regulation of Production Processes in the European Union: A Political-Economy Approach", *Aussenwirtschaft*, 51: 559–82. Reprinted in Bommer (1998), ch. 5.

——(1998). *Economic Integration and the Environment: A Political-Economic Perspective*. Cheltenham: Edward Elgar.

——(1999). "Environmental Policy and Industrial Competitiveness: The Pollution-Haven Hypothesis Reconsidered", *Review of International Economics*, 7: 342–55.

Bommer, R., and Schulze, G. G. (1999). "Environmental Improvement with Trade Liberalization", *European Journal of Political Economy*, 15: 639–61.

Botteon, M., and Carraro, C. (1997). "Burden-Sharing and Coalition Stability in Environmental Negotiations with Asymmentric Countries", in Carraro, C. (ed.), *International Environmental Negotiations: Strategic Policy Issues*. Cheltenham: Edward Elgar, ch. 3, 26–55.

Bourke, I. J. (1988). *Trade in Forest Products: A Study of the Barriers Faced by Developing Countries*. FAO Forestry Paper 83. Rome: FAO.

——(1995). "International Trade in Forest Products and the Environment", *Unasylva*, 46/183: 11–17.

——(1998). "Prospects and Challenges in the Supply and Demand of Timber in the Global Market." Paper presented to the International Timber Conference '98, Kuala Lumpur, Malaysia, 5–7 March.

Bovenberg, L. A. (1995). "Environmental Taxation and Employment", *De Economist*, 143: 111–40.

——(1999). "Green Tax Reforms and the Double Dividend: An Updated Reader's Guide", *International Tax and Public Finance*, 6: 421–43.

Bovenberg, L. A., and de Mooij, R. A. (1994). "Environmental Levies and Distortionary Taxation", *American Economic Review*, 94: 1085–9.

——(1996). "Environmental Taxation and the Double-dividend: The Role of Factor Substitution and Capital Mobility", in Carraro, C. and Siniscalco, D. (eds.), *Environmental Fiscal Reform and Unemployment*. Dordrecht: Kluwer Academic publishers, 3–52.

——(1997). "Environmental Tax Reform and Endogenous Growth", *Journal of Public Economics*, 63/2: 207–37.

Bovenberg, L. A., and Goulder, L. H. (1996). "Optimal Environmental Taxation in the Presence of Other Taxes: General Equilibrium Analyses", *American Economic Review*, 86: 985–1000.

Bovenberg, L. A., and Ploeg, F. van der (1994). "Green Policies and Public Finance in a Small Open Economy", *Scandinavian Journal of Economics*, 96/3: 343–63.

——(1998). "Tax Reform, Structural Unemployment and the Environment", *Scandinavian Journal of Economics*, 100/3: 593–610.

Bovenberg, L. A., and Smulders, S. (1995). "Environmental Quality and Pollution Augmenting Technical Change in a Two-Sector Endogenous Growth Model", *Journal of Public Economics*, 57/3: 369–91.

——(1996). "Transitional Impacts of Environmental Policy in an Endogenous Growth Model", *International Economic Review*, 37/4: 861–93.

Brander, J. A., and Spencer, B. (1985). "Export Subsidies and International Market Share Rivalry", *Journal of International Economics*, 18: 83–100.

Brander, J. A., and Taylor, M. S. (1997a). "International Trade and Open Access Renewable Resources: The Small Open Economy Case", *Canadian Journal of Economics*, 30/3: 526–552.

—— (1997b). "International Trade between Consumer and Conservationist Countries", *Resource and Energy Economics*, 19/4: 267–97.

Bredahl, M., and Holleran, E. (1997). *Technical Regulations and Food Safety in NAFTA*. Columbia, Mo.: University of Missouri, Dept. of Agricultural Economics.

Brennan, G., and Buchanan, J. (1980). *The Power to Tax: Analytical Foundations of Fiscal Constitution*. Cambridge UK: Cambridge University Press.

Bretschger, L. (1997). "International Trade, Knowledge Diffusion, and Growth", *International Trade Journal*, 11/3: 327–48.

—— (1998a). "The Sustainability Paradigm: A Macroeconomic Perspective", *Revue Région et Développement*, 7: 73–103.

—— (1998b). "How to Substitute in Order to Sustain: Knowledge Driven Growth under Environmental Restrictions", *Environment and Development Economics*, 3: 425–42.

—— (1998c). "Nachhaltige Entwicklung der Weltwirtschaft: Ein Nord/Süd-Ansatz", *Schweizerische Zeitschrift für Volkswirtschaft und Statistik*, 134/3: 369–90.

—— (1999). *Growth Theory and Sustainable Development*. Aldershot UK: Edward Elgar.

Bromley, D. (1996). *The Environmental Implications of Agriculture*. Madison: University of Wisconsin, Dept. of Agricultural and Applied Economics. Staff Paper no. 401. Prepared for OECD Seminar on Environmental Benefits from Agriculture: Issues and Policies, Helsinki, Finland, 10–13 September 1996. (Available AgEcon Search Web site, http://agecon.lib.umn.edu/wis/stpap401.pdf.)

Browder, J. O., Matricardi, E. A. T., and Abdala, W. S. (1996). "Is Sustainable Tropical Timber Production Financially Viable? A Comparative Analysis of Mahogany Silviculture among Small Farmers in the Brazilian Amazon.", *Ecological Economics*, 16: 147–59.

Brown, D., Deardorff, A., and Stern, R. (1992). "A North American Free Trade Agreement: Analytical Issues and a Computational Assessment", *World Economy*, 15: 11–29.

Buchanan, J. (1994). "Lagged Implementation as an Element in Constitutional Strategy", *European Journal of Political Economy*, 10: 11–26.

Buchanan, J., and Tullock, G. (1975). "Polluters' Profits and Political Response, Direct Control Versus Taxes", *American Economic Review*, 65: 139–47.

Buchholz, W., and Konrad, K. A. (1994). "Global Environmental Problems and the Strategic Choice of Technology", *Journal of Economics/Zeitschrift für Nationalökonomie*, 60/3: 299–321.

Bucovetsky, S. (1991). "Asymmetric Tax Competition", *Journal of Urban Economics*, 30: 167–81.

Buitrón, X., and Mulliken, T. (1997). *CITES Appendix III and the Trade in Big-Leasfed Mahogany. A Traffic Network Report*. Cambridge, UK: Traffic.

Buongiorno, J., and Manurung, T. (1992). "Predicted Effects of an Import Tax in the European Community on the International Trade in Tropical Timbers", *mimeo.*, Department of Forestry. Madison: University of Wisconsin.

Burfisher, M. E., House, R. M., and Langley, S. V. (1992). "Free Trade Impacts on U.S. and Southern Agriculture", *Southern Journal of Agricultural Economics*, 24/1: 61–78.

Caldwell, L. K. (1990). *International Environmental Policy*, second edn. Durham NC: Duke University Press.

Carraro, C. (ed.) (1994). *Trade, Innovation, Environment*. Dordrecht *et al.*: Kluwer.

Carraro, C., and Siniscalco, D. (1992). "The International Dimension of Environmental Policy", *European Economic Review*, 36: 379–87.

—— (1993). "Strategies for the International Protection of the Environment", *Journal of Public Economics*, 52: 309–28.

—— (1995). "Policy Coordination for Sustainability: Commitments, Transfers, and Linked Negotiations", in Goldwin, I. and Winters, A. (eds.), *The Economics of Sustainable Development*. Cambridge: Cambridge University Press.

—— (1997). "RD Cooperation and the Stability of International Environmental Agreements", in Carraro, C. (ed.), *International Environmental Agreements: Strategic Policy Issues*. Cheltenham: E. Elgar, ch. 5, 71–96.

Cesar, H., and de Zeeuw, A. (1996). "Issue Linkage in Global Environmental Problems", in Xepapadeas, A. (ed.), *Economic Policy for the Environment and Natural Resources: Techniques for the Management and Control of Pollution*. Cheltenham: E. Elgar, 158–73.

Chander, P., and Tulkens, H. (1995). "A Core-Theoretic Solution for the Design of Cooperative Agreements on Transfrontier Pollution", *International Tax and Public Finance*, 2: 279–93.

—— (1997). "The Core of an Economy with Multilateral Environmental Externalities", *International Journal of Game Theory*, 26: 379–401.

Chapman, D. (1991). "Environmental Standards and International Trade in Automobiles and Copper: The Case for a Social Tariff", *Natural Resources Journal*, 31: 449–61.

Chapman, D., Agras, J., and Suri, V. (1999). "Industrial and Resource Location, Trade, and Pollution", in Dore, H., and Mount, T. (eds.), *Global Environmental Economics*. Malden/MA and Oxford/UK: Blackwell, 267–84.

Charnovitz, S. (1993). "Taxonomy of Environmental Trade Measures", *Georgetown International Environmental Law Review*, 6/1: 1–46.

—— (1997). "A Critical Guide to the WTO's Report on Trade and Environment: A Commentary", *Arizona Law Review*, 14/2: 341–79.

Chayes, A., and Chayes, H. (1993). "On Compliance", *International Organization*, 47: 175–205.

Chichilnisky, G. (1977). "Economic Development and Efficiency Criteria in the Satisfaction of Basic Needs", *Applied Mathematical Modelling*, 1.

—— (1994). "North-South Trade and the Global Environment", *American Economic Review*, 84: 851–74.

—— (1996). "Property Rights and the Dynamics of North-South Trade", in Bredahl, M. E., Ballenger, N., Dunmore, J. C., and Roe, T. L. (eds.), *Agriculture, Trade and the Environment: Discovering and Measuring the Critical Linkages*. Boulder, Colo.: Westview Press, 97–110.

Chichilnisky, G., and Di Matteo, M. (1996). *Trade, Migration and Environment: A General Equilibrium Analysis*. Milan: Fondazione Eni Enrico Mattei, Nota di lavoro 71.96.

Clarke, R., Boero, G., and Winters, A. (1996). "Controlling Greenhouse Gases: A Survey of Global Macroeconomic Studies", *Bulletin of Economic Research*, 48: 269–308.

Cline, W. R. (1992). "The Economics of Global Warming", Washington, DC: Institute for International Economics.

Clower, R. (1993). "The State of Economics: Hopeless but not Serious?", in Colander, D. (ed.), *The Spread of Economic Ideas*. Cambridge: Cambridge University Press.

Coase, R. H. (1960). "The Problem of Social Cost", *Journal of Law and Economics* 3: 1–44.

Coates, D. (1996). "Jobs Versus Wilderness Areas: The Role of Campaign Contributions", in Cole, M., Rayner, A., and Bates, J. (1998). "Trade Liberalisation and the Environment: The Case of the Uruguay Round", *World Economy*, 21: 33–347.

Congleton, R. D. (1992). "Political Regimes and Pollution Control". *Review of Economics and Statistics* (August), 74: 412–421.

—— (1995). "Toward a Transactions Cost Theory of Environmental Treaties", *Economia della Scelte Pubbliche*, 13: 119–39.

—— (ed.) (1996). *The Political Economy of Environmental Policy*. Ann Arbor: University of Michigan Press.

Congleton, R. D. (ed.) (1996). *The Political Economy of Environmental Protection: Analysis and Evidence*. Ann Arbor: University of Michigan Press, 69–96.

Conrad, K. (1993). "Taxes and Subsidies for Pollution-Intensive Industries as Trade Policy", *Journal of Environmental Economics and Management*; 25: 121–35.

—— (1996). "Choosing Emission Taxes under International Price Competition", in Carraro, C., Katsoulacos, Y., and Xepapadeas, A. (eds.), *Environmental Policy and Market Structure*. Dordrecht: Kluwer, 85–98.

—— (1999). Voluntary Environmental Agreements, Emission Taxes and International Trade: The Importance of Timing Strategies, University of Mannheim, Department of Economics and Statistics, Discussion Paper 562–99.

—— (2000). "Strategic Trade Policy under International Price and Quantity Competition", *Review of International Economics*, forthcoming.

Copeland, B. R., (1991). "International Trade in Waste Products in the Presence of Illegal Disposal", *Journal of Environmental Economics and Management*, 20: 143–62.

—— (1994). "International Trade and the Environment: Policy Reform in a Polluted Small Economy", *Journal of Environmental Economics and Management*, 26: 44–65.

Copeland, B. R., and Taylor, M. S. (1994). "North-South Trade and the Environment", *Quarterly Journal of Economics*, 109: 755–87.

—— (1995). "Trade and Transboundary Pollution", *American Economic Review*, 85: 716–37.

—— (1997). *A Simple Model of Trade, Capital Mobility, and the Environment*. NBER Working Paper Series, no. 5898.

Coughlin, P., and Nitzan, S. (1981). "Electoral Outcomes with Probabilistic Voting and Nash Social Welfare Maxima". *Journal of Public Economics* 15: 113–22.

Cropper, M. L., and W. E. Oates (1992). "Environmental Economics: A Survey", *Journal of Economic Literature*, 30: 675–740.

Daly, H. E. (1990). "Toward some Operational Principles of Sustainable Development", *Ecological Economics*, 2: 1–6.

Daly, H. E., and R. Goodland (1994). "An Ecological-Economic Assessment of Deregulation of International Commerce under GATT", *Ecological Economics*, 9: 73–92.

Dasgupta, P. S. (1995). "Optimal Development and NNP", in Goldin, I., and Winters, L. A. (eds.), *The Economics of Sustainable Development*. Cambridge: Cambridge University Press, 111–43.

D'Aspremont, C. A., and Gabszewicz, J. J. (1986). "On the Stability of Collusion", in Matthewson, G. F., and Stiglitz, J. E. (eds.), *New Developments in the Analysis of Market Structure*. New York: MacMillan, 243–64.

de Bruyn, S., and Heintz, R. (1999). "The Environmental Kuznets Curve Hypothesis", in van den Bergh, J. (ed.), *Handbook of Environmental and Resource Economics*. Cheltenham: Edward Elgar, ch. 46: 656–77.

Dean, J. (1992). "Trade and Environment: A Survey of the Literature", in Low, P. (ed.), *International Trade and the Environment*. World Bank Discussion Paper 159. Washington, DC: World Bank, 15–28.

—— (1999). "Testing the Impact of Trade Liberalization on the Environment", in Fredriksson, P. (ed.), *Trade, Global Policy, and the Environment*. World Bank Discussion Paper no. 402. Washington, DC: The World Bank, 55–63.

Demsetz, H. (1967). "Towards a Theory of Property Rights", *American Economic Review* 57: 347–60.

Dijkstra, B. R. (1999). *The Political Economy of Environmental Policy*. Cheltenham: Edward Elgar.

Diwan, I., and Shafik, N. (1992). "Investment, Technology, and the Global Environment: Towards International Agreement in a World of Dispute", in Low, P. (ed.), *International Trade and the Environment*. World Bank Discussion Paper 159. Washington, DC: World Bank, 263–87.

Dixit, A. (1985). "Tax Policy in Open Economies", in A. Auerbach and M. Feldstein (eds.), *Handbook of Public Economics, Vol. 1*. Amsterdam: North-Holland.

—— (1996). "Special-Interest Lobbying and Endogenous Commodity Taxation", *Eastern Economic Journal*, 22: 375–88.

Dockner, E., and Long, N. V. (1993). "International Pollution Control: Cooperative vs. Non-Cooperative Strategies", *Journal of Environmental Economics and Management*, 25: 13–29.

Donsimoni, M. P., Economides, N. S., and Polimarchakis, H. M. (1986). "Stable Cartels", *International Economic Review*, 27: 317–27.

Douma, W. T., (1991). *International Regulations on the Export of Hazardous Waste: The Evolution of a New Piece of International Environmental Law*, Groningen: Rijksuniversiteit Papers on Development and Security No. 33.

Downs, A. (1957). *An Economic Theory of Democracy*. New York: Harper and Row.

Drake, P., Haid, D., Hammons, E. J., and Williams, D. (1995). *Analysis of Macroeconomic Trends in the Supply and Demand of Sustainably Produced Tropical Timber from the Asia-Pacific Region Phase II*. Report to the International Tropical Timber Organization, Yokohoma.

Dubois, O., Robins, N., and Bass, S. (1995). *Forest Certification. Report to the European Commission*. London: International Institute for Environment and Development.

Duerksen, C., and Leonard, H. (1980). "Environmental Regulations and the Location of Industries an International Perspective", *Columbia Journal of World Business*, 15: 52–68.

Eaton, J., and G. Grossman (1986). "Optimal Trade and Industrial Policy under Oligopoly", *Quarterly Journal of Economics*, 101: 383–406.

Economist, The (1992). "Let Them Eat Pollution", 8 February 1992, 66.

EIA (Environmental Investigation Agency) (1996). *Corporate Power, Corruption and the Destruction of the World's Forests: The Case for a New Global Forest Agreement.* London: EIA.

Elbasha, E. H., and Roe, T. (1996). "On Endogenous Growth: The Implications of Environmental Externalities", *Journal of Environmental Economics and Management*, 31/2: 240–68.

Eliste, P., and P. Fredriksson (1999). "The Political Economy of Environmental Regulations, Government Assistance, and Foreign Trade", in Fredriksson, P. (ed.), *Trade, Global Policy, and the Environment.* World Bank Discussion Paper no. 402, Washington, DC, ch. 9, 129–39.

Elliot, G. (1994). "The Trade Implications of Recycled Content in Newsprint: The Canadian Experience", in Organization for Economic Cooperation and Development (OECD), *Life-Cycle Management and Trade.* Paris: OECD.

Elster, J. (1989). "Social Norms and Economic Theory", *Journal of Economic Perspectives*, 3: 99–117.

Endres, A. (1993). "Internationale Vereinbarungen zum Schutz der globalen Umweltressourcen: Der Fall porportionaler Emissionsreduktion", *Aussenwirtschaft*, 48/1: 51–76.

Endres, A., and Finus, M. (1998). "Playing a Better Global Emission Game: Does it Help to be Green?", *Swiss Journal of Economics and Statistics*, 134/1: 21–40.

Ervin, D. E. (1997a). *Agriculture, Trade and the Environment: Anticipating the Policy Challenges.* Paris: Organization for Economic Cooperation and Development. OECD/GD(97)171.

——(1997b). *Interactions Between Trade and the Environment in the Fruit and Vegetable Sector of OECD Countries.* Paris: Organization for Economic Cooperation and Development. Unpublished report.

——(1999). "Toward GATT-Proofing Environmental Programmes for Agriculture", *Journal of World Trade*, 33/2: 63–82.

Ervin, D. E., and Fox, G. (1999). "Environmental Policy Considerations in the Grain-Livestock Subsectors in Canada, Mexico, and the United States", in Lyons, R., Knutson, R., and Meilke, K. (eds.), *Economic Harmonization in the Canadian, U.S., and Mexican Grain-Livestock Subsector.* Winnipeg: University of Manitoba, 275–303.

Ervin, D. E., and Keller, V. N. (1996). "Key Questions", in Bredahl, M. E., Ballenger, N., Dunmore, J. C., and Roe, T. L. (eds.), *Agriculture, Trade and the Environment: Discovering and Measuring the Critical Linkages.* Boulder, Colo.: Westview Press, 281–94.

Ervin, D. E., and Schmitz, A. (1996). "A New Era of Environmental Management in Agriculture?", *American Journal of Agricultural Economics*, 78: 1198–206.

Ervin, D. E., Runge, C., Runge, F., Graffy, E. A., Anthony, W. E., Batie, S. S., Faeth, P., Penny, T., and Warman, T. (1998). "Agriculture and the Environment: A New Strategic Vision", *Environment*, 40/6: 8–15, 35–40.

Esty, D. (1994). *Greening the GATT: Trade, Environment, and the Future.* Washington, DC: Institute for International Economics.

——(1996). "Revitalizing Environmental Federalism", *Michigan Law Review*, 95: 570–653.

Ethier, W. (1984). "Higher Dimensional Issues in Trade Theory", in Jones, R., and Kenen, P. (eds.), *Handbook of International Economics, Vol.1 (International Trade)*. Amsterdam: North Holland, 131–84.

Ewijk, C. van, and Wijnbergen, S. van (1995). "Can Abatement Overcome the Conflict between Environment and Economic Growth?", *De Economist*, 143/2: 197–216.

Eyckmans, J. (1997). "Nash Implementation of a Proportional Solution to International Pollution Control Problems", *Journal of Environmental Economics and Management*, 33/3: 314–30.

Faeth, P. (1995). *Growing Green: Enhancing the Economic and Environmental Performance of U.S. Agriculture*. Washington, DC: World Resources Institute.

FAO (Food and Agricultural Organization of the United Nations) (1995). *FAO Forest Products Yearbook 1993*. Rome: FAO.

——(1997). *State of the World's Forests 1997*. Rome: FAO.

——(1998). FAOSTAT Agriculture Data. Web site http://apps.fao.org/cgi-bin/nph-db.pl?subset = agriculture (accessed June 1999).

Farzin, Y. H. (1996). "Optimal Pricing of Environmental and Natural Resource Use with Stock Externalities", *Journal of Public Economics*, 62: 31–57.

Feld, L. P., Pommerehne, W. W., and Hart, A. (1996). "Private Provision of a Public Good" in Congleton, R. D. (ed.), *The Political Economy of Environmental Protection: Analysis and Evidence*. Ann Arbor: University of Michigan Press, 227–50.

Felder, S. T., and Rutherford, F. (1993). "Unilateral CO_2 Reductions and Carbon Leakage: The Consequences of International Trade in Oil and Basic Materials", *Journal of Environmental Economics and Management*, 25/2: 162–76.

Felder, S., and Schleiniger, R. (1995). "Domestic Environmental Policy and International Factor Mobility: A General Equilibrium Analysis", *Swiss Journal of Economics and Statistics*, 131: 547–58.

——(2000). "Optimal Differentiation of International Environmental Taxes in the Presence of National Labor Market Distortions", *Environmental and Resource Economics*, 15: 89–102.

Felke, R. (1998). *European Environmental Regulations and International Competitiveness: The Political Economy of Environmental Barriers to Trade*. Baden-Baden: Nomos.

Ferguson, W., and Padula, A. (1994). *Economic Effects of Banning Methyl Bromide for Soil Fumigation*. Washington, DC: US Dept. of Agriculture, Economic Research Service. Agricultural Economic Report no. 677.

Ferrantino, M. (1997). "International Trade, Environmental Quality and Public Policy", *World Economy*, 20: 43–72.

Finus, M., and Rundshagen, B. (1998). "Toward a Positive Theory of Coalition Formation and Endogenous Instrumental Choice in Global Pollution Control", *Public Choice*, 96: 145–86.

Flora, D. F., and McGinnis, W. J. (1991). *Effects of Spotted-Owl Reservations, The State Log Embargo, Forest Replanning and Recession on Timber Flows and Prices in the Pacific Northwest and Abroad*, unpublished review draft, Trade Research, Pacific Northwest Research Station, USDA Forest Service, Seattle.

Folmer, H., Mouche, P. van, and Ragland, S. E. (1993). "Interconnected Games and International Environmental Problems", *Environmental and Resource Economics*, 3/4: 313–35.

Forster, B. (1977). "Pollution Control in a Two-Sector Dynamic General Equilibrium Model", *Journal of Environmental Economics and Management*, 4: 305–12.

Forsythe, K., and Evangelou, P. (1994). "Costs and Benefits of Irradiation Quarantine Treatments for U.S. Fruit and Vegetable Imports", in Sullivan, J. (ed.), *Environmental Policies: Implications for Agricultural Trade*. Washington, DC: US Dept. of Agriculture, Economic Research Service, 82–90. Foreign Agricultural Economic Report no. 252.

Francois, J., McDonald, B., and Nordstrom, H. (1995). "Assessing the Uruguay Round", in Martin, W., and Winters, A. (eds.), *The Uruguay Round and the Developing Countries*. World Bank Discussion Paper no. 307. Washington, DC: World Bank.

Fredriksson, P. (1997a). "The Political Economy of Pollution Taxes in a Small Open Economy", *Journal of Environmental Economics and Management*, 33: 44–58.

——(1997b). "Environmental Policy Choice: Pollution Abatement Subsidies", *Resource and Energy Economics*, 20: 51–63.

——(1999a). "The Political Economy of Trade Liberalization and Environmental Policy", *Southern Economic Journal*, 65: 513–25.

——(ed.) (1999b). "Trade, Global Policy, and the Environment." World Bank Discussion Paper no. 402. Washington, DC: World Bank.

Fredriksson, P., and Gaston, N. (1999). "The Importance of Trade for the Ratification of the 1992 Climate Change Convention", in Fredriksson, P. G. (ed.), *Trade, Global Policy, and the Environment*. World Bank Discussion Papers. Washington, DC: The World Bank, ch. 12.

——(2000). "Environmental Governance in Federal Systems: The Effects of Capital Competition and Lobby Groups", *Economic Inquiry*, 38: 501–14.

Friedman, J, Gerlowski, D., and Silberman, J. (1992). "What Attracts Foreign Multinational Corporations? Evidence From Branch Plant Location in the United States", *Journal of Regional Science*, 32: 403–18.

Fullerton, D. (1997). "Environmental Levies and Distortionary Taxation: Comment", *American Economic Review*, 87: 245–51.

Fullerton, D., and Metcalf, G. E. (1997). "Environmental Taxes and the Double Dividend Hypothesis: Did you Really Expect Something for Nothing?", NBER working paper 6199.

Gabel, H. L. (1994). "The Environmental Effects of Trade in the Transport Sector", in *The Environmental Effects of Trade*. Paris: Organization for Economic Cooperation and Development, 153–73.

Gardner, B. L. (1996). "Environmental Regulation and the Competitiveness of U.S. Agriculture", in Bredahl, M. E., Ballenger, N., Dunmore, J. C., and Roe, T. L. (eds.), *Agriculture, Trade and the Environment: Discovering and Measuring the Critical Linkages*. Boulder, Colo.: Westview Press, 215–30.

Germain, M., Toint, P., and Tulkens, H. (1997). *Financial Tranfers to Ensure Cooperative International Optimality in Stock Pollutant Abatement*, CORE Discussion Paper no. 9701, Center for Operations Research Econometrics (CORE), Université catholique de Louvain, Belgium.

Gillis, M. (1990). *Forest Incentive Policies*. Paper prepared for the World Bank Forest Policy Paper. Washington, DC: World Bank.

Giordano, M. (1994). "Tropical Forest Policy and Trade: A Case Study of Malaysia", in Sullivan, J. (ed.), *Environmental Policies: Implications for Agricultural Trade*.

Washington, DC: US Dept. of Agriculture, Economic Research Service, 113–21. Foreign Agricultural Economic Report no. 252.

Golombek, R., Hagem, C., and Hoel, M. (1994). "The Design of a Carbon Tax in an Incomplete International Climate Agreement", in Carraro, C. (ed.), *Trade, Innovation, Environment*, Dordrecht: Kluwer, ch. 3.3, 323–61.

Gould, D., and G. Woodbridge (1998). "The Political Economy of Retaliation, Liberalization and Trade Wars", *European Journal of Political Economy*, 14: 115–37.

Goulder, L. H. (1995). "Environmental Taxation and the Double Dividend: A Reader's Guide", *International Tax and Public Finance*, 2: 157–83.

Goulder, L. H., Parry, I. W. H., and Burtraw, D. (1997). "Revenue-Raising versus other Approaches to Environmental Protection: the Critical Significance of Preexisting Tax Distortions", *Rand Journal of Economics*, 28: 708–31.

Gradus, R., and Smulders, S. (1993). "The Trade-off between Environmental Care and Long-Term Growth: Pollution in Three Prototype Growth Models", *Journal of Economics*, 58/1: 25–51.

Gray, H., and Walter, I. (1983). "Investment Related Trade Distortions", *Journal of World Trade Law*, 17: 283–307.

Grossman, G. M. (1995). "Pollution and Growth: What Do we Know?", in Goldin, I., and Winters, L. A. (eds.), *The Economics of Sustainable Development*, Cambridge: Cambridge University Press.

Grossman, G. M., and Helpman, E. (1991). *Innovation and Growth in the Global Economy*, Cambridge Mass.: MIT Press.

——(1994). "Protection for Sale", *American Economic Review*, 84: 833–50.

——(1995). "Trade Wars and Trade Talks", *Journal of Political Economy*, 103: 675–708.

Grossman, G. M., and Krueger, A. B. (1993). "Environmental Impacts of a North American Free Trade Agreement", in Garber, P. (ed.), *The US–Mexico Free Trade Agreement*. Cambridge Mass.: MIT Press.

——(1995). "Economic Growth and the Environment", *Quarterly Journal of Economics*, 110: 353–77.

Grut, M., Gray, J. A., and Egli, N. (1991). *Forest Pricing and Concession Policies: Managing the High Forests of West and Central Africa*. World Bank Technical Paper no. 143, Africa Technical Department Series. Washington, DC: The World Bank.

Gürtzgen, N., and M. Rauscher (1999). "Environmental Policy, Intra-Industry Trade, and Transfrontier Pollution", University of Rostock, Dept. of Economics, mimeo.

Gylfason, T., Herbertsson, T. T., and Zoega, G. (1997). "A Mixed Blessing: Natural Resources and Economic Growth", *Centre of Economic Policy Research*, Discussion Paper no. 1668.

Haupt, A. (1999). "Environmental Product Standards, International Trade and Monopolistic Competition", Dept. of Economics, European University Viandrina, Frankfurt (Oder), mimeo.

Heal, G. (1994). "Formation of International Environmental Agreements", in Carraro, C. (ed.), *Trade, Innovation, Environment*, Dordrecht: Kluwer, ch. 3.2, 301–22.

Heister, J., Mohr, E., Plesmann, W., Stähler, F., Stoll, T., and Wolfrum, R. (1995). "Economic and Legal Aspects of International Environmental Agreements: The Case of Enforcing and Stabilizing an International CO_2 Agreement", Kiel Working Paper no. 711, Kiel Institute of World Economics, Kiel.

Helfand, G. E. (1991). "Standards Versus Standards: The Effects of Different Pollution Restrictions", *American Economic Review*, 81: 622–34.

Henderson, V. (1996). "Effects of Air Quality Regulation", *American Economic Review*, 86: 789–813.

Hettich, F. (1998). "Growth Effects of a Revenue Neutral Environmental Tax Reform", *Journal of Economics*, 67/3: 287–316.

Hettich, F., and Svane, M. S. (1998). "Environmental Policy in a Two Sector Endogenous Growth Model", EPRU Working Paper Series 1998-04, Copenhagen Business School.

Hettige, H., Martin, P., Sing, M., and Wheeler, D. (1994). *The Industrial Pollution Projection System*. Washington, DC: World Bank.

Hillman, A. (1998). "Political Economy and Political Correctness", *Public Choice*, 96: 219–39.

Hillman, A., and Ursprung, H. (1988). "Domestic Politics, Foreign Interests, and International Trade Policy", *American Economic Review*, 78: 729–45.

—— (1992). "The Influence of Environmental Concerns on the Political Determination of Trade Policy", in Anderson, K., and Blackhurst, R. (eds.), *The Greening of World Trade Issues*. London: Harverster Wheatsheaf.

—— (1994a). "Greens, Supergreens, and International Trade Policy: Environmental Concerns and Protectionism", in Carraro, C. (ed.), *Trade, Innovation, Environment*. Dordrecht: Kluwer, 75–108.

—— (1994b). "Domestic Politics, Foreign Interests, and International Trade Policy: Reply", *American Economic Review*, 84: 1476–78.

Hoekman, B., and Leidy, M. (1992). "Environmental Policy Formation in a Trading Economy: A Public Choice Perspective", in Anderson, K. and Blackhurst, R. (eds.), *The Greening of World Trade Issues*. New York: Harvester-Wheatsheaf.

Hoel, M. (1991). "Global Environmental Problems: The Effects of Unilateral Actions Taken by One Country", *Journal of Environmental Economics and Management*, 21: 55–70.

—— (1992). "International Environment Conventions: The Case of Uniform Reductions of Emissions", *Environmental and Resource Economics*, 2: 141–59.

—— (1994). "Efficient Climate Policy in the Presence of Free Riders", *Journal of Environmental Economics and Management*, 27: 259–74.

—— (1997a). Environmental Policy with Endogenous Plant Locations, *Scandinavian Journal of Economics*, 99: 241–59.

—— (1997b). "International Coordination of Environmental Taxes", in Carraro, C. (ed.), *New Directions in the Economic Theory of the Environment*. Cambridge: Cambridge University Press, ch. 5, 105–46.

Hoel, M., and Schneider, K. (1997). "Incentives to Participate in an International Environmental Agreement", *Environmental and Resource Economics*, 9: 153–70.

Holländer, H. (1990). "A Social Exchange Approach to Voluntary Cooperation", *American Economic Review*, 80: 1157–67.

Holmlund, B., and Kolm, A-S. (1997). "Environmental Tax Reform in a Small Open Economy with Structural Unemployment". Department of Economics, Uppsala University, working paper 1997: 2.

Howarth, R. B. (1995). "Sustainability under Uncertainty: A Deontological Approach", *Land Economics*, 71/4: 417–27.

Howarth, R. B., and Norgaard, R. B. (1992). "Environmental Valuation under Sustainable Development", *American Economic Review, Papers and Proceedings*, 82/2: 473–7.

Howitt, R. (1991). "Water Policy Effects on Crop Production and Vice-Versa: An Empirical Approach", in Just, R. E. and Bockstael, N. (eds.), *Commodity and Resource Policies in Agricultural Systems*. New York: Springer-Verlag, 234–53.

Hyde, W. F., and Sedjo, R. A. (1992). "Managing Tropical Forests: Reflections on the Rent Distribution Discussion", *Land Economics*, 68/3: 343–50.

Jaffe, A., and Palmer, K. (1997). "Environmental Regulation and Innovation: A Panel Data Study", *Review of Economics and Statistics*, 79: 610–19.

Jaffe, A., Peterson, S., Rotney, P., and Stavins, R. (1995). "Environmental Regulation and the Competitiveness of U.S. Manufacturing: What Does the Evidence Tell Us?", *Journal of Economic Literature*, 33: 132–63.

Johnson, H. (1953). "Optimum Tariffs and Retaliation", *Review of Economic Studies*, 21: 142–53.

Jones, T., and Wibe, S. (1992). *Forests: Market and Intervention Failures: Five Case Studies*. London: Earthscan.

Just, R. E., Lichtenberg, E., and Zilberman, D. (1991). "Effects of Feed Grain and Wheat Programs on Irrigation and Groundwater Depletion in Nebraska", in Just, R. E. and Bockstael, N. (eds.), *Commodity and Resource Policies in Agricultural Systems*. New York: Springer-Verlag, 215–33.

Kahane, L. (1996). "Congressional Voting Patterns on NAFTA: An Empirical Analysis", *American Journal of Economics and Sociology*, 55: 395–409.

Kalt, J. (1988). "The Impact of Domestic Environmental Regulatory Policies on U.S. International Competitiveness", in Spence, M., and Hazard, H. (eds.), *International Competitiveness*. Cambridge Mass.: Ballinger, 221–62.

Kemp, M. (1964). *The Pure Theory of International Trade*. Englewood Cliffs NJ: Prentice-Hall.

Kemp, R. H., and Phantumvanit, D. (1995). "1995 Mid-Term Review of Progress Towards the Achievement of the Year 2000; Objective. Report to the International Tropical Timber Council XIX Session, 8–16 November", Yokohoma.

Kennedy, P. W. (1994). "Equilibrium Pollution Taxes in Open Economies with Imperfect Competition", *Journal of Environmental Economics and Management*, 27: 49–63.

Kerr, S. (1995). *Markets versus International Funds for Implementing International Environmental Agreements: Ozone Depletion and the Montreal Protocol*. Working Paper No. 95–12, The Center for International Affairs, Harvard University, Harvard.

Kiekens, J.-P. (1995). "Timber Certification: A Critique", *Unasylva*, 46/183: 27–8.

—— (1997). *Eco-Certification: Tendances Internationales et Implications Forstières et Commerciales*, Étude réalisée pour le Ministère de l'Environment, des Ressource Naturelles et de l'Agriculture de la Région Wallonne (Belgique), Brussels.

Killinger, S. (1996). "Indirect Internalization of International Environmental Externalities", *Finanzarchiv*, 53/3 + 4: 332–68.

Kirchgässner, G., and Mohr, E. (1996). "Trade Restrictions as Viable Means of Enforcing Compliance with International Environmental Law: An Economic

Assessment", in Wolfrum, R. (ed.), *Enforcing Environmental Standards: Economic Mechanisms as Viable Means?*, Berlin: Springer, 199–226.

Koch, K.-J., and Schulze, G. (1998). "Equilibria in Tax Competition Models", in K. Koch and K. Jaeger (eds.), *Trade, Growth, and Economic Policy in Open Economies, Essays in Honor of Hans-Jürgen Vosgerau*. Berlin: Springer, 281–311.

Koopmans, T. (1965). "On the Concept of Optimal Economic Growth", *Pontificiae Academiae Scientorum Scripta Varia*, 28, reprint in "The Econometric Approach to Development Planning". Amsterdam: North-Holland, 1966.

Körber, A. (1998). "Why Everybody Loves Flipper: The Political Economy of the U.S. Dolphin-safe Laws", *European Journal of Political Economy*, 14: 457–509.

Koskela, E., and Schöb, R. (1999). "Alleviating Unemployment: The Case for Green Tax Reforms", *European Economic Review*, 43: 1723–46.

Kox, H., and Tak, C. van der (1996). "Nontransboundary Pollution and the Efficiency of International Environmental Cooperation", *Ecological Economics*, 19: 247–59.

Kreps, D. M., Milgrom, P., Roberts, J., and Wilson, R. (1982). "Rational Cooperation in the Finitely Repeated Prisoner's Dilemma", *Journal of Economic Theory*, 27: 245–52.

Krissoff, B., Ballenger, N., Dunmore, J. C., and Gray, D. (1996). *Exploring Linkages Among Agriculture, Trade and Environment: Issues for the Next Century*. Washington, DC: US Dept. of Agriculture, Economic Research Service. Agricultural Economic Report no. 738. (Available at Web site http://www.ers.usda.gov/epubs/pdf/aer738/index.htm.)

Krugman, P. (1980). "Scale Economies, Product Differentiation, and the Pattern of Trade", *American Economic Review*, 70: 950–9.

Krutilla, K. (1991). "Environmental Regulation in an Open Economy", *Journal of Environmental Economics and Management*, 20: 127–42.

Kuch, P., and Reichelderfer, K. (1992). "The Environmental Implications of Agricultural Support Programs: A United States Perspective", in Becker, T., Gray, R., Schmitz, A. (eds.), *Improving Agricultural Trade Performance Under the GATT*. Kiel, Germany: Wissenschaftsverlag Vauk Kiel KG, 215–31.

Kuik, O., and Verbruggen, H. (1995). *International Trade and Nature Conservation: The Polluter-Pays Principle*. Amsterdam: Institute for Environmental Studies. R-95/02.

Kverndokk, S. (1993). "Global CO_2 Agreements: A Cost-Effective Approach", *The Energy Journal*, 14/2: 91–112.

—— (1994). "Coalitions and Side Payments in International CO_2 Treaties", in van Ierland, E. C. (ed.), *International Environmental Economics: Theories, Models and Applications to Climate Change, International Trade and Acidification*. Amsterdam: Elsevier, 45–76.

Laffont, J.-J., and Tirole, J. (1996). "Pollution Permits and Compliance Strategies", *Journal of Public Economics*, 62: 85–125.

Larson, B. A., and Tobey, J. A. (1994). "Uncertain Climate Change and the International Policy Response", *Ecological Economics*, 11: 77–84.

Lee, D. (1999). "Lowering the Cost of Pollution Control Versus Controlling Pollution", *Public Choice*, 100: 123–34.

Lee, H., and Roland-Holst, D. (1993). *International Trade and the Transfer of Environmental Costs and Benefits*. Paris: Organization for Economic Cooperation and Development. Technical Paper no. 91.

Leetmaa, S., and Smith, M. (1996). *The Conservation Reserve Program: Implications of a Reduced Program on U.S. Grain Trade*. Washington, DC: US Dept. of Agriculture, Economic Research Service. Working Paper.

Leidy, M., and Hoekman, B. (1994). "'Cleaning Up' while Cleaning Up? Pollution Abatement, Interest Groups and Contingent Trade Policies", *Public Choice*, 78: 241–58.

—— (1996). "Pollution Abatement, Interest Groups, and Contingent Trade Policies", in Congleton, R. (ed.), *The Political Economy of Environmental Protection*. Ann Arbor: University of Michigan Press, 43–68.

Leonard, J. (1988). *Pollution and the Struggle for World Product*. Cambridge: Cambridge University Press.

Levinson, A. (1995). "Environmental Regulations and Industry Location: International and Domestic Evidence", in Bhagwati, J., and Hurdec, R. (eds.), *Fair Trade and Harmonization: Prerequisite for Free Trade?, Volume I: Economic Analysis*. Cambridge Mass.: MIT Press, ch. 11. 429–57.

—— (1996). "Environmental Regulations and Manufacturers' Location Choices: Evidence from the Census of Manufactures", *Journal of Public Economics*, 62: 5–29.

—— (1999a). "State Taxes and Interstate Hazardous Waste Shipments", *American Economic Review*, 89: 666–77.

—— (1999b). "NIMBY Taxes Matter: The Case of State Hazardous Waste Disposal Taxes", *Journal of Public Economics*, 74: 31–51.

Long, V. N., and Siebert, H. (1991). "Institutional Competition Versus ex-ante Harmonization: The Case of Environmental Policy", *Journal of Institutional and Theoretical Economics*, 147: 296–311.

Low, P. (1992a). "Trade Measures and Environmental Quality: The Implications for Mexico's Exports", in Low, P. (ed.), *International Trade and the Environment*. World Bank Discussion Paper 159. Washington, DC: World Bank, 105–20.

—— (ed.) (1992b). "International trade and environment", World Bank Discussion Papers 159, The World Bank, Washington, DC.

Low, P., and Yeats, A. (1992). "Do 'Dirty' Industries Migrate?", in Low, P. (ed.), *International Trade and the Environment*. World Bank Discussion Paper 159. Washington, DC: World Bank, 89–103.

Lucas, R. E., Wheeler, D., and Hettige, H. (1992). "Economic Development, Environmental Regulation, and the International Migration of Toxic Industrial Pollution: 1960–1988", in Low, P. (ed.), *International Trade and the Environment*. World Bank Discussion Paper 159. Washington, DC: World Bank, 67–86.

Ludema, R. D., and Wooton, I. (1994). "Cross-border Externalities and Trade Liberalization: The Strategic Control of Pollution", *Canadian Journal of Economics*, 27: 950–66.

Lutz, E. (1992). "Agricultural Trade Liberalization, Price Changes, and Environmental Effects", *Environmental and Resource Economics*.

MacDougall, G. D. A. (1960). "The Benefits and Costs of Private Investment from Abroad: A Theoretical Approach", *Economic Record*, 36: 13–35.

Maestad, O., (1998). "On the Efficiency of Green Trade Policy", *Environmental and Resource Economics*, 11: 1–18.

Magee, S. P., Brock, W., and Young, L. (1989). *Black Hole Tariffs and Endogenous Political Theory*. Cambridge: Cambridge University Press.

Mäler, K.-G. (1990). "International Environmental Problems", *Oxford Review of Economic Policy*, 6/1: 80–108.

—— (1991). "Incentives in International Environmental Problems", in Siebert, H. (ed.), *Environmental Scarcity: The International Dimension*. Tübingen: Mohr, 75–93.

Maloney, M. T., and McCormick, R. E. (1982). "A Positive Theory of Environmental Quality Regulation", *Journal of Law and Economics*, 25: 99–123.

Mani, M., and Wheeler, D. (1999). "In Search of Pollution Havens? Dirty Industry in the World Economy", in Fredriksson, P. (ed.), *Trade, Global Policy, and the Environment*. World Bank Discussion Paper no. 402. Washington, DC: The World Bank, 115–28.

Mani, M., Pargal, S., and Huq, M. (1996). "Does Environmental Regulation Matter? Determinants of the Location of New Manufacturing Plants in India in 1994." Washington, DC: Worldbank, Policy Research Working Papers no. 1718. (Available also online at http://www.worldbank.org/html/dec/Publications/Workpapers/WPS1700series/wps1718/ wps1718.pdf.)

Marchant, M. A., and Ballenger, N. (1994). "The Trade and Environment Debate: Relevant for Southern Agriculture?", *Journal of Agricultural and Applied Economics*, 26/1: 108–28.

Marks, St., and McArthur, J. (1990). "Empirical Analyses of the Determination of Protection: A Survey and Some New Results", in Odell, J., and Willett, T. (eds.), *International Trade Policies*. Ann Arbor: University of Michigan Press, 105–39.

Markusen, J. (1975a). "International Externalities and Optimal Tax Structures", *Journal of International Economics*, 5: 15–29.

—— (1975b). "Cooperative Control of International Pollution and Common Property Resources", *Quarterly Journal of Economics*, 89: 618–32.

—— (1997). "Costly Pollution Abatement, Competitiveness and Plant Location Decisions", *Resource and Energy Economics*, 19: 29–320.

Markusen, J., and Venables, A. (1998). "Multinational Firms and the New Trade Theory", *Journal of International Economics*, 46: 183–203.

Markusen, J., Morey, E., and Olewiler, N. (1993). "Environmental Policy when Market Structure and Plant Locations are Endogenous", *Journal of Environmental Economics and Management*, 24: 69–86.

—— (1995). "Competition in Regional Environmental Policies when Plant Locations are Endogenous", *Journal of Public Economics*, 56: 55–77.

McCarl, B., Nayda, W., and Houshmand, Z. (1994). *Environmental and Economic Effects of Trade and Agricultural Policy Alterations*. Washington, DC: US Congress, Office of Technology Assessment.

McConnell, K. E. (1997). "Income and the Demand for Environmental Quality", *Environment and Development Economics*, 2/4: 383–99.

McGuire, M. (1982). "Regulation, Factor Rewards, and International Trade", *Journal of Public Economics*, 17: 335–54.

Meadows D. M., Meadows D., Randers J., and Behrens, W. (1972). *The Limits to Growth: A Report for The Club of Rome's Project on the Predicament of Mankind*. New York: Universe Books.

Merrifield, J. D. (1988). "The Impact of Selected Abatement Strategies on Transnational Pollution, the Terms of Trade, and Factor Rewards: A General Equilibrium Approach", *Journal of Environmental Economics and Management*, 15: 259–84.

M'Gonigle, R. M., and Zacher, M. W. (1979). *Pollution, Politics, and International Law*. Berkeley: University of California Press.

Miranowski, J., Hrubovcak, J., and Sutton, J. (1991). "The Effects of Commodity Programs on Resource Use", in Just, R. E. and Bockstael, N. (eds.), *Commodity and Resource Policies in Agricultural Systems*. New York: Springer.

Mohr, E. (1991). "Global Warming: Economic Policy in the Face of Positive and Negative Spillovers", in Siebert, H. (ed.), *Environmental Scarcity: The International Dimension*, Tübingen: Mohr, 187–212.

——(1995). "International Environmental Permit Trade and Debt: The Consequences of Country Sovereignty and Cross-Default Policies", *Review of International Economics*, 3/1: 1–19.

Mohr, E., and Thomas, J. (1998). "Pooling Sovereign Risks: The Case of Environmental Treaties and International Debt", *Journal of Development Economics*, 55/1: 153–69.

Mooij, R. A. de (2000). *Environmental Taxation and the Double Dividend*. Contributions to Economic Analysis 246, Amsterdam: North-Holland.

Moomaw, W. R., and Unruh, G. C. (1997). "Are Environmental Kuznets Curves Misleading us? The Case of CO_2 Emissions", *Environment and Development Economics*, 2/4: 451–63.

Moyers, B. (1990). *Global Dumping Ground: The International Traffic in Hazardous Waste*. Washington: Seven Locks Press.

Mueller, D. C. (1989). *Public Choice II*. Cambridge: Cambridge University Press.

Murdoch, J. C., and Sandler, T. (1997). "The Voluntary Provision of a Pure Public Good: The Case of Reduced CFCs Emissions and the Montreal Protocol", *Journal of Public Economics*, 63: 331–49.

Murdoch, J. C., Sandler, T., and Sargent, K. (1997). "A Tale of Two Collectives: Sulfur versus Nitrogen Oxides Emission Reduction in Europe", *Economica*, 64: 281–301.

Murrel, P. O., and Ryterman, R. (1991). "A Methodology for Testing Comparative Economic Theories: Theory and Application to East-West Environmental Problems", *Journal of Comparative Economics*, 15: 582–601.

NEI (Netherlands Economic Institute) (1989). *An Import Surcharge on the Import of Tropical Timber in the European Community: An Evaluation*. Rotterdam: NEI.

Nielsen S. B., Pedersen, L. H., and Sørensen, P. B. (1995). "Environmental Policy, Pollution, Unemployment, and Endogenous Growth", *International Tax and Public Finance*, 2/2: 185–205.

Nunez-Muller, M. (1990). "The Schoenberg Case: Transfrontier Movements of Hazardous Waste", *Natural Resources Journal*, 30/1: 153–61.

Oates, W. E., and Schwab, R. M. (1988). "Economic Competition Among Jurisdictions: Efficiency Enhancing or Distortion Inducing?", *Journal of Public Economics*, 35: 333–54.

OECD (Organization for Economic Cooperation and Development) (1978). *Macroeconomic Evaluation of Environmental Programmes*. Paris: Organization for Economic Cooperation and Development.

——(1993). *Agricultural and Environmental Policy Integration: Recent Progress and New Directions*. Paris.

——(1994a). *Methodologies for Environmental and Trade Reviews*. Paris. OECD/GD(94)103.

——(1994b). *The Environmental Effects of Trade*. Paris.

OFI (Oxford Forestry Institute) in association with the Timber Research and Development Association (TRADA) (1991). *Incentives in Producer and Consumer Countries to Promote Sustainable Development of Tropical Forests*, ITTO Pre-Project Report, PCM, PCF, PCI(IV)/1/Rev. 3, OFI, Oxford.

Olson, M. (1965). *The Logic of Collective Action. Public Goods and the Theory of Groups*. Cambridge, Mass.: Harvard University Press.

Orden, D., and Roberts, D. H. (1997). *Understanding Technical Barriers to Agricultural Trade*. St Paul, Minn.: International Agricultural Trade Research Consortium. Conference proceedings.

Osborn, T. (1993). "The Conservation Reserve Program: Status, Future, and Policy Options", *Journal of Soil and Water Conservation*, 48/4: 271–8.

Osteen, C. D., and Szedmra, P. I. (1989). *Agricultural Pesticide Use Trends and Policy Issues*. Washington, DC: US Dept. of Agriculture, Economic Research Service. Agricultural Economic Report no. 622.

OTA (US Congress, Office of Technology Assessment) (1992). *Trade and the Environment: Conflicts and Opportunities*. Washington, DC: US Government Printing Office. OTA-BP-ITE–94.

——(1993). *Harmful Nonindigenous Species in the United States: Summary*. Washington, DC: US Government Printing Office. OTA-F–566.

——(1995a). *Agriculture, Trade, and the Environment: Achieving Complementary Policies*. Washington, DC: US Government Printing Office. OTA-ENV–617.

——(1995b). *Targeting Environmental Priorities in Agriculture: Reforming Program Strategies*. Washington, DC: US Goverment Printing Office. OTA-ENV–640.

Paden, R. (1994). "Free Trade and Environmental Economics", *Agriculture and Human Values*, 11/1: 47–54.

Pajuoja, H. (1995). "The Outlook for the European Forest Resources and Roundwood Supply." *UN/ECE? FAO Timber And Forest Discussion Papers*. ECE/TIM/DP/4. Geneva: United Nations Economic Commission for Europe.

Palmer, K., Oates, W. E., and Portney, P. R. (1995). "Tightening Environmental Standards: The Benefit-Cost or the No-Cost Paradigm?", *Journal of Economic Perspectives*, 9/4: 119–32.

Paris, R., and Ruzicka, I. (1991). *Barking Up the Wrong Tree: The Role of Rent Appropriation in Sustainable Forest Management*. Asian Development Bank Environment Office Occasional Paper no.1.

Parry, I. W. H. (1995). "Pollution Taxes and Revenue Recycling", *Journal of Environmental Economics and Management*, 29: 64–77.

Pearce, D. (1990). "Economics and the Global Environmental Challenge", *Journal of International Studies*, 19/3: 365–87.

Pearce, D., and Atkinson, G. (1998). "The Concept of Sustainable Development: An Evaluation of its Usefulness Ten Years after Brundtland", *Schweizerische Zeitschrift für Volkswirtschaft und Statistik*, 134/3: 251–69.

Pearce, D. W., Barbier, E., and Markandya, A. (1988). "Sustainable Development and Cost Benefit Analysis", Canadian Environmental Assessment Research Council Workshop on Integrating Economic and Environmental Assessment, Vancouver, Canada, November 17–18.

——(1990). *Sustainable Development, Economics and Environment in the Third World*. London: Edward Elgar.

Pearson, C. (ed.) (1987). *Multinational Corporations, Environment, and the Third World*. NC: Duke University Press and World Resources Institute.

Peltzman, S. (1976). "Towards a More General Theory of Regulation", *Journal of Law and Economics*, 19: 211–40.

Perez-Garcia, J. M. (1991). *An Assessment of the Impacts of Recent Environmental and Trade Restrictions on Timber Harvests and Exports*. Working Paper no. 33, Center for International Trade in Forest Products. Seattle: University of Washington.

Perez-Garcia, J. M., and Lippke, B. R. (1993). *Measuring the Impacts of Tropical Timber Supply Constraints, Tropical Timber Trade Constraints and Global Trade Liberalization*. LEEC Discussion Paper 93-03, London Environmental Economics Centre, UK.

Perroni, C., and Wigle, R. (1994). "International Trade and Environmental Quality: How Important are the Linkages?", *Canadian Journal of Economics*, 27: 551–67.

Pethig, R. (1976). "Pollution, Welfare and Environmental Policy in the Theory of Comparative Advantage", *Journal of Environmental Economics and Management*, 2: 160–9.

Petrakis, E., and Xepapadeas, A. (1996). "Environmental Consciousness and Moral Hazard in International Agreements to Protect the Environment", *Journal of Public Economics*, 60/1: 95–110.

Pezzey, J. (1989). *Economic Analysis of Sustainable Growth and Sustainable Development*. World Bank, Environment Department Working Paper no. 15.

Pezzey, J. C. V., and Park, A. (1998). "Reflections on the Double Dividend Debate", *Environmental and Resource Economics*, 11: 539–55.

Pflüger, M. (1996). "Ecological Dumping in a General Equilibrium Model with Regional Externalities and Monopolistic Competition," University of Freiburg, Institut für Allgemeine Wirtschaftsforschung, mimeo.

Ploeg, F. van der, and Ligthart, J. E. (1994). "Sustainable Growth and Renewable Resources in the Global Economy", in Carraro, C. (ed.), *Trade, Innovation, Environment*, Dordrecht: Kluwer, ch. 2.5, 259–80.

Ploeg, F. van der, and Zeeuw, A. de (1991). "A Differential Game of International Pollution Control", *Systems Control Letters*, 17: 409–14.

——(1992). "International Aspects of Pollution Control", *Environmental and Resource Economics*, 2: 117–39.

Porter, M. (1991). "America's Green Strategy", *Scientific American*, April 1991: 96.

Porter, M. E., and Linde, C. van der (1995). "Toward a New Conception of the Environment-Competitiveness Relationship", *Journal of Economic Perspectives*, 9/4: 97–118.

Proost, S., and Regemorter, D. van (1995). "The Double Dividend and the Role of Inequality Aversion and Macroeconomic Regimes", *International Tax and Public Finance*, 2: 207–19.

Radetzki, M. (1992). "Economic Growth and the Environment", in Low, P. (ed.), *International Trade and the Environment*. World Bank Discussion Paper 159. Washington, DC: World Bank, 121–34.

Rauscher, M. (1991a). "Foreign Trade and the Environment", in Siebert, H. (ed.), *Environmental Scarcity: The International Dimension*. Mohr: Tübingen.

——(1991b). "National Environmental Policies and the Effects of Economic Integration", *European Journal of Political Economy*, 7: 313–29.

Rauscher, M. (1992). "Economic Integration and the Environment: Effects on Members and Non-Members", *Environmental and Resource Economics*, 2: 221–36.

—— (1993). "Environmental Regulation and the International Capital Allocation", Milan: Fondazione ENI Enrico Matthei, Nota di Lavoro 79.93. Forthcoming: Mäyler, K.-G. (ed.), *International Environmental Problems: An Economic Perspective*. Kluwer: Amsterdam.

—— (1994). "On Ecological Dumping", *Oxford Economic Papers*, 46: 822–40.

—— (1995). Environmental Regulation and the Location of Polluting Industries, *International Tax and Public Finance*, 2: 229–44.

—— (1997). *International Trade, Factor Movements, and the Environment*. Oxford: Clarendon Press.

Ray, E. (1995). "Global Pollution Effects of US Protectionism", *The International Trade Journal*, 9: 475–93.

Rege, V. (1994). "GATT Law and Environment-Related Issues Affecting the Trade of Developing Countries", *Journal of World Trade*, 10/3: 95–169.

Repetto, R., and Gillis, M. (1988). *Public Policies and the Misuse of Forest Resources*. Cambridge: Cambridge University Press.

Reppelin-Hill, V. (1999). "Trade and Environment: An Empirical Analysis of the Technology Effect in the Steel Industry", *Journal of Environmental Economics and Management*, 38: 283–301.

Resource Assessment Commission (1991). *Forest and Timber Inquiry*, Draft Report. Volumes 1 and 2, Australian Government Publishing Service.

Reynolds, R., Moore, W., Arthur-Worsop, M., and Storey, M. (1993). *Impacts on the Environment of Reduced Agricultural Subsidies: A Case Study of New Zealand*. Wellington, NZ: Ministry of Agriculture and Fisheries. MAF Policy Technical Paper 93/12.

Ribaudo, M. O. (1997). "Managing Agricultural Water Pollution: Are States Doing the Job?", in D. D. Southgate, J. Hammonds, and M. Livingston (eds.), *Devolution in Environmental Policy: Proceedings of the Policy Consortium Symposium*, Washington, DC, 5 June 1997. Washington, DC: World Bank, 78–102.

Ribaudo, M. O., and Shoemaker, R. A. (1995). "The Effect of Feedgrain Program Participation on Chemical Use", *Agricultural and Resource Economics Review*, 24/2: 211–20.

Richardson, J., and Mutti, J. (1976). "Industrial Development through Environmental Controls: The International Competitive Aspect", in Walter, I. (ed.), *Studies in International Environmental Economics*. New York: Wiley.

—— (1977). "International Competitive Displacement from Environmental Control", *Journal of Environmental Economics and Management*, 4: 135–52.

Roberts, D. H., and DeRemer, K. (1997). *Overview of Foreign Technical Barriers to U.S. Agricultural Exports*. Washington, DC: US Dept. of Agriculture, Economic Research Service, Commercial Agriculture Division. Staff Paper no. AGES–9705.

Roberts, D. H., and Orden, D. (1997). "Determinants of Technical Barriers to Trade: The Case of U.S. Phytosanitary Restrictions on Mexican Avocados, 1972–1995", in Orden, D. and Roberts, D. H. (eds.), *Understanding Technical Barriers to Agricultural Trade*. St Paul, Minn.: International Agricultural Trade Research Consortium. Conference proceedings.

Robison, D. (1988). "Industrial Pollution Abatement: The Impact on the Balance of Trade", *Canadian Journal of Economics*, 21: 187–99.

Romer, P. M. (1990). "Endogenous Technological Change", *Journal of Political Economy*, 98/5, part 2, S71–S102.

Rothman, D. (1998). "Environmental Kuznets Curves: Real Progress or Passing the Buck? A Case for Consumption Based Approaches", *Ecological Economics*, 25/2: 177–94.

Rowley, C. K., Tollison, R. D., and Tullock, G. (1988). *The Political Economy of Rent Seeking*. Boston: Kluwer Academic Publishers.

Runge, C. F. (1990). "Trade Protectionism and Environmental Regulations: The New Nontariff Barriers", *Northwestern Journal of International Law and Business*, 11/1: 47–61.

—— (1994). *Freer Trade, Protected Environment: Balancing Trade Liberalization and Environmental Interests*. New York: Council on Foreign Relations.

—— (1995). "Trade, Pollution, and Environmental Protection", in Bromley, D. W. (ed.), *Handbook of Environmental Economics*. Oxford: Basil Blackwell.

Sachs, J., and Warner, A. M. (1995). *Natural Resource Abundance and Economic Growth*. NBER Working Paper no. 5398.

Samuelson, P. (1954). "The Pure Theory of Public Expenditure", *Review of Economic Studies*, 36: 387–9.

Sandler, T. (1996). "A Game Theoretic Analysis of Carbon Emissions", in Congleton, R. D. (ed.), *The Political Economy of Environmental Protection*.

Schleich, J. (1999). "Environmental Quality with Endogenous Domestic and Trade Policies", *European Journal of Political Economy*, 15: 53–71.

Schleich, J., and Orden, D. (1999). "Environmental Quality and Industry Protection with Noncooperative Versus Cooperative Domestic and Trade Policies", Department of Economics, Stanford University, mimeo.

Schmidt, C. (1997). "Enforcement and Cost-effectiveness of International Environmental Agreements: The Role of Side Payments", Diskussionsbeiträge des Sonderforschungsbereich 178, Serie II, Nr. 350. Universität Konstanz.

Schneeweis, T., and Schulze, G. (2001). "Environmental Investment and International Competitiveness: Empirical Evidence from Firm Panel Data", University of Konstanz, mimeo.

Schneider, K. (1997). "Involuntary Unemployment and Environmental Policy: the Double Dividend Hypothesis", *Scandinavian Journal of Economics*, 99: 45–60.

Schöb, R. (1996) "Evaluating Tax Reforms in the Presence of Externalities", *Oxford Economic Papers*, 48: 537–55.

—— (1997). "Environmental Taxes and Preexisting Distortions: The Normalization Trap", *International Tax and Public Finance*, 4: 167–76.

Schulze G., and Ursprung, H. (1999). "Globalization of the Economy and the Nation State", *World Economy*, 22: 295–352.

—— (2000). "Economic Integration and Environmental Economics: A Survey of the Political-economic Literature", in Hans-Jürgen Vosgerau (ed.), *Institutional Arrangements for Global Economic Integration*. London: MacMillan.

Schweinberger, A. (1997). "Environmental Policies, Comparative Advantage and the Gains/Losses from International Trade", *The Japanese Economic Review*, 48: 199–212.

Sedjo, R. A., and Lyon, K. S. (1990). *The Long-Term Adequacy of World Timber Supply. Resources for the Future.* Washington DC.

Sedjo, R. A., Bowes, M., and Wiseman, C. (1991). *Toward a Worldwide System of Tradeable Forest Protection and Management Obligations.* Washington DC.

Seldon, T., and Song, D. (1994). "Environmental Quality and Development: Is There a Kuznets Curve for Air Pollution?", *Journal of Environmental Economics and Management*, 27: 147–62.

Sen, P., and Smulders, S. (2000). "The Double Dividend Hypothesis and Trade Liberalization", Discussion paper, Center for Economic Research, Tilburg University: 2000–06.

Siebert, H. (1974). "Environmental Protection and International Specialization", *Weltwirtschaftliches Archiv*, 110: 494–508.

——(1977). "Environmental Quality and the Gains from Trade", *Kyklos*, 30: 657–73.

Sinner, J. (1994). "Trade and the Environment: Efficiency, Equity and Sovereignty Considerations", *Australian Journal of Agricultural Economics*, 38/2: 171–87.

Sinner, J., Cairns, I., Storey, M., and Warmington, B. (1995). *Agri-Environmental Programmes in New Zealand: A Report to OECD.* Wellington, NZ: Ministry of Agriculture and Fisheries. MAF Policy Technical Paper 95/3.

Smith, T. T. Jr., and Kromarek, P. (1989). *Understanding US and European Environmental Law.* Boston: Martinus Nijhoff.

Smulders, S. (1995a). "Environmental Policy and Sustainable Economic Growth, An Endogenous Growth Perspective", *De Economist*, 143/2: 163–95.

——(1995b). "Entropy, Environment, and Endogenous Economic Growth", *International Tax and Public Finance*, 2/2: 319–40.

Smulders, S., and Gradus, R. (1996). "Pollution Abatement and Long-term Growth", *European Journal of Political Economy*, 12/3: 505–32.

Spencer, B., and Brander, J. (1983). "International R&D Rivalry and Industrial Strategy", *Review of Economic Studies*, 50: 707–22.

Stähler, F. (1992). "Pareto Improvements by In-kind Transfers", Kiel Working Paper no. 541, Kiel Institute of World Economics, Kiel.

——(1993). "On The Economics of International Environmental Agreements", Kiel Working Paper no. 600, Kiel Institute of World Economics, Kiel.

Stanton, G. (1997). "Implications of the WTO Agreement on Sanitary and Phytosanitary Measures", in Orden, D. and Roberts, D. H. (eds.), *Understanding Technical Barriers to Agricultural Trade.* St Paul, Minn.: International Agricultural Trade Research Consortium. Conference proceedings.

Stigler, G. (1971). "The Theory of Economic Regulation", *Bell Journal of Economics*, 2: 3–21.

Stiglitz, J. (1974). "Growth with Exhaustible Natural Resource: Efficient and Optimal Growth Paths", *Review of Economic Studies*, Symposium on the Economics of Exhaustible Natural Resource, 123–37.

Sugden, R. (1984). "Reciprocity: The Supply of Public Goods through Voluntary Contributions", *The Economic Journal*, 94: 772–87.

Suri, V., and Chapman, D. (1998). "Economic Growth, Trade and Energy: The Implica-tions for the Environmental Kuznets Curve", *Ecological Economics*, 25/2: 195–208.

Sutton, J. D. (1989). *Resource Policy Subsidies and the GATT Negotiations.* Washington, DC: US Dept. of Agriculture, Economic Research Service. Agricultural Economic Report no. 616.

Taylor, T. (1997). "Relationship Between Production and the Environment: The Case of Tomato Production in Florida", Paris: Organization for Economic Cooperation and Development. Report prepared for OECD Group Meeting on Fruits and Vegetables, Paris, 8–9 October 1997.

Telser, L. (1980). "A Theory of Self-Enforcing Agreements", *Journal of Business*, 53: 27–44.

Thornsbury, S., Roberts, D., DeRemer, K., and Orden, D. (1997). "A First Step in Understanding Technical Barriers to Agricultural Trade", Paper presented at XXIII International Conference of Agricultural Economists, Sacramento, California, August 1997.

Tobey, J. A. (1990). "The Effects of Domestic Environmental Policies on Patterns of World Trade: An Empirical Test", *Kyklos*, 43/2: 191–209.

——(1991). "The Effects of Environmental Policy towards Agriculture on Trade: Some Considerations", *Food Policy*, 16/2: 90–4.

Tobey, J. A., and Reinert, K. A. (1991). "The Effects of Domestic Agricultural Policy Reform on Environmental Quality", *Journal of Agricultural Economics Research*, 43/2: 20–8.

Tobey, J. A., and Smets, H. (1996). "The Polluter-Pays Principle in the Context of Agriculture and the Environment", *World Economy*, 19/1: 63–87.

Tollison R. D., and Willett, T. D. (1979). "An Economic Theory of Mutually Advantageous Issue Linkages in International Negotiations", *International Organization*, 33: 309–46.

Toman, M. A. (1994). "Economics and Sustainability: Balancing Trade-offs and Imperatives", *Land Economics*, 70/4: 393–413.

Turner, R. K. (1988). "Sustainability, Resource Conservation and Pollution Control: an Overview", in Turner, R. K. (ed.), *Sustainable Environmental Management: Principles and Practice*. London: Belhaven Press.

Ugelow, J. (1982). "A Survey of Recent Studies on Cost of Pollution Control and the Effects on Trade", in Rubin, S. (ed.), *Environment and Trade*. New Jersey: Allanheld, Osmun and Co.

Ulph, A. (1993). "Environmental Policy, Plant Location and Government Protection", in Carraro, C. (ed.), *Trade, Innovation, and the Environment*. Dordrecht: Kluwer: 123–63.

——(1994). "Environmental Policy and International Trade: A Survey of Recent Economic Analysis", Milan: Fondazione ENI Enrico Matthei, Nota di Lavoro 53.94.

——(1996). "Environmental Policy and International Trade When Governments and Producer Act Strategically", *Journal of Environmental Economics and Management*, 30: 265–81.

——(1997). "Harmonisation, Minimum Standards and Optimal International Environmental Policy Under Asymmetric Information", University of Southampton Discussion Paper in Economics and Econometrics, no. 9701.

Ulph, A., and Ulph, D. (1994). "The Optimal Time Path of a Carbon Tax", *Oxford Economic Papers*, 46: 857–68.

Ulph, A., and Valentini, L. (1997). "Plant Location and Strategic Environmental Policy with Inter-sectoral Linkages", *Resource and Energy Economics*, 19: 363–83.

UNCTAD (1993). *Program on Transnational Corporations, Environmental Management in Transnational Corporations: Report on the Benchmark Environmental Survey*. New York: United Nations Conference on Trade and Development.

US EPA (US Environmental Protection Agency) (1988). *Regulatory Impact Analysis: Protection of Stratospheric Ozone, Volume I: Regulatory Impact Analysis Document*. Washington, DC.

USDA (US Dept of Agriculture) (1994). *Agricultural Resources and Environmental Indicators*. Washington, DC: US Dept of Agriculture, Economic Research Service. Agricultural Handbook no. 705.

——(1997). *Agricultural Resources and Environmental Indicators, 1996–97*. Washington, DC: US Dept of Agriculture, Economic Research Service. Agricultural Handbook no. 712.

van Beers, C., and van den Bergh, J. (1997). "An Empirical Multi-Country Analysis of the Impact of Environmental Regulations on Foreign Trade Flows", *Kyklos*, 50: 29–46.

van Grasstek, C. (1992). "The Political Economy of Trade and the Environment in the United States Senate", in P. Low (ed.), *International Trade and the Environment*. World Bank Discussion Papers, no. 159. Washington, DC, 227–44.

Vanberg, V., and Buchanan, J. (1989). "Interests and Theories in Constitutional Choice", *Journal of Theoretical Politics*, 1: 49–62.

Varangis, P. N., Crossley, R., and Primo Braga, C. A. (1995). *Is There a Commercial Case for Tropical Timber Certification?* Policy Research Working Paper 1479. Washington, DC: The World Bank.

Vaubel, R., and Willett, T. D., (eds.) (1991). *The Political Economy of International Organizations*. Boulder Colo.: Westview Press.

Verbruggen, H., and Jansen, M. A. (1995). "International Coordination of Environmental Policies", in Folmer, H., Gabel, H. L., and Opschoor, H. (eds.), *Principles of Environmental and Resource Economics*, Aldershot: E. Elgar, ch. 10, 228–52.

Verbruggen, H., Kuik, O., and Bennis, M. (1995). *Environmental Regulations as Trade Barriers for Developing Countries: Eco-Labelling and the Dutch Cut Flower Industry*. London: International Institute for Environment and Development. CREED Working Paper no. 2.

Vincent, J. R. (1992). "The Tropical Timber Trade and Sustainable Development", *Science*, 256: 1651–55.

Vincent, J. R., and Binkley, C. S. (1991). *Forest Based Industrialization: A Dynamic Perspective*. Development Discussion Paper no. 389. Cambridge, Mass.: Harvard Institute for International Development (HIID).

Vogel, D. (1992). "Disscussant's Comment, to van Grassteck (1992)", in P. Low (ed.), *International Trade and the Environment*. World Bank Discussion Papers, no. 159. Washington, DC, 245–6.

Vousden, N. (1990). *The Economics of Trade Protection*. Cambridge: Cambridge University Press.

Walter, I. (1973). "The Pollution Content of American Trade", *Western Economic Journal*, 11: 61–70.

—— (1975). *The International Economics of Pollution*. London: Macmillan.

—— (1976). *Studies in International Environmental Economics*. New York: Wiley.

—— (1982). "Environmentally Induced Relocation to Developing Countries", in Rubin, S. (ed.), *Environment and Trade*. New Jersey: Allanheld, Osmun and Co.

Weaver, P. M., Gabel, H. L., Bloemhof-Ruwaard, J. M., and Wassenhove, L. N. (1995). *Optimising Environmental Product Life Cycles: A Case Study of the European Pulp and Paper Sector*. CMER Working Papers 95/29/EPS/TM. Centre for the Management of Environmental Resources, INSEAD, Fountainebleau, France.

Weingast, B., and Moran, M. (1983). "Bureaucratic Discretion or Congressional Control", *Journal of Political Economy*, 91: 765–800.

Welsch, H. (1992). "Equity and Efficiency in International CO_2 Agreements", in E. Hope and S. Strøm (eds.), *Energy Markets and Environmental Issues: A European Perspective*, Proceedings of a German–Norwegian Energy Conference, Bergen, 5–7 June 1991, Oslo: Scandinavian University Press, ch. 12, 211–25.

Whalley, J. (1996). "Quantifying Trade and Environment Linkages Through Economywide Modeling", in Bredahl, M. E., Ballenger, N., Dunmore, J. C., and Roe, T. L. (eds.), *Agriculture, Trade and the Environment: Discovering and Measuring the Critical Linkages*. Boulder, Colo.: Westview Press, 137–48.

Wheeler, D., and Martin, P. (1992). "Prices, Policies and the International Diffusion of Clean Technology: The Case of Wood Pulp Production", in Low, P. (ed.), *International Trade and the Environment*. World Bank Discussion Paper 159. Washington, DC: World Bank, 197–224.

Wibe, S. (1991). *Market and Intervention Failures in the Management of Forests, Report to the Environment Committee*. Paris: OECD.

Williamson, O. E. (1983). "Credible Commitments: Using Hostages to Support Exchange", *American Economic Review*, 73: 519–40.

Wilson, J. (1999). "Theories of Tax Competition", *National Tax Journal*, 52: 269–304.

World Commission on Environment and Development (1987): *Our Common Future*. New York: Oxford University Press.

WRI (World Resources Institute) (1992). *World Resources 1992–93*. New York: Oxford University Press.

WTO (World Trade Organization) (1994). *The Results of the Uruguay Round of Multilateral Trade Negotiations. Market Access for Goods and Services: Overview of the Results*. Geneva: GATT.

—— (1998). *WTO Annual Report 1998: Special Topic: Globalization and Trade*. Geneva.

Yang, C. (1995). "Endogenous Tariff Formation under Representative Democracy: A Probabilistic Voting Model", *American Economic Review*, 85: 956–63.

Yarkin, C., Sunding, D., Zilberman, D., and Siebert, J. (1994). "Cancelling Methyl Bromide for Postharvest Use to Trigger Mixed Economic Results", *California Agriculture*, 48/3: 16–21.

Yezer, A., and Philipson, A. (1974). *Influence of Environmental Considerations on Agriculture and Industrial to Locate Outside the Continental US*. Public Interest Economic Center.

Yohe, G. W. (1979). "The Backward Incidence of Pollution Control: Some Comparative Statics in General Equilibrium", *Journal of Environmental Economics and Management*, 6: 187–98.

References

Young, A. (1991). "Learning by Doing and Dynamic Effects of International Trade", *Quarterly Journal of Economics*, 106/2: 369–405.

Young, O. R. (1989). "The Politics of International Regime Formation: Managing Natural Resources and the Environment", *International Organization*, 43: 349–75.

Young, C. E., and Osborn, C. T. (1990). *The Conservation Reserve Program: An Economic Assessment*. Washington, DC: US Dept. of Agriculture, Economic Research Service. Agricultural Economic Report no. 626.

Zhang, Z., and Folmer, H. (1995). "The Choice of Policy Instruments for the Control of Carbon Dioxide Emissions", *Intereconomics*, 30/3: 133–42.

Index